T0201305

WIRELESS COMMUNICATIONS SYSTEMS DESIGN

WIRELESS COMMUNICATIONS SYSTEMS DESIGN

Haesik Kim
VTT Technical Research Centre of Finland

WILEY

Library of Congress Cataloging-in-Publication Data

A catalogue record for this book is available from the British Library.

ISBN: 9781118610152

Set in 10/12pt Times by SPi Global, Pondicherry, India

1   2015

To my wife Hyeeun,
daughter Naul,
son Hanul,
and
mother Hyungsuk.

Contents

Preface

The main purpose of this book is to help senior undergraduate students, graduate students, or young engineers design wireless communications systems by themselves. The basic architecture of wireless communications has not changed much in recent decades. Especially, most of the broadband wireless communications (WiMAX, LTE, WLAN, UWB, etc.) adopt similar techniques (OFDM/OFDMA, turbo codes/LDPC, MIMO, AMC, etc.). Therefore, this book is intended to provide the basic knowledge and methodology for wireless communications systems design. This book is based on my experience in various areas (standards, research and development) of wireless communications as a researcher, developer, and consultant.

Many different interests are entangled in the wireless communications systems design. The designed wireless communication system should satisfy various interested parties such as vendors, mobile operators, regulators, and mobile users. For example, in a wireless communication system, an officer of a regulatory agency will look for efficient spectrum usage and for violation of the regulation for the frequency band. A mobile operator will look out for business opportunities, service quality, and the cost of the infrastructure. A mobile phone vendor or a network equipment vendor looks for the cost-effective design of the wireless communication system. An end user is interested in whether or not it can use a network with enough data rates and seamless connection. In order to build wireless communication systems, we consider many requirements such as spectrum, system capacity, service area, radio resource allocation, cost, and QoS. Though it is not easy to find an optimal solution, we should find a good trade-off. Wireless communication system designers require a wide range of data because one decision in one design step is closely related to another decision in the next design step. This process is very complicated and it is very difficult to find an optimal solution. Thus, they often make a decision subjectively and empirically. This book provides the reader with a big picture of wireless communications systems design in terms of a mobile device designer. It covers from fundamentals of wireless communication theory to wireless communications block implementation and system integration.

A brief design flow of wireless communications systems is as follows: Firstly, international regulator such as the International Telecommunication Union (ITU) defines the spectrum allocation for different services. The regional regulatory bodies such as European Conference of Postal and Telecommunications Administrations (CEPT) in Europe and Federal

Communications Commission (FCC) in the United States play quite an important role in decision making of ITU meetings. They are responsible for spectrum licensing and protection. After spectrum allocation for a certain service, ITU may label certain frequency allocations for a certain technology and determine the requirements for the technology. For the requirements, some standardization bodies such as the 3rd Generation Partnership Project (3GPP) and the Institute of Electrical and Electronics Engineers (IEEE) make their contributions to ITU to guarantee that their technology will meet the requirements. The actual technology is specified in the internal standardization such as 3GPP and IEEE and some standards are established at regional level. Secondly, a wireless channel is defined. Its mathematical model is developed to define the test scenario for the evaluation of the proposed technical solutions or algorithms. For example, short-range wireless channel models are different from cellular channel models. In cellular systems, we widely use empirical channel models such as ITU channel models. Thirdly, each block of the wireless communication system is selected to satisfy the requirement of the wireless communication system. This step is discussed and defined in wireless communication standards such as 3GPP and IEEE. The next steps are wireless communication system implementation. They are carried out by vendors. Fourthly, receiver algorithms under the given standard are selected and their performances are confirmed by a floating point design. Basically, standard bodies define interfaces, signaling protocol, and data structures and also give transmitter and receiver requirements. However, many freedoms of actual design and implementation are left. Fifthly, a fixed point design is performed. In this step, the architecture of a target wireless communication system is defined and the complexity can be roughly calculated. Lastly, conventional chip design processes as the remaining steps (RTL design and back-end design) are performed. This book focuses on each wireless communication algorithm design and deals with architecture design and implementation.

This book considers a broadband wireless communication system based on OFDM/OFDMA system because it is widely used in the modern wireless communication systems. The organization of the book is as follows: In Part I, the overall wireless communication theories are introduced. The basic structure of wireless communications systems is composed of a transmitter, a wireless channel, and a receiver. Their mathematical model is defined and their important theories are introduced. In addition, several physical layer techniques for mitigating wireless channel impairments are discussed. Chapter 5 "Wireless Channel Impairment Mitigation Techniques" provides a reader with fundamentals of Parts II and III. In Part II, key wireless communication blocks (FEC, MIMO, OFDM, channel estimation, equalization, and synchronization) are designed. In wireless communications systems, each block is deployed symmetrically. For example, if turbo encoder and QPSK modulator are used in a transmitter, turbo decoder and QPSK demodulator should be used in a receiver. Therefore, each block should be designed and verified pairwise. In Part III, wireless communications systems design is introduced from radio planning to system integration. Chapter 11 "Radio Planning" introduces a reader to wireless network design. In Chapter 12 "Wireless Communications System Design and Considerations," we discuss the design methodology and implementation techniques. In Chapter 13, we roughly design a 4G mobile device.

I am pleased to acknowledge the support of VTT Technical Research Centre of Finland and John Wiley & Sons and the valuable discussion of my colleagues and experts in Lancaster University, University of Leeds, Nokia, Samsung, NEC, NICT, Cadence, and so on. I am grateful for the support of my family and friends.

Haesik Kim
VTT Oulu, Finland

List of Abbreviations

CDMA	Code Division Multiple Access
CEPT	European Conference of Postal and Telecommunications Administrations
CID	Connection Identifier
CINR	Carrier to Interference-plus-Noise Ratio
CIR	Channel Impulse Response
CMOS	Complementary Metal Oxide Semiconductor
CORDIC	COordinate Rotation DIgital Computer
CP	Cyclic Prefix
CPS	Common Part Sublayer
CPU	Central Processing Unit
CQI	Channel Quality Indicators
CQICH	Channel Quality Information Channel
CRC	Cyclic Redundancy Check
CS	Convergence Sublayer
CSI	Channel State Information
CTC	Convolutional Turbo Code
DAC	Digital to- analogue Converter
DC	Direct Current
DDCE	Decision-Directed Channel Estimation
DFG	Data Flow Graph
DFT	Discrete Fourier Transform
DMA	Direct Memory Access
DNS	Domain Name System
DPSK	Differential Phase Shift Keying
DRS	Demodulation Reference Signal
DS	Direct-Sequence
DSL	Digital Subscriber Line
DSP	Digital Signal Processor
DVB-T	Digital Video Broadcasting-Terrestrial
DVFS	Dynamic Voltage and Frequency Scaling
DwPTS	Downlink Pilot Time Slot
EDA	Electronic Design Automation
EGC	Equal Gain Combining
EIRP	Effective (or Equivalent) Isotropic Radiated Power
eNB/eNodeB	Evolved Node B
ESD	Energy Spectral Density
ETSI	European Telecommunications Standards Institute
FCC	Federal Communications Commission
FCH	Frame Control Head
FDD	Frequency Division Duplexing
FDM	Frequency Division Multiplexing
FDMA	Frequency Division Multiple Access
FEC	Forward Error Correction
FEQ	Frequency Domain eQualization
FFT	Fast Fourier Transform
FIR	Finite Impulse Response

FPGA	Field Programmable Gate Array
FSD	Fixed Sphere Decoding
FSM	Finite State Machine
FTP	File Transfer Protocol
FUSC	Fully Used Subchannels
GF	Galois Field
GOS	Grade of Service
GP	Guard Period
GPIO	General-Purpose Input/Output
GSM	Global System for Mobile Communications
HDL	Hardware Description Language
HSPA	High Speed Packet Access
HTTP	Hypertext Transfer Protocol
HW	Hardware
IC	Integrated Circuit
ICI	Inter-Carrier Interference
ICMP	Internet Control Message Protocol
IDFT	Inverse Discrete Fourier Transform
IEEE	Institute of Electrical and Electronics Engineers
IFFT	Inverse Fast Fourier Transform
IGMP	Internet Group Management Protocol
IP	Internet Protocol
ISI	Inter Symbol Interference
ISO	International Standard Organisation
ITU	International Telecommunication Union
ITU-R	International Telecommunication Union Radio communications sector
JPEG	Joint Photographic Experts Group
LDPC	Low Density Parity Check
LFSR	Linear Feedback Shift Register
LLC	Logical Link Control
LLR	Log Likelihood Ratio
LMMSE	Linear MMSE
LNA	Low Noise Amplifier
LO	Local Oscillator
LPF	Low Pass Filter
LS	Least Square
LT	Luby Transform
LTE	Long Term Evolution
LTE-A	Long Term Evolution-Advanced
LUT	Look-Up Table
MAC	Media Access Control
MAP	Maximum a Posteriori
MAPL	Maximum Allowable Path Loss
MBSFN	Multicast Broadcast Signal Frequency Network
MIMO	Multiple Input Multiple Output
ML	Maximum Likelihood

MMSE	Minimum Mean Squared Error
MPEG	Moving Picture Experts Group
MPSK	M-ary Phase Shift Keying
MPSoC	MultiProcessor System-on-Chip
MRC	Maximal Ratio Combining
MS	Mobile Station
MSE	Mean Squared Error
NMT	Nordic Mobile Telephone
NP	Nondeterministic Polynomial
NTT	Nippon Telegraph and Telephone
Ofcom	Office of Communications
OFDM	Orthogonal Frequency Division Multiplexing
OFDMA	Orthogonal Frequency Division Multiple Access
OSI	Open Systems Interconnection
PA	Power Amplifier
PAPR	Peak-to-Average Power Ratio
PBCH	Physical Broadcast Channel
PCFICH	Physical Control Format Indicator Channel
PCI	Protocol Control Information
PDCCH	Physical Downlink Control Channel
PDF	Probability Density Function
PDSCH	Physical Downlink Shared Channel
PDU	Protocol Data Unit
PED	Partial Euclidean Distance
PEP	Pairwise Error Probability
PHICH	Physical Hybrid ARQ Indicator Channel
PHY	PHYsical Layer
PLL	Phase Locked Loop
PMCH	Physical Multicast Channel
PN	Pseudo-Noise
PPP	Point-to-Point Protocol
PRACH	Physical Random Access Channel
PSD	Power Spectral Density
PSS	Primary Synchronization Signal
PUCCH	Physical Uplink Control Channel
PUSC	Partially Used Subchannels
PUSCH	Physical Uplink Shared Channel
QC	Quasi Cyclic
QoS	Quality of Service
QPSK	Quadrature Phase Shift Keying
R2MDC	Radix-2 Multi-path Delay Commutator
R2SDF	Radix-2 Single-path Delay Feedback
RACH	Random Access Channel
RB	Resource Block
RF	Radio Frequency
RM	Reed-Muller

rms	Root Mean Square
RS	Reed-Solomon
RSC	Recursive Systematic Convolutional
RSSI	Received Signal Strength Indication
RTG	Receive Transition Gap
RTL	Register-Transfer Level
SAP	Service Access Points
SC	Selection Combining
SD	Sphere Decoding
SDMA	Space-Division Multiple Access
SDR	Software Defined Radio
SDU	Service Data Unit
SFID	Service Flow Identifier
SIC	Successive Interference Cancellation
SINR	Signal to Interference plus Noise Ratio
SISO	Soft Input Soft Output
SMB	Server Message Block
SNR	Signal to Noise Ratio
SoC	System-on-Chip
SOVA	Soft Output Viterbi Algorithm
SPC	Single Parity Check
SRAM	Static Random Access Memory
SRS	Sounding Reference Signal
SS	Synchronization Signal
SSH	Secure Shell
SSL	Secure Sockets Layer
SSS	Secondary Synchronization Signal
STBC	Space Time Block Code
STC	Space Time Coding
STTC	Space Time Trellis Code
SVD	Singular Value Decomposition
SW	Software
TC	Turbo Code
TCM	Trellis Coded Modulation
TCP	Transmission Control Protocol
TDD	Time Division Duplexing
TDMA	Time Division Multiple Access
TEQ	Time Domain eQualization
TTG	Transmit Transition Gap
UART	Universal Asynchronous Receiver/Transmitter
UDP	User Datagram Protocol
UE	User Equipment
UMTS	Universal Mobile Telecommunication System
UpPTS	Uplink Pilot Time Slot
UWB	Ultra-WideBand
VHDL	VHSIC Hardware Description Language

VLSI	Very-Large-Scale Integration
W-CDMA	Wideband Code Division Multiple Access
WiMAX	Worldwide Interoperability for Microwave Access
WLAN	Wireless Local Area Network
ZC	Zadoff-Chu
ZF	Zero Forcing
ZP	Zero Padding

Part I

Wireless Communications Theory

1

Historical Sketch of Wireless Communications

The reason why humans have become the most advanced species is that they produce information, store it on paper or in electronic devices, and exchange it among them. Especially, the exchange and diffusion of information has changed people's lifestyle significantly. For example, let's assume a traveler is visiting a place. Before cellular phones were in use, the traveler had to plan his visit carefully. He should book the hotel and flight and collect the information about the location manually beforehand. To locate the hotel or attraction points, he should make use of a map. Some people find this difficult as they may be disorientated. After cellular phones have come into use, a traveler can book a hotel and flight on the website using his smart phone. Once he reaches the place, the phone can guide him to the attraction points and provide useful information such as about a nice restaurant or a nearby bargain sale shop. In addition, he can make use of his phone to check email or stock price anytime and at anyplace. This drastic change in lifestyle is due to high-speed wireless communication. In this chapter, we will trace back through successive stages of wireless communications development in technical and economical aspects.

1.1 Advancement of Wireless Communications Technologies

Smoke signals used by Indian tribes are considered to be the start of wireless communication systems. Transmitting and receiving, and sending a message from one place to another place are pre-planned by them. However, the transmission range is limited to visual distance and can be carried out only in good weather. There are similar alternatives such as communication drums, signal lamps, carrier pigeons and semaphore flags. All the above have been used for thousands of years and semaphore flags are still being used in maritime communications.

The innovative paper *On Physical Lines of Force* was published by Scottish physicist and mathematician J. C. Maxwell between 1861 and 1862 [1]. This paper mathematically describes

Wireless Communications Systems Design, First Edition. Haesik Kim.
© 2015 John Wiley & Sons, Ltd. Published 2015 by John Wiley & Sons, Ltd.

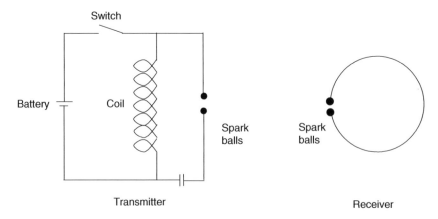

Figure 1.1   Hertz's experiment

how electromagnetic waves propagate. He predicted that the speed of electromagnetic wave is the same as that of the light waves. In 1880s, many scientists tried to prove the existence of electromagnetic waves. H. R. Hertz built an experimental apparatus to prove Maxwell's theory in 1887. The apparatus consists of simple transmitter and receiver with a small gap through which sparks could leap as shown in Figure 1.1. The transmitter can generate a spark and the receiver was placed several yards away from the transmitter. If the second spark appears in the receiver after the transmitter generates the first spark, it means the electromagnetic wave was transmitted and Maxwell's theory is correct. He published his work in the book *Electric Waves: Being Researches on the Propagation of Electric Action with Finite Velocity through Space* [2].

In the 1890s, many scientists continued Hertz's experiments. French scientist, E. Branly, invented the metal filings coherer which consists of a tube containing two electrodes. This device could detect the electromagnetic waves. Russian scientist, A. S. Popov, built a controllable electromagnetic system. On March 24, 1896, he demonstrated a radio transmission between two buildings in St. Petersburg. His paper "Apparatus for the detection and recording of electrical oscillations" [3] was published in the *Journal of the Russian Physical Chemical Society*. G. Marconi known as the father of long-distance radio transmission began his experiment in Italy contemporaneously. His experiment was nothing new but he focused on developing a practical radio system. He kept doing experiment with extending the communication distance. In 1901, he built a wireless transmission station in Cornwall, England, and successfully transmitted a radio signal to Newfoundland (it is now a part of Canada) across the Atlantic Ocean. His radio system was huge and expensive equipment with 150 m antenna, high power, and low frequency. In 1906, L. D. Forest invented a vacuum tube which made the radio system to become smaller. This radio system was used by the US government and purchased by many other countries before the Great War. After the end of the Great War, there were many efforts to find alternatives to fragile vacuum tubes. American physicist W. Shockley and chemist S. Morgan in Bell Labs established a group that worked on solid-state physics and developed a transistor. This device opened a new era of electronics. This revolution made the field of wireless communication systems to become narrower and closer to the public. A transistor radio developed in 1954 was a small portable wireless receiver and the most popular wireless communication device.

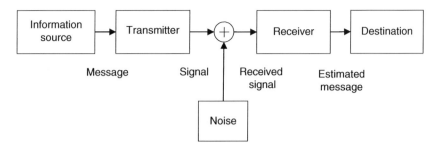

Figure 1.2   Shannon's communication architecture

Another revolution came from Bell Labs at the same time. Bell Labs scientist, C. E. Shannon, proposed information theory and published the landmark paper "A mathematical theory of communication" [4] in *Bell System Technical Journal*. Scientists at that time wanted to know how to measure information and how much information can be sent in a channel. Shannon adopted the concept of entropy to measure information, which was used in thermodynamics. Entropy of information theory means a level of the uncertainty of a random variable. He defined the channel capacity as the maximum rate of reliable communications over a noisy channel. In addition, he designed the communication architecture and is shown in Figure 1.2.

All of the current communication systems are based on Shannon's communication architecture. This architecture was innovative because a communication system designer can treat each component of the communication system separately. The time information theory was proposed became the golden age for the communication society. Many scientists developed new communication theories and implemented a new communication system. Another driving force of wireless communication systems came from the evolution of electronics. In 1958, engineer J. Kilby from Texas Instruments invented the Integrated Circuit (IC) and another engineer R. Noyce from Fairchild developed it independently a half year later. Noyce's IC chip was made of silicon while Kilby's IC chip was made of germanium. Noyce's IC chip was close to practical solutions and became an industry standard of the modern IC chips because silicon is much cheaper and easier to handle than germanium. As electronic devices evolve, wireless communication systems could be portable. The weight of the world's first mobile phone was over 30 kg. However, wireless communication systems reached greater levels due to IC technology and gradually the weights of mobile phones were significantly reduced.

A cellular system which has hexagonal cells covering a whole area without overlaps was introduced in the paper "The cellular concept" by V. H. MacDonald [5]. This paper produced another landmark concept and overcame many problems in wireless communication system such as power consumption, coverage, user capacity, spectral efficiency, and interference. The frequency reuse is one of the key concepts in the cellular network. The coverage of the cellular radio system is divided into hexagonal cells which are assigned different frequencies (F1–F4). Each cell does not have adjacent neighboring cells with same frequency as shown in Figure 1.3. Thus, cochannel interferences can be reduced, cell capacity can be increased, and cell coverage can be extended.

In each cell, it is necessary to have a multiple access scheme that enables many users to access a cellular network. Several multiple access schemes such as Frequency Division Multiple Access

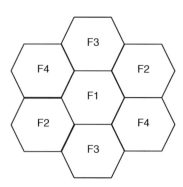

Figure 1.3   Example of frequency reuse

(FDMA), Time Division Multiple Access (TDMA), Code Division Multiple Access (CDMA), and Orthogonal Frequency Division Multiple Access (OFDMA) are widely used in cellular systems. This new concept opened another new era of wireless communications. Based on this concept, many commercial wireless communication systems were deployed. In 1979, Nippon Telegraph and Telephone (NTT) Corporation deployed the first commercial cellular network in Tokyo. Nordic Mobile Telephone (NMT) was launched in Finland, Denmark, Norway, and Sweden in 1981. Advanced Mobile Phone System (AMPS) developed by Bell Labs was deployed in Chicago, USA, on October 13, 1983, and Motorola mobile phones were used. These cellular networks were analogue-based systems. The analogue cellular phones were not popular due to high cost. After the digital Global System for Mobile Communications (GSM) called the 2nd Generation (2G) was launched in Finland in 1991, the mobile phone finally became an essential device. The huge success of GSM attracted many people to wireless communications. With wireless communication technologies advancing at a fast-growing rate, many new wireless communications were developed in not only long ranges but also short ranges. The 2G system provided voice service to users but the 3rd Generation (3G) focused on data service. The Universal Mobile Telecommunication System (UMTS) was one of the 3G systems standardized by the 3rd Generation Partnership Project (3GPP). The UMTS based on Wideband Code Division Multiple Access (W-CDMA) provided a high data rate service by including many new technologies such as turbo codes and adaptive modulation coding. High-Speed Packet Access (HSPA), Long-Term Evolution (LTE), and Long-Term Evolution-Advanced (LTE-A) kept achieving a high data rate because the volume of data service in wireless communication systems was getting bigger. In addition, the advent of smart phone brought the upheavals in wireless communication industry. The voice call is no longer the main feature of a mobile phone. Data services such as web browsing, video call, location service, internet games, and email service have become more important. Thus, the data rate has become the key metric to evaluate wireless communication systems. Table 1.1 shows us the evolution of 3GPP standards.

1.2   Wireless Communications, Lifestyles, and Economics

Let's imagine we have to send a message (100 alphabets) across the Atlantic Ocean. Before the advent of wireless communication systems, we had to deliver it by ship and it took about three weeks. The data rate was (100 alphabets $\times$ 8bits)/(3 weeks $\times$ 7 days $\times$ 24 hours $\times$ 60

Table 1.1 Evolution of 3GPP standards

	Data rate	Key features	Release date
GSM	DL: 9.6 kbps UL: 9.6 kbps	Digital, TDMA	1991
GPRS	DL: 14.4–115.2 kbps UL: 14.4–115.2 kbps	TDMA, GMSK, convolutional coding	1999
UMTS (Release 99)	DL: 384 kbps UL: 384 kbps	WCDMA, turbo coding	March 2000
UMTS (Release 4)	DL: 384 kbps UL: 384 kbps	Higher chip rate than release 99	March 2001
HSDPA (Release 5)	DL: Up to 14 Mbps UL: 384 kbps	HARQ, fast scheduling, channel quality feedback, AMC	June 2002
HSUPA (Release 6)	DL: Up to 14 Mbps UL: Up to 5 Mbps	Multimedia broadcast multicast service (MBMS), integration with WiFi	March 2005
HSPA+ (Release 7)	DL: Up to 28 Mbps UL: Up to 11.5 Mbps	MIMO, higher order modulation, latency reduction	December 2007
LTE (Release 8 and 9)	DL: 140 Mbps (10 MHz)/300 Mbps (20 MHz) UL: 25 Mbps (10 MHz)/75 Mbps (20 MHz)	OFDMA, dual carrier HSPA, SON, femtocell	December 2008 (Release 8) December 2009 (Release 9)
LTE-A (Release 10)	DL: 1 Gbps (peak download) UL: 500 Mbps (peak upload)	Carrier aggregation (CA), coordinated multiple point transmission and reception (CoMP), relay	March 2011

minutes × 60 seconds) = 0.00044 bps. After wireless telegraph was invented, the transmission time was reduced to about 2 minutes and the data rate was reduced to (100 alphabets × 8 bits)/(2 minutes × 60 seconds) = 6.67 bps. Now, let us compare these with the modern wireless communication technology such as GSM. The data rate of GSM is 9.6 kbps. Thus, it was raised by a factor of about 20 million times and about 1440 times, respectively. When comparing GSM with LTE-A, the data rate of LTE-A is 1 Gbps and LTE-A was raised by a factor of about 104,000 times. In terms of the transmission rate, we made great strides in the wireless communications technologies. It took 150 years to build the current cellular system from telegraph. Especially, it took only 20 years from GSM to LTE-A. The data rate improvement of wireless communications is summarized in Figure 1.4.

How does the improvement of wireless communication technologies affect people's life? If we consider the cost of delivery, it must be a significant impact. When we sail across the Atlantic Ocean in order to deliver a short message, we should spend for labour, fuel, ship maintenance, and so on. Besides the cost of delivery, we already have experienced the big change caused by the developments in wireless communication. The invention of a transistor radio made people listen to brand new music and the latest news in real time. Especially, when a weather centre issues a storm warning, radio is the most efficient way to distribute

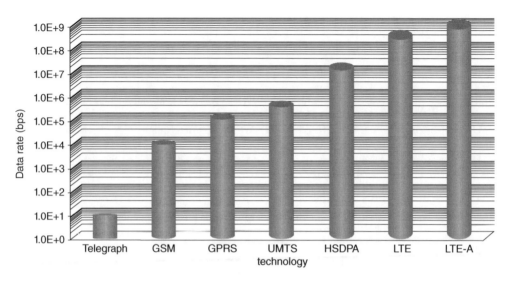

Figure 1.4 Data rates of wireless communications

information. The advent of the point-to-point communication brought another drastic change in our life. The first popular personal wireless communication device is a pager. This device can receive a short message consisting of a few digits that could be telephone numbers or some codes. After receiving the digits, people should find a landline phone to make a call. This device enables people to connect more freely. However, this device is one-way communication and does not support voice calls and relies on a landline phone. Today, the digital 2G system has become more popular and useful personal wireless communication device. The voice quality it provided is almost the same as a landline phone service. People were connected wirelessly and could work inside and outside. The 3G system made a revolution in the usage of Internet and the advent of smartphones, thereby drastically changing lifestyle. It supports broadband service and can access mobile web. Smartphone is mobile terminal supporting voice call and mobile computing. This device enables mobile users to trade stocks, browse webs, download files, exchange emails, trace locations, play video games, and so on. Regardless of time and place, people can access Internet.

Now, let's take a look at commercial usages and economics of wireless communications. When the electromagnetic waves are actively researched, many scientists didn't realize their commercial value. H. Hertz who proved the existence of electromagnetic waves was one of them. He said *It is of no use whatsoever this is just an experiment that proves Maestro Maxwell was right. We just have these mysterious electromagnetic waves that we cannot see with the naked eye. But they are there* [6]. When he was questioned about their commercial importance, he said "*Nothing*." However, G. Marconi was different. He applied for a patent for his invention and was awarded famous British patent No. 7777 "Improvements in apparatus for wireless telegraphy" [7]. He established his company, Wireless Telegraph and Signal Company, in 1897, and provided telegraphic service. Until a transistor radio was invented, wireless communication systems were of limited usage because it was bulky and expensive, and therefore, it was operated among wireless stations. The invention of transistor radio made advertisers more fascinating about the device. Radio commercial is still one of the most important modes of marketing. Their interest in wireless communication systems has been sped-up since the advent of television. The wireless communication device having become a personal device, spectrum shortage problem started to occur. Therefore, the International Telecommunication Union (ITU) started coordinating the use of radio spectrums and building communication standard. The role of ITU is to allocate frequencies to some wireless communications in overall point of view, and national regulators such as the Office of Communications (Ofcom) of the United Kingdom allocates frequencies to specific uses with complying with ITU guidelines. The usage of frequencies differs across counties. For example, Britain operates frequency bands from 88 MHz to 1 GHz for TV broadcasting (40%), defence (22%), GSM (10%), and maritime communication (1%) [8]. As the number of different wireless communication systems is rapidly rising, the price of the frequency band is getting higher and there is an increased shortage of frequency band. Basically, a government sells them to telecommunication operators by spectrum auction. The Britain sold the 3G frequency bands to telecommunication operators in 2000 and the total winning bid was €38.3 billion. Thus, wireless resources have become one of the most valuable natural resources in the world.

References

[1] J. C. Maxwell, "On Physical Lines of Force," *Philosophical Magazine*, 1861.

[2] H. Hertz, *Electric Waves: Being Researches on the Propagation of Electric Action with Finite Velocity through Space*, Authorized English translation by D. E. Jones, Macmillan and Company, New York, 1893.

[3] A. S. Popov, "Apparatus for the Detection and Recording of Electrical Oscillations" (in Russian), *Zhurnal Russkag Fizicheskoi Khimii Obshchestva (Physics, Pt. I)*, vol. **28**, pp. 1–14, 1896.

[4] C. E. Shannon, "A Mathematical Theory of Communication," *Bell System Technical Journal*, vol. **27**, pp. 379–423 & 623–656, 1948.

[5] V. H. MacDonald, "The Cellular Concept," *Bell System Technical Journal*, vol. **58**, no. 1, pp. 15–42, 1979.

[6] A. Norton, *Dynamic Fields and Waves: The Physical World*, Institute of Physics in Association with the Open University, Bristol, p. 38, 2000.

[7] G. Marconi, "Improvements in Apparatus for Wireless Telegraphy," British patent No. 7,777, April 26, 1900 (Application) and April 13, 1901 (Accepted).

[8] M. Cave, "*Independent Review of Radio Spectrum Management*," Consultation Paper, HM Treasury and Department of Trade and Industry, London, 2001.

2

Probability Theory

The reason why many wireless communication books start from probability theory is that wireless communications deal with uncertainty. If there are no channel impairments by nature, we can receive the transmitted messages without any distortion and don't need to care about probability theory. However, nature distorts and interferes when electromagnetic waves propagate. In wireless communication systems, the received messages over a wireless channel include many channel impairments such as thermal noises, interferences, frequency and timing offsets, fading, and shadowing. Thus, wireless communication systems should overcome these impairments.

2.1   Random Signals

Wireless communication systems are designed to deliver a message over a wireless channel. It is assumed that the transmitted messages include a random source and the received messages cannot be predicted with certainty. In addition, wireless channel impairments including thermal noises are expressed as random factors. Therefore, we need to know mathematical expression and characteristics of random signals.

We can divide signals into *deterministic signals* and *non-deterministic signals*. The deterministic signals are predictable for arbitrary time. It is possible to reproduce identical signals. The deterministic signals can be expressed by a simple mathematical equation and each value of the deterministic signal can be fixed as shown in Figure 2.1. We know each value with certainty at any time through mathematical calculation.

On the other hand, non-deterministic signals are either *random signals* or *irregular signals*. The random signals cannot be expressed by a simple mathematical equation and each value of the random signal cannot be predicted with certainty as shown in Figure 2.2.

Therefore, we use probability to express and analyze a random signal. The irregular signals are not describable by probability theory. It occasionally occurs in wireless communications.

Wireless Communications Systems Design, First Edition. Haesik Kim.
© 2015 John Wiley & Sons, Ltd. Published 2015 by John Wiley & Sons, Ltd.

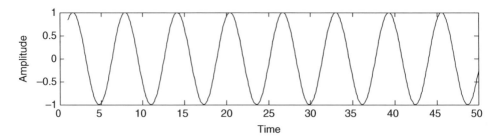

Figure 2.1   Example of a deterministic signal

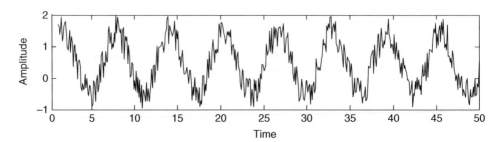

Figure 2.2   Example of a random signal

The statistic metrics such as average and variance are useful tool to understand random signals. Now, we look into not one random signal but a collection or ensemble of random signals and define useful terms. A *random variable* is useful to express unpredictable values. It is defined as follows:

Definition 2.1   Random variable

A random variable, X, is a possible sample set of a certain signal source.

There are two types of random variables: *discrete* random variable when X has discrete values and *continuous* random variable when X has continuous values. The *probability distribution* (or *probability function*) of a random variable is probabilities corresponding to each possible random value. If we deal with continuous random variables, it is difficult to express the probabilities of all possible events. Thus, we define the *Probability Density Function* (PDF) which is the probability distribution of a continuous random variable. The probability distribution, $P(X)$, of the discrete random variable, $X = x_i$, is defined as follows:

$$P(X = x_i) = p_i \tag{2.1}$$

where X takes n values and the probability, p_i, has the following properties:

$$0 \le p_i \le 1 \tag{2.2}$$

$$p_1 + p_2 + \cdots + p_n = 1 \tag{2.3}$$

$$P(X = x_i \text{ or } X = x_j) = P(X = x_i) + P(X = x_j) \tag{2.4}$$

where $i \neq j$ and $0 \leq i, j \leq n$.

Example 2.1 Discrete random variable and probability distribution

Let a discrete random variable, X, and corresponding probability, p_i, have the following values:

X	1	2	3	4	5
p_i	0.1	0.2	0.4	0.1	0.2

Check the properties of the discrete random variable and probability distribution.

Solution

Each probability $\left(p_i = \{0.1, 0.2, 0.4, 0.1, 0.2\} \right)$ of the random variable, X, satisfies (2.2), (2.3), and (2.4) as follows:

$$0 \leq p_i = \{0.1, 0.2, 0.4, 0.1, 0.2\} \leq 1$$

$$0.1 + 0.2 + 0.4 + 0.1 + 0.2 = 1.$$

When we pick up two random values, x_1 and x_2, (2.4) is satisfied as follows:

$$P(X = x_1 \text{ or } X = x_2) = P(x_1 = 1) + P(x_2 = 2) = p_1 + p_2 = 0.1 + 0.2 = 0.3.$$

In addition, another combination of random values has same results.   ∎

The *Cumulative Distribution Function* (CDF) (or *distribution function*), $F_X(x_i)$, of the discrete random variable, X, is defined as follows:

$$F_X(x_i) = P(X \leq x_i). \tag{2.5}$$

This means the probability that the random variable, X, is less than or equal to x_i. When the random variable, X, has interval $(x_a, x_b]$, the probability distribution and the cumulative distribution function can be represented as follows:

$$P(x_a < X \leq x_b) = F_X(x_b) - F_X(x_a) \tag{2.6}$$

where the notation ( ] denotes a semi-closed interval and $x_a < x_b$. The properties of the distribution function are as follows:

$$\lim_{x_i \to -\infty} F_X(x_i) = 0 \tag{2.7}$$

$$\lim_{x_i \to +\infty} F_X(x_i) = 1 \tag{2.8}$$

$$0 \leq F_X(x_i) \leq 1 \tag{2.9}$$

$$F_X(x_a) \leq F_X(x_b) \text{ if } x_a \leq x_b. \tag{2.10}$$

We will meet those two important functions when dealing with wireless communication systems. The most important probability distribution in wireless communications is *Gaussian* (or *Normal*) *distribution*. It is defined as follows:

$$P(X) = \frac{1}{\sigma\sqrt{2\pi}} e^{-\frac{1}{2}\left(\frac{X-\mu}{\sigma}\right)^2} \tag{2.11}$$

where σ and μ are standard deviation and mean of the distribution, respectively. The cumulative distribution function of Gaussian distribution is as follows:

$$F_X(x_i) = \frac{1}{2}\left[1 + \mathrm{erf}\left(\frac{x_i - \mu}{\sqrt{2\sigma^2}}\right)\right] \tag{2.12}$$

where error function, erf( ), is defined as follows:

$$\mathrm{erf}(x) = \frac{2}{\sqrt{\pi}} \int_0^x e^{-t^2}\, dt. \tag{2.13}$$

Wireless communication systems basically overcome many types of channel impairments. Thus, we must describe and analyze the noise mathematically and the natural noise such as thermal noises is expressed by Gaussian distribution.

Example 2.2   Gaussian distribution and its cumulative distribution function
Let a random variable, X, have the following parameters:

	Gaussian 1	Gaussian 2	Gaussian 3
Standard deviation (σ)	0.2	0.5	1
Mean (μ)	0	0.1	−0.2

Plot the Gaussian distributions and its cumulative distribution functions.

Solution
From (2.11), we plot the Gaussian distributions as shown in Figure 2.3.
From (2.12), we plot their cumulative distribution functions as shown in Figure 2.4.   ∎

When a system is composed of a sample signal $(s \in S)$ and a collection of time function $(t \in (-\infty, \infty))$, we define a random process as follows:

Definition 2.2   Random process

A random process, $X(s, t)$, is a collection or an ensemble of random variables from a certain source. It usually represents a random value of time.

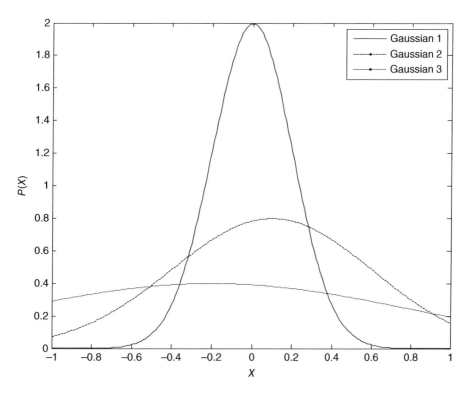

Figure 2.3 Gaussian distributions

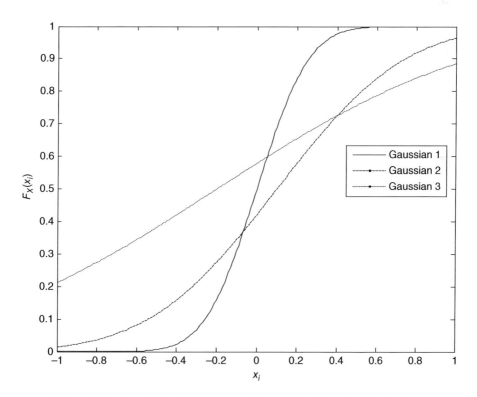

Figure 2.4 Cumulative distribution functions of three Gaussian distributions

A random signal cannot be predicted but we may forecast future values from previous events using probability theory. Thus, we can deal with wireless channel impairments and distorted signals (thermal noise, nonlinear distortion of electronic devices, etc.) randomly but not irregularly. The random process is very useful model to describe an information source, transmission, and noise. When we consider a random noise generator, its waveforms are as shown in Figure 2.5.

The random noise generator creates n waveforms. Each sample signal, s_i, for a specific time (t_1) is a random variable. When we observe only random variable, s_i, the probability distribution is expressed as follows:

$$P(s_i) = p_i \quad \text{where} \quad 0 \le p_i \le 1 \quad \text{and} \quad i = 1, 2, \ldots, n \tag{2.14}$$

and

$$\sum_{i=1}^{n} p_i = 1. \tag{2.15}$$

When we observe one sample signal, s_1, we can have a sample time function, $X(t)$. The random process, $X(s, t)$, becomes a real number, $X(s_1, t_1)$, when we fix a specific sample signal (s_1) and time (t_1).

2.2   Spectral Density

An electromagnetic wave transmits information and transfers a power in the air. When we analyze the distribution of the signal strength in frequency domain, the spectral density of the electromagnetic wave is very useful concept. When we consider a signal over time, $s(t)$, we find the *Energy Spectral Density* (ESD) and *Power Spectral Density* (PSD). If $s(t)$ is a voltage across a resistor, R, to be $1\,\Omega$, the instantaneous power, $p(t)$, is as follows:

$$p(t) = \frac{(s(t)^2)}{R} = s(t)^2 \tag{2.16}$$

and the total energy, E_s, of $s(t)$ is as follows:

$$E_s = \int_{-\infty}^{\infty} s(t)^2 \, dt = \int_{-\infty}^{\infty} |S(\omega)|^2 \, df \tag{2.17}$$

where $S(\omega)$ is the Fourier transform of $s(t)$, ω is the angular frequency $(\omega = 2\pi f)$, and f is the ordinary frequency. We can obtain this relationship between time domain energy and frequency domain energy from Parseval's theorem [1]. The energy spectral density of the signal can be found as follows:

$$E_s = \int_{-\infty}^{\infty} |S(\omega)|^2 \, df = \int_{-\infty}^{\infty} |S(2\pi f)|^2 \, df = \int_{-\infty}^{\infty} E(f) df \tag{2.18}$$

where $E(f)$ denotes the squared magnitude and energy spectral density of the signal $s(t)$. It means the signal energy per unit bandwidth and its dimension is joule per hertz. Basically, the

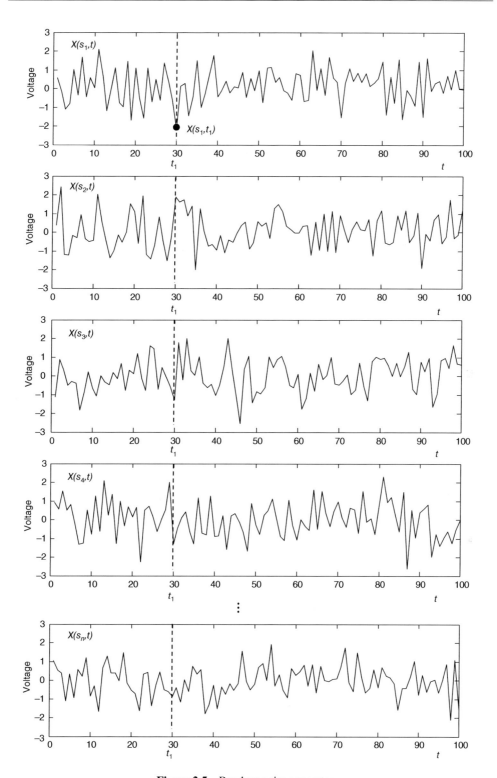

Figure 2.5   Random noise generator

energy is infinite but the power is finite. Therefore, the power spectral density is more useful to describe a periodic signal. We define an average power, P_s, as follows:

$$P_s = \lim_{T \to \infty} \frac{1}{T} \int_{-T/2}^{T/2} s(t)^2 \, dt \tag{2.19}$$

$$= \lim_{T \to \infty} \int_{-T/2}^{T/2} \frac{1}{T} |S(2\pi f)|^2 \, df \tag{2.20}$$

$$= \int_{-T/2}^{T/2} \lim_{T \to \infty} \frac{1}{T} |S(2\pi f)|^2 \, df \tag{2.21}$$

$$= \int_{-T/2}^{T/2} P(f) \, df \tag{2.22}$$

where $P(f)$ is the power spectral density of the periodic signal $s(t)$. It means the signal power per unit bandwidth and its dimension is watts per hertz. If we have a non-periodic signal, we can observe a certain time interval of the signal and have its power spectral density.

Example 2.3   Power spectral density
Consider a periodic signal, $s(t)$, and Fourier transform having the following parameters:

Periodic signal	$s(t) = \cos(2\pi t \cdot 10)$
Sample frequency	32 Hz
Time interval	0–0.32 seconds
FFT size	128

Plot the normalized single-side power spectral density of the signal, $s(t)$.

Solution
We find the normalized power spectral density using Fourier transform and plot them as shown in Figures 2.6 and 2.7.   ∎

2.3   Correlation Functions

A correlation function is used to know the relationship between random variables. In wireless communication systems, synchronization blocks are based on this correlation function. In a random process, we assume to deal with a whole signal. However, we may sometime have a part of the signal and need to describe it. In this case, we can derive some parameter from the ensemble. If we obtain an average from this ensemble, we call this ensemble average.

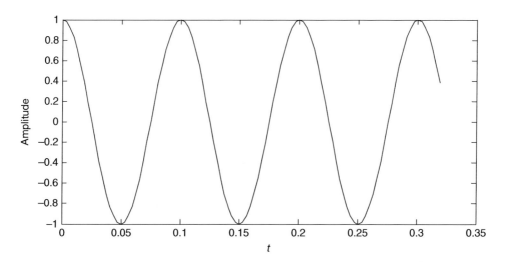

Figure 2.6   Periodic signal, $s(t)$

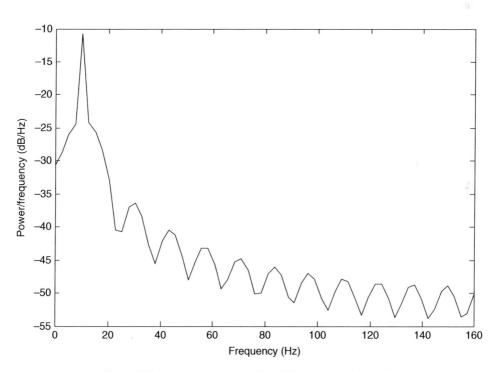

Figure 2.7   Power spectral density of the periodic signal, $s(t)$

When a discrete random variable, X, has values, x_i, and probabilities, p_i, where $0 \le i \le n$, the *expectation* (or *mean function*) of this random variable is as follows:

$$E[X] = \sum_{i=1}^{n} x_i p_i. \qquad (2.23)$$

For a continuous random process, we define it as follows:

Definition 2.3 Expectation

The expectation of a random process, $X(s, t)$, is

$$E[X(t_k)] = \overline{X(t_k)} = \int_{-\infty}^{\infty} x p_{X(t_k)}(x) dx$$

where $X(t_k)$ and $p_{X(t_k)}(x)$ are the random variable and the probability distribution when observing the random process at a certain time, t_k, respectively.

An autocorrelation is used to measure a similarity of signals and means how much a signal matches with a time lag version. We define the autocorrelation function using expectation and Figure 2.8 shows us its visual description.

Definition 2.4 Autocorrelation function

The autocorrelation of a random process, $X(s, t)$, is

$$R_X(t_i, t_j) = E\left[X(t_i)X(t_j)\right]$$

where $X(t_i)$ and $X(t_j)$ are the random variables when we observe the random process at a certain time t_i and t_j, respectively.

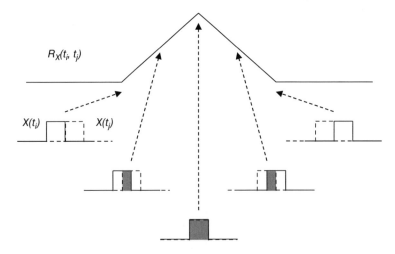

Figure 2.8   Example of an autocorrelation

When a random process is not varying with the time origin and all statistic parameters are not affected, we call it *strict sense stationary process*. If two statistic parameters, expectation and variance, are not varying with respect to time shift, we call it *wide sense stationary process*. For a wide sense stationary process, we express expectation and autocorrelation function as follows:

$$E\left[X(t_k)\right] = m_X \quad \text{where } m_X \text{ is a constant,} \tag{2.24}$$

and

$$R_X(t_i, t_j) = R_X(t_i - t_j). \tag{2.25}$$

This equation means that the autocorrelation depends on the time difference $(\tau = t_i - t_j)$. Now, we express the autocorrelation as follows:

$$R_X(\tau) = \int_{-\infty}^{\infty} X(t)X^*(t-\tau)dt \tag{2.26}$$

$$= \int_{-\infty}^{\infty} X^*(t)X(t+\tau)dt \tag{2.27}$$

where the operation "*" represents the complex conjugate.

When we need to know a similarity of two different random processes, cross-correlation is useful and measures a similarity between a transmitted signal and a stored signal in the receiver. The cross-correlation function is defined as follows:

Definition 2.5   Cross-correlation function

The cross-correlation of two random process, $X(s, t)$ and $Y(s, t)$, is

$$R_{X,Y}(t_i, t_j) = E\left[X(t_i)Y(t_j)\right]$$

where $X(t_i)$ and $Y(t_j)$ are the random variables when observing the random process at a certain time t_i and t_j, respectively.

The visual description of cross-correlation is shown in Figure 2.9.

A matched filter based on the cross correlation function is very useful for detecting a known signal from a signal that has been contaminated by wireless impairments. The input, $x(t)$, of the matched filer can be expressed as follows:

$$x(t) = As(t - t_d) + n(t) \tag{2.28}$$

where $s(t)$, A, t_d, and $n(t)$ are a known transmitted signal, an unknown scaling factor, an unknown time delay, and additive noise, respectively. The output, $y(t)$, of the matched filter is expressed as follows:

$$y(t) = x(t) \circledast h(t) \qquad (2.29)$$

$$y_0(t) = As(t - t_d) \circledast h(t) \qquad (2.30)$$

$$n_0(t) = n(t) \circledast h(t) \qquad (2.31)$$

where $h(t)$ and the operation "$\circledast$" denote a matched filter and convolution, respectively. The system model of the matched filter is illustrated in Figure 2.10.

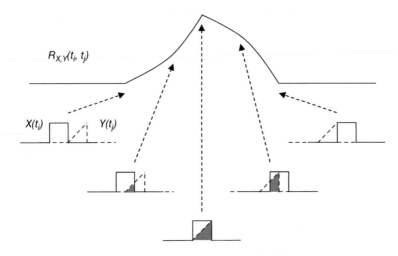

Figure 2.9   Example of a cross-correlation

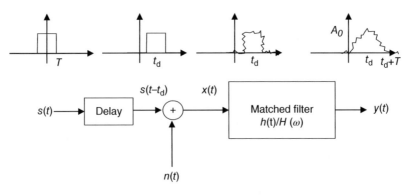

Figure 2.10   Matched filter system model

Basically, the matched filter is a linear filter and has a sharp peak output. We know information of the transmitted signal by a time $(t_d + T)$ and amplitude (A_0) of the matched filter output. The impulse response of the matched filter with additive white noise is

$$h(t) = s^*(-t + t_d). \tag{2.32}$$

We represent the matched filter output using autocorrelation function as follows:

$$y(t) = \int_{-\infty}^{\infty} x(\tau)h(t - \tau)d\tau \tag{2.33}$$

$$= \int_{-\infty}^{\infty} x(\tau)s^*(-t + \tau + t_d)d\tau \tag{2.34}$$

$$= \int_{-\infty}^{\infty} s^*(u)x(u + (t - t_d))du \tag{2.35}$$

$$= R_{sx}(t - t_d) \tag{2.36}$$

where $u = -t + \tau + t_d$. If the input, $x(t)$, is perfectly matched, the output of the matched filter is identical to autocorrelation of $s(t)$ as following:

$$y(t) = R_s(t - t_d). \tag{2.37}$$

Example 2.4   Matched filter
Let $s(t)$ be the transmitted sinc function signal as follows:

$$s(t) = \frac{\sin(\pi t)}{\pi t} \quad \text{where} -5 \leq t \leq 5$$

and the received signal, $r(t)$, is $r(t) = s(t) + n(t)$ where $-5 \leq t \leq 5$. Describe the matched filter process.

Solution
The transmitted signal, $s(t)$, is illustrated as shown in Figure 2.11.
    The received signal, $r(t)$, is illustrated as shown in Figure 2.12 and the impulse response of the matched filter is

$$h(t) = s^*(-t) = \frac{\sin(-\pi t)}{-\pi t} = \frac{\sin(\pi t)}{\pi t} \quad \text{where} -5 \leq t \leq 5$$

and this is identical to the transmitted signal, $s(t)$.

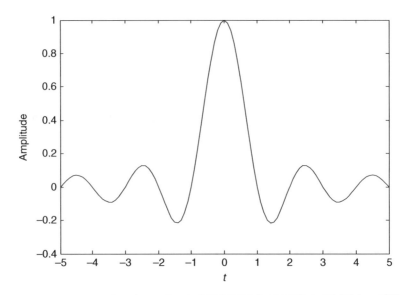

Figure 2.11   Transmitted sinc function signal, $s(t)$

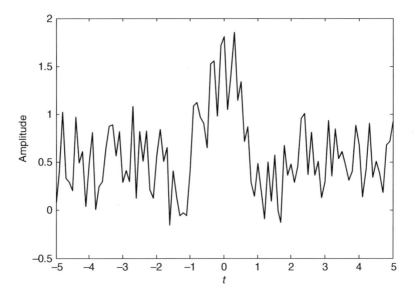

Figure 2.12   Received sinc function signal, $r(t)$

Therefore, the output, $y(t)$, of the matched filter is

$$y(t) = \int_{-\infty}^{\infty} r(\tau)h(t-\tau)d\tau$$

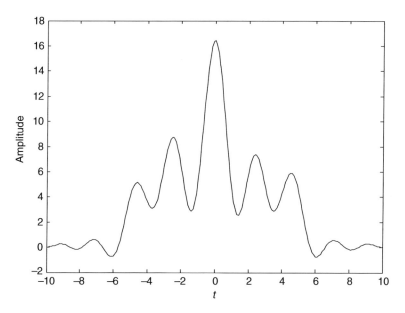

Figure 2.13   Matched filter output, $y(t)$

$$= \int_{-\infty}^{\infty} \left(\frac{\sin(\pi\tau)}{\pi\tau} + n_0(\tau) \right) \left(\frac{\sin(\pi(t-\tau))}{\pi(t-\tau)} \right) d\tau$$

where $n_0(\tau)$ is additive noise. The output, $y(t)$, is illustrated in Figure 2.13.   ∎

2.4   Central Limit Theorem

When we face a noise that the nature has made electrical component noises or thermal noises, we assume it follows Gaussian distribution because of central limit theorems. In this section, we investigate central limit theorem and its characteristic. The central limit theorem is defined as follows:

Definition 2.6   Central limit theorem

Let random variables $\{X_1, X_2, \ldots, X_n\}$ be an independent and identically distributed sequence with expectation, $E[X_i] = \mu$, and variance, $\mathrm{Var}(X_i) = \sigma^2$. The sample average, $\bar{X}$, and the sample sum, S_n, of the random variables are approximately normally distributed when n is large enough and have following characteristics:

$E(\bar{X}) = \mu$ and $\mathrm{Var}(\bar{X}) = \sigma^2/n$

$E(S_n) = n\mu$ and $\mathrm{Var}(S_n) = n\sigma^2.$

This theorem means that the sample average and sum have Gaussian distribution regardless of distribution of each random variable. Now, we can find many examples from our daily life. Example 2.5 shows us that a random number generation converges to normal distribution.

Example 2.5   Central limit theorem

Consider n random source generators whose output is 0 or 1. When each random source generator produces 10 000 random samples, observe the samples sum (X value) while the number of random source generators increases.

Solution

Using simple computer simulation, we can find the samples sum as shown in Figure 2.14.

As we can observe Figure 2.14, the frequency of occurrences is almost even when the number of the random source generator is 1. As the number of the random source generators increases, the distribution of sample sum goes to normal distribution.   ■

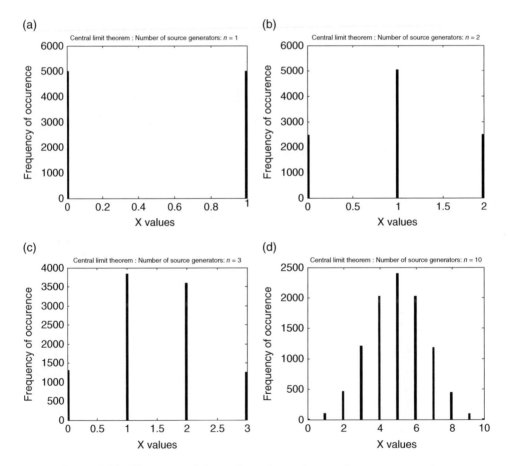

Figure 2.14   Histograms of observed sample sum by n random source generators

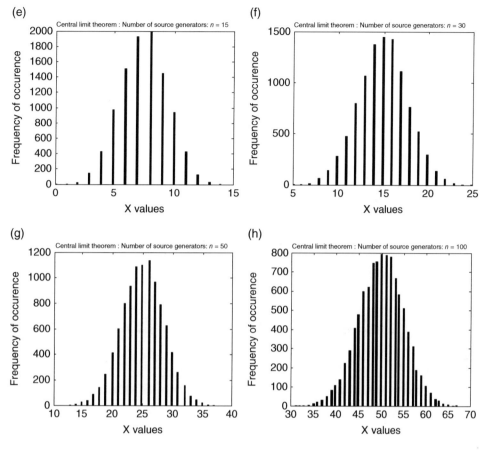

Figure 2.14 *(Continued)*

A Gaussian distribution with average value, μ, and variance, σ^2, is denoted as $N(\mu, \sigma^2)$. It has been expressed as (2.11). The standard Gaussian distribution ($N(0, 1)$) with $\mu = 0$ and $\sigma^2 = 1$ is

$$P(X) = \frac{1}{\sqrt{2\pi}} e^{-\frac{x^2}{2}} \tag{2.38}$$

The cumulative distribution function of the standard Gaussian distribution is

$$F_X(x_i) = \frac{1}{2}\left[1 + \mathrm{erf}\left(\frac{x_i}{\sqrt{2}}\right)\right]. \tag{2.39}$$

Summary 2.1   Probability theory

1. The deterministic signals can be expressed by a simple mathematical equation. However, the random signals cannot be expressed by a simple mathematical equation and each value of a random signal cannot be predicted with certainty.
2. The Gaussian (or Normal) distribution is defined as follows:

$$P(X) = \frac{1}{\sigma\sqrt{2\pi}} e^{-\frac{1}{2}\left(\frac{X-\mu}{\sigma}\right)^2}$$

where σ and μ are standard deviation and mean of the distribution, respectively.
3. The average power, P_s, is defined as follows:

$$P_s = \lim_{T \to \infty} \frac{1}{T} \int_{-T/2}^{T/2} s(t)^2 \, dt = \int_{-T/2}^{T/2} P(f) \, df$$

where $P(f)$ is power spectral density of the periodic signal $s(t)$.
When a random process is not varying with the time origin and all statistic parameters are not affected, we call it strict sense stationary process.

2.5   Problems

2.1. Let a random variable, X, and corresponding probabilities, p_i, have the following values:

X	1	2	3	4	5	6	7
p_i	0.05	0.25	0.15	0.1	0.2	0.15	0.1

Check the properties of random variable and probability distribution.

2.2. Let a random variable, X, have the following parameters:

	Gaussian 1	Gaussian 2	Gaussian 3	Gaussian 4	Gaussian 5
σ	0.1	0.3	0.7	0.5	0.2
μ	0.2	0.5	−0.8	−0.2	0.1

Plot the Gaussian distribution and its cumulative distribution function.

2.3. Consider a non-periodic signal, $s(t)$, and Fourier transform having the following parameters:

Non-periodic signal	$s(t) = \cos(2\pi t \cdot 10) + \sin(2\pi t \cdot 10^2) + N(t)$ where $N(t)$ is a random noise
Sample frequency	32 Hz
Time interval	0–0.32 seconds
FFT size	128

Plot the normalized single side power spectral density of the signal, $s(t)$.

2.4. Let $s(t)$ be the transmitted signal with a rectangular shape pulse as following:

$$s(t) = A \operatorname{rect}\left(\frac{t - T/2}{T}\right)$$

where

$$\operatorname{rect}(t) = \begin{cases} 0 & \text{if } |t| > \dfrac{1}{2} \\ \dfrac{1}{2} & \text{if } |t| = \dfrac{1}{2} \\ 1 & \text{if } |t| < \dfrac{1}{2} \end{cases}$$

and the received signal, $r(t)$, is $r(t) = s(t) + n(t)$. Describe the matched filter process.

2.5. Consider a dice tossing with n dices whose output is from 1 to 6. When we toss 10 000 times with n dices, observe the samples sum (X value) while the number of dice increases.

2.6. Consider a coin tossing. The random process, $X(t)$, is defined as follows:

$$X(t) = \begin{cases} \cos(\pi t) & \text{when head shows up} \\ 3t & \text{when tail shows up} \end{cases}.$$

Plot the sample functions and find the PDF of the random variables at $t = 1$.

2.7. Consider $X(t) = A \sin(2\pi t + \theta)$ where A is constant and θ is a random variable uniformly distributed in the range $0 \le \theta \le \pi$. Check whether or not the process is wide sense stationary process.

2.8. Show that the cross-correlation of $f(t)$ and $g(t)$ is equivalent to the convolution of $f^*(-t)$ and $g(t)$.

2.9. Consider a Gaussian noise with the *psd* of $N_0/2$. Find the autocorrelation function of the output process when a low pass filter with bandwidth B receives Gaussian noise as an input.

2.10. The number of phone call per each hour in one base station follows Gaussian distribution with $\mu = 500$ and $\sigma = 300$. Suppose that a random sample of 30 hours is selected and the sample mean $\bar{x}$ of the incoming phone calls is computed. Describe the distribution of the sample mean $\bar{x}$ and find the probability that the sample mean $\bar{x}$ of the incoming phone calls for 30 hours is larger than 650.

2.11.   A dice is tossed 50 times. Using the central limit theorem, find that probability that (i) the sum is greater than 250, (ii) the sum is equal to 200, and (iii) the sum is less than 150.

Reference

[1] S. Haykin, *Communication Systems*, John Wiley & Sons, Inc., Now York, 1983.

3

Wireless Channels

Wireless communication systems are designed to recover received signals that are damaged by wireless channel impairments. In classical communication theory, wireless channels were unknown and people think unpredictable noises occur in wireless channels. However, many scientists and engineers investigated wireless channels and developed different types of wireless channel models. They can be categorized as additive white Gaussian noise, jitter, phase shift, path loss, shadowing, multipath fading, interference, and so on. They give the transmitted signals different types of damages. In addition, transmission distance, mobility, geographical feature, weather, antenna position, and signal waveform affect the parameters of wireless channel models. For example, the channel model of short range wireless communications is different from the one of cellular systems. An urban area channel model is different from that of a rural one. A mobile user has more severe frequency offset due to the Doppler effect. Therefore, it is important to understand wireless channels when designing wireless communication systems. In this chapter, we look into wireless channels and their characteristics.

3.1 Additive White Gaussian Noise

In Chapter 2, we have defined noise as Gaussian distribution due to central limit theorem. When the noise power has a flat power spectral density, we call it *white noise* and its single-sided power spectral density ($P_{wn}(f)$ (W/Hz)) can be denoted as follows:

$$P_{wn}(f) = N_0 \tag{3.1}$$

where N_0 is constant. This equation means the white noise has an equal power spectral density for all frequency bands. This white noise affects most types of signals independently so that we usually call it *Additive White Gaussian Noise* (AWGN). If a wireless channel is given, we

Wireless Communications Systems Design, First Edition. Haesik Kim.
© 2015 John Wiley & Sons, Ltd. Published 2015 by John Wiley & Sons, Ltd.

can calculate *capacity* of the wireless channel. The channel capacity of single antenna systems for AWGN channel is defined as follows:

Definition 3.1   Channel capacity for AWGN

$$C_{awgn} = W \log_2 \left(1 + \frac{P_r}{N_0 W} \right)$$

where C_{awgn}, P_r, and W are the channel capacity (bits/s), average received power (W), and bandwidth (Hz), respectively.

Channel capacity, C_{awgn}, means the amount of information we can send through the given channel. We call this equation *Shannon-Hartley theorem* [1]. The average received signal power over the channel noise ($P_r/N_0 W$) is simply the signal to noise ratio (SNR) and C_{awgn}/W is simply the normalized channel capacity. Figure 3.1 illustrates their relationship.

In Figure 3.1, we can observe two areas: theoretically impossible area and theoretically possible area. When we want to transmit a signal with data rate, R, it is not possible to

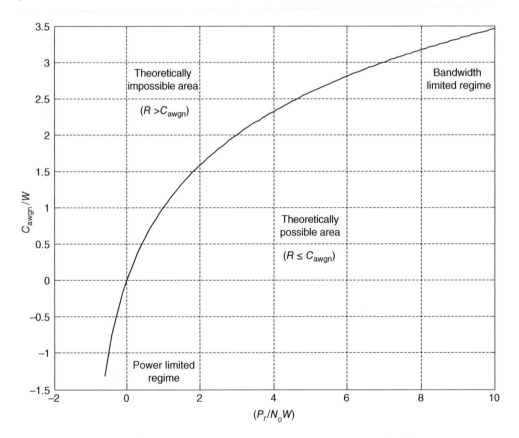

Figure 3.1   Channel capacity of single antenna system for AWGN

communicate reliably in this area ($R > C_{\text{awgn}}$). The curve represents the limitation of liable transmission under the given channel. When the average received power is large enough (SNR $\gg$ 1), the capacity reaches the saturation point in the power and regards as a linear function in the bandwidth. Therefore, we call this area *bandwidth limited regime*. The capacity equation can be expressed approximately as follows:

$$C_{\text{awgn}} \approx 0.332W\left(10\log_{10}\text{SNR}\right). \tag{3.2}$$

On the contrary, when the average received power is small enough (SNR $\ll$ 1), the capacity reaches the saturation point in the bandwidth and regards as a linear function in the power. Therefore, we call this area *power limited regime*. The capacity equation can be expressed approximately as follows:

$$C_{\text{awgn}} \approx 1.44\frac{S}{N_0}. \tag{3.3}$$

For infinite bandwidth, $S = E_b C_{\text{awgn}}$. (3.3) can be rewritten as follows:

$$C_{\text{awgn}} \approx 1.44\frac{E_b C_{\text{awgn}}}{N_0} \tag{3.4}$$

where E_b is a bit energy. We can obtain the following equation:

$$\frac{E_b}{N_0} \approx \frac{1}{1.44} = 0.694 = -1.6 \text{ in dB.} \tag{3.5}$$

We call this value (0.694 or −1.6 in dB) *Shannon limit*. This value means it is not possible to communicate reliably below this value.

Example 3.1   Channel capacity for AWGN

Consider a communication system with the following parameters:

$$\text{SNR} = 20\,\text{dB and Band width} = 10 \text{ kHz}$$

Calculate the channel capacity C_{awgn} for AWGN.

Solution

$$\text{SNR} = 20\,\text{dB} = 100\,\text{W}$$

From Definition 3.1, the channel capacity is

$$C_{\text{awgn}} = 10\log_2\left(1+100\right) = 66.58 \text{ kbps.} \qquad\blacksquare$$

3.2   Large-Scale Path Loss Models

An electromagnetic wave propagates from a transmitter to a receiver over the air. During this propagation, the energy of the electromagnetic wave is reduced. Basically, we can express this process as path loss models. One of simple channel models is *Friis transmission model* [2]. It describes a simple free space channel without obstructions between a transmitter and a receiver. In this model, the electromagnetic waves spread outward in a sphere, and the received power, P_r, is given as follows:

$$P_r = P_t G_t G_r \left(\frac{\lambda}{4\pi R} \right)^2 \tag{3.6}$$

where P_t is the transmitted power, G_t and G_r are the transmitter antenna gain and the receiver antenna gain, respectively, λ is the wavelength, and R is the distance between a transmitter and a receiver. The Friis transmission model is illustrated in Figure 3.2.

The isotropic antenna of the transmitter radiates the Effective Isotropic Radiated Power (EIRP) uniformly. It is represented by

$$\text{EIRP} = P_t G_t \tag{3.7}$$

and the antenna of the receiver collects the transmitted power as much as the antenna size (or the effective area) of the receiver. We call it the antenna aperture, A_e. The received power is related to this antenna aperture and the surface area of the sphere, $4\pi R^2$.

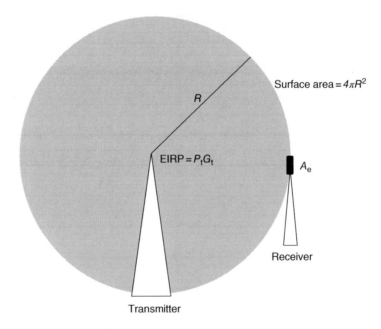

Figure 3.2   Friis transmission model

The received power is proportional to $A_e/4\pi R^2$. Therefore, we can present the received power, P_r, as follows:

$$P_r = P_t G_t \frac{A_e}{4\pi R^2}. \tag{3.8}$$

The relationship between the receiver antenna gain and effective aperture is

$$G_r = \frac{4\pi}{\lambda^2} A_e. \tag{3.9}$$

Combining (3.8) and (3.9), we can obtain (3.6). This equation means the received power depends on the square of the distance proportionally because the other parameters are basically given. In practical systems, the received power decreases more quickly due to reflection, diffraction, and scattering. These mechanisms are governed by the Maxwell equations. Reflection is the change in the direction of an electromagnetic wave when it encounters an object. The behavior of the reflection is explained by *Fresnel equations*. Diffraction is the bending of light when it passes the edge of an object. The behavior of diffraction is explained by *Huygen's principle*. Scattering is the production of other small electromagnetic waves when it encounters an object with a rough surface. The process of the scattering is quite complex and some types of scattering are Rayleigh scattering, Mie scattering, Raman scattering, Tyndall scattering, and Brillouin scattering.

Example 3.2   Friis transmission equation
Consider a wireless communication system with the following parameters:

Transmission power	10 W
Transmitter antenna gain	1
Receiver antenna gain	0.8
Carrier frequency	20 MHz
Distance between transmitter and receiver	100 m

Calculate the received power using the Friis transmission equation.

Solution
The wavelength of the carrier frequency 20 MHz is $\lambda = c/f = 3 \times 10^8/20 \times 10^6 = 15$ m
From (3.6), the received power is

$$P_r = P_t G_t G_r \left(\frac{\lambda}{4\pi R} \right)^2 = 10 \cdot 1 \cdot 0.8 \cdot \left(\frac{15}{4\pi \cdot 100} \right)^2 = 0.0011 \, \text{W} = 1.1 \, \text{mW}. \qquad \blacksquare$$

Figure 3.3 illustrates the two-ray ground reflection model.

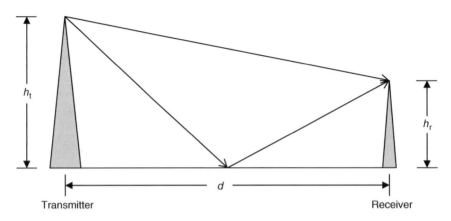

Figure 3.3   Two-ray ground reflection model

When considering the reflection from ground, we can use the two-ray ground reflection model as follows:

$$P_r = P_t G_t G_r \left(\frac{h_t h_r}{d^2} \right)^2 \qquad (3.10)$$

where h_t and h_r denote the transmitter antenna height and receiver antenna height, respectively. Unlike Friis transmission equation, the received power is affected by the antenna heights. The fourth power of the distance, d^4, is included in the equation so that the received power decreases more quickly according to the distance. This equation does not depend on the wavelength.

Example 3.3   Two-ray ground reflection model
Consider a wireless communication system with the following parameters:

Transmission power	10 W
Transmitter antenna gain	1
Receiver antenna gain	0.8
Transmitter antenna height	10 m
Receiver antenna height	1.5 m
Distance between transmitter and receiver	100 m

Calculate the received power using the two-ray ground reflection model.

Solution
From (3.7), the received power is

$$P_r = P_t G_t G_r \left(\frac{h_t h_r}{d^2} \right)^2 = 10 \cdot 1 \cdot 0.8 \cdot \left(\frac{10 \cdot 1.5}{100^2} \right)^2 = 0.018 \, \text{mW}. \qquad \blacksquare$$

In practical systems, we use empirical path loss models. The received power of the empirical models is expressed as follows:

$$P_r = P_t P_0 \left(\frac{d_0}{d} \right)^{\gamma} \tag{3.11}$$

where P_0 is the power at the reference distance, d_0, and γ is the path loss exponent which typically has 2–8 according to carrier frequency, environment, and so on. The path loss is defined as the difference between the transmitted power and the received power as follows:

Definition 3.2   Path loss

$$PL = P_t - P_r = 10 \log_{10} \frac{P_t}{1 \text{mW}} - 10 \log_{10} \frac{P_r}{1 \text{mW}} = 10 \log_{10} \frac{P_t}{P_r}$$

where PL, P_t, and P_r are the path loss (dB), transmitted power (W), and received power (W), respectively.

From Definition 3.2, the path loss of (3.11) is

$$PL(d) = PL_0 + 10\gamma \log_{10} \frac{d}{d_0} \tag{3.12}$$

where PL_0 is the path loss at the reference distance d_0 (dB). When a propagating electromagnetic wave encounters some obstruction such as wall, building, and tree, reflection, diffraction, and scattering happen and the electromagnetic wave is distorted. We call this effect *shadowing* or *slow fading* and use the modified path loss model as follows:

$$PL(d) = PL_0 + 10\gamma \log_{10} \frac{d}{d_0} + \chi \tag{3.13}$$

where χ denotes a Gaussian distributed random variable with standard deviation, σ, and represents shadowing effect.

Summary 3.1   Wireless channels and path loss models

1. The white noise affects most types of signals independently so that we usually call it Additive White Gaussian Noise (AWGN).
2. The channel capacity, C_{awgn}, means how much information we can send through the given channel.
3. In the Friis transmission model, the received power, P_r, is given as follows:

$$P_r = P_t G_t \frac{A_e}{4\pi R^2} = P_t G_t G_r \left(\frac{\lambda}{4\pi R} \right)^2$$

where P_t is the transmitted power, G_t and G_r are the transmitter and receiver antenna gain, respectively, λ is the wavelength, R is the distance between a transmitter and a receiver, and A_e is the antenna aperture.

4. In the two-ray ground reflection model, the received power, P_r, is given as follows:

$$P_r = P_t G_t G_r \left(\frac{h_t h_r}{d^2} \right)^2$$

where h_t and h_r denote the transmitter antenna height and receiver antenna height, respectively.

5. The received power of the empirical model is

$$P_r = P_t P_0 \left(\frac{d_0}{d} \right)^\gamma$$

where P_0 is the power at the reference distance d_0 and γ is the path loss exponent which typically has 2–8 according to carrier frequency, environment, and so on.

3.3 Multipath Channels

When an electromagnetic wave propagates in the air, reflection happens and it produces some reflected waves as shown in Figure 3.4. Although the receiver wants to have the original signal only, it receives the reflected waves together with the original signal. The reflected signals travel different paths and they have different amplitudes and phase noises. Thus, many signals with different delays and amplitudes arrive at the receiver.

The power delay profile is widely used to analyze multipath wireless channels. It is described by several parameters such as the mean excess delay, maximum excess delay, root mean square (rms) delay spread, and maximum delay spread. Among them, the maximum delay spread, τ_{max}, and the rms delay spread, τ_{rms}, are important multipath channel parameters. The maximum delay spread means the total time interval we can receive the reflected signals with high energy. A high value of τ_{max} means a highly dispersive channel. However, the maximum delay spread can be defined in several ways. Therefore, the rms delay spread is more widely used and indicates the rms value of the delay profile. We use this parameter to decide coherence bandwidth. One example of the power delay profile is shown in Figure 3.5.

The multipath channel is modeled as a linear Finite Impulse Response (FIR) filter. The impulse response, $h(t, \tau)$, of the multipath channel is

$$h(t,\tau) = \sum_{i=1}^{N} a_i(t)\delta(\tau - \tau_i) \tag{3.14}$$

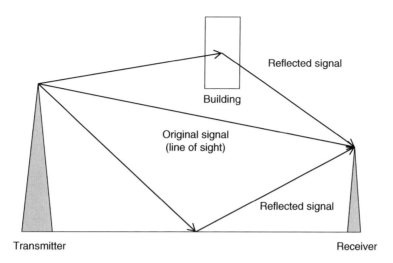

Figure 3.4   Original signal and reflected signals

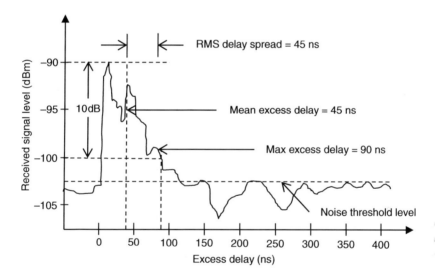

Figure 3.5   Power delay profile example

where $a_i(t)$ and $\delta(\tau - \tau_i)$ denote amplitude of a multipath and the Dirac delta function, respectively. N means N different reflected paths. One example is illustrated in Figure 3.6.

We can observe two different reflected paths in the delay profile and represent this using a three-tap FIR filter as shown in Figure 3.7. In this figure, delay, τ_1 and τ_2, can be implemented by shift-registers and amplitude, a_0, a_1, and a_2, can be implemented by multiplications (or amplifiers).

From the power delay profile, we can analyze the multipath channel.

If a wireless channel is not constant over a transmit bandwidth and severely distorted, it would be very difficult to establish a reliable radio link. Thus, we define the coherence bandwidth (B_c) as follows:

Definition 3.3   Coherence bandwidth

The coherence bandwidth is a statistical measurement of the bandwidth where the channel is regarded as the *flat* channel, which means two signals passing through the channel experience similar gain and phase rotation.

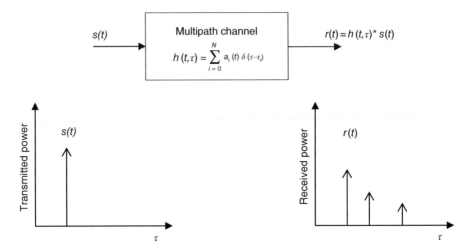

Figure 3.6   Multipath channel model

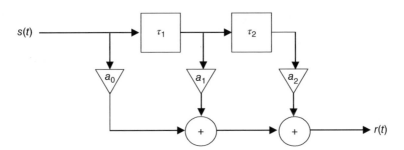

Figure 3.7   FIR filter for multipath channel model

It simply means the range of frequencies we use to send a signal without serious channel distortions. The rms delay spread, τ_{rms}, is inversely proportional to the coherence bandwidth as follows:

$$B_c \propto \frac{1}{\tau_{rms}}. \tag{3.15}$$

Therefore, the larger τ_{rms} means the wireless channel becomes more frequency selective channel. The more accurate definition of B_c is related to the frequency correlation function. For correlation greater than 0.9 [3],

$$B_c \approx \frac{1}{50\tau_{rms}}. \tag{3.16}$$

For correlation greater than 0.5 [3],

$$B_c \approx \frac{1}{5\tau_{rms}}. \tag{3.17}$$

When we design a symbol structure of wireless communication systems, the symbol length should be defined according to τ_{rms}. For example, if the symbol length is greater than $10\,\tau_{rms}$, we assume the wireless communication system does not need to mitigate Inter Symbol Interference (ISI). If it is less than $10\,\tau_{rms}$, we must avoid ISI using several techniques such as an equalizer. If it is much less than $10\,\tau_{rms}$, a reliable communication is not possible.

Example 3.4   Coherence bandwidth
Consider a wireless channel with the rms delay spread $(\tau_{rms}) = 1.5\,\mu s$. Find the coherence bandwidth (B_c) for correlation greater than 0.5 and possible data rate to avoid ISI. Check whether or not the wireless communication system with signal bandwidth (200 kHz) needs an equalizer.

Solution
From (3.17), the coherence bandwidth is

$$B_c \approx \frac{1}{5\tau_{rms}} = \frac{1}{5 \times 1.5\,\mu s} = 133\,\text{kHz}.$$

We can simply represent the relationship between data rate (R bps) and bandwidth (W Hz) as $W \approx 2R$. Since we want to send a signal without ISI, it is a safe estimation that the signal should be transmitted over the coherent bandwidth. Thus, the data rate should be less than 66.5 kbps (=133/2). The coherence bandwidth (133 kHz) is less than the signal bandwidth (200 kHz). Therefore, the wireless channel is not reliable and an equalizer is needed.   ■

Typically, the rms delay spread of cellular systems is less than 8 μs. The urban or mountain area has longer than 8 μs due to more reflections. The rms delay spread of indoor small cells is less than 50 ns.

A wireless channel is varying according to time, movement, environment, and so on. If a wireless channel changes every moment, we will not know wireless channel characteristics and cannot design wireless communication systems. Therefore, we assume a wireless channel is constant during certain time. The coherence time means the certain time. It is defined as follows:

Definition 3.4   Coherence time

The coherence time is the time interval over which a channel impulse response is regarded as not varying.

Similar to the power delay profile, the Doppler power spectrum provides statistical information about the coherence time. The Doppler power spectrum is affected by a user mobility while the power delay profile is affected by a multipath. A moving wireless communication device produces the Doppler shift (Δf). The Doppler shift is expressed as follows:

$$\Delta f = \frac{v}{\lambda}\cos\theta = \frac{vf}{c}\cos\theta \qquad (3.18)$$

where v, λ, θ, and c denote the speed of a moving wireless communication device, the wavelength of the carrier electromagnetic wave, the angle of arrival with respect to a direction of the moving wireless device, and the speed of light, respectively. When a receiver moves in the opposite direction of the electromagnetic wave ($\theta=0°$), the Doppler shift is

$$\Delta f_{\theta=0°} = \frac{v}{\lambda}. \qquad (3.19)$$

When a receiver moves in the same direction of the electromagnetic wave ($\theta=180°$), the Doppler shift is

$$\Delta f_{\theta=180°} = -\frac{v}{\lambda}. \qquad (3.20)$$

When a receiver moves in perpendicular to the electromagnetic wave ($\theta=90°$), the Doppler shift is

$$\Delta f_{\theta=90°} = 0. \qquad (3.21)$$

Therefore, the maximum Doppler shift (f_m) means $\cos\theta=1$ and is represented as follows:

$$f_m = \frac{v}{\lambda} \qquad (3.22)$$

and the coherence time (T_c) is inversely proportional to the Doppler spread as follows:

$$T_c \propto \frac{1}{f_m} \qquad (3.23)$$

For correlation greater than 0.5 [3],

$$T_c \approx \sqrt{\frac{9}{16\pi f_m^2}} = \frac{0.423}{f_m}. \qquad (3.24)$$

If the symbol duration is greater than the coherence time, the wireless channel changes during symbol transmission. We call this *fast fading*. On the other hands, if the symbol duration is less than the coherence time, the wireless channel does not change during symbol transmission. We call this *slow fading*.

Example 3.5   Coherence time

Consider a mobile receiver with $v = 80$ km/h and $f_c = 2$ GHz. Find the Doppler shift and the coherence time for the correlation greater than 0.5.

Solution

From (3.22), the maximum Doppler shift is

$$f_m = \frac{v f_c}{c} = \frac{(80 \text{ km/h})(2 \times 10^9 \text{ Hz})}{3 \times 10^8 \text{ m/s}} = \frac{(22.222)(2 \times 10^9)}{3 \times 10^8} = 148.1 \text{ Hz}.$$

From (3.24), the coherence time is

$$T_c = \frac{0.423}{f_m} = \frac{0.423}{148.1} = 2.856 \text{ ms}.$$ ∎

The Rayleigh fading distribution is widely used to model a small-scale fading and describes a statistical time varying model for the propagation of electromagnetic waves. When the time variant channel impulse response in a flat fading channel is given by

$$c(t) = a(t) e^{j\varphi(t)} \tag{3.25}$$

where $a(t)$ and $\varphi(t)$ denote the magnitude and phase rotation, respectively, the probability density function of the Rayleigh distribution is represented by

$$p_{\text{Rayleigh}}(a) = \frac{a}{\sigma^2} e^{-(a^2/2\sigma^2)} \tag{3.26}$$

where σ^2 denotes the time average power of the received signal. One example of the Rayleigh fading is illustrated in Figure 3.8.

When there is one strong multipath signal such as a line-of-sight signal, we can use the Rice distribution. It is expressed as follows:

$$p_{\text{Rice}}(a) = \frac{a}{\sigma^2} e^{-\frac{\left(a^2 + a_0^2\right)}{2\sigma^2}} I_0 \left(\frac{a a_0}{\sigma^2} \right) \tag{3.27}$$

where a_0 and I_0 denote the peak magnitude of one strong multipath and the modified Bessel function of the first kind. Every natural phenomenon in time domain has corresponding phenomenon in frequency domain. Sometimes we explain the power delay profile in frequency domain through the Fourier transform. The multipath fading results in significant power losses, frequency and phase offsets, and so on. Therefore, a wireless communication system designer should understand fading environments clearly and select suitable mitigation techniques. The diversity techniques are very important techniques to overcome fading channels. Basically, this technique uses the fact that each wireless channel

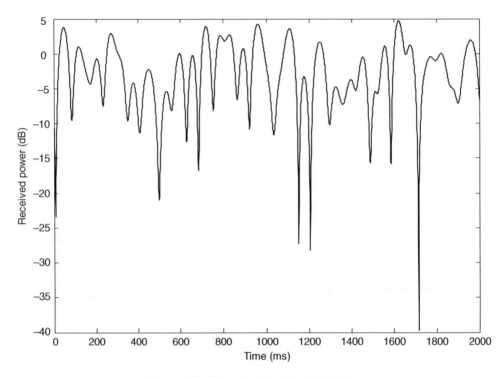

Figure 3.8   Example of the Rayleigh fading

experiences different fading. For example, some part of a signal passing through channel 1 can be damaged seriously but same part of a signal passing through channel 2 may not be damaged. A receiver can recover the transmitted signal through observing multiple signals passing through different channels. More detail will be dealt in Chapter 5. The type of small-scale fading is illustrated in Figure 3.9 where B_s and T_s denote the signal bandwidth and time interval, respectively.

The capacity, $C_{\text{freq sel}}$ (bits/s/Hz), of the frequency selective fading channel [4] is given by

$$C_{\text{freq sel}} = \sum_{n=0}^{N-1} \log_2 \left(1 + \frac{P_n^* |\overline{h}_n|^2}{N_0} \right) \qquad (3.28)$$

where $|\overline{h}_n|^2$ is the sub-channel gain and P_n^* can be found by Lagrangian method as follows:

$$P_n^* = \max\left(\left(\frac{1}{\lambda_L} - \frac{N_0}{|\overline{h}_n|^2} \right), 0 \right) \qquad (3.29)$$

where λ_L is the Lagrangian multiplier. We call this power allocation *water-filling*. One example of the water-filling algorithm is illustrated in Figure 3.10.

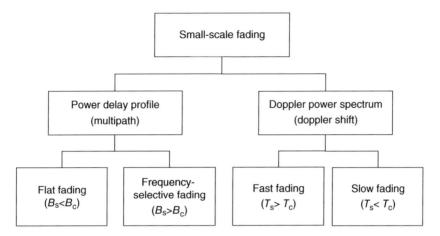

Figure 3.9   Type of small-scale fading

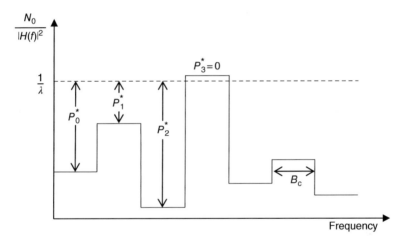

Figure 3.10   Water-filling algorithm

In the fast fading channel, we transmit a signal in time interval LT_c and have approximately same channel gain for lth coherence time period. Therefore, the average capacity, C_{fast} (bits/s/Hz), is

$$C_{\text{fast}} = \frac{1}{L} \sum_{l=1}^{L} \log_2 \left(1 + \left| \overline{h_l} \right|^2 \text{SNR} \right). \tag{3.30}$$

When L is large, the capacity is

$$C_{\text{fast}} = E \left[\log_2 \left(1 + \left| \overline{h_l} \right|^2 \text{SNR} \right) \right] \tag{3.31}$$

The outage capacity is used for slow-fading channels because the capacity is a random variable and there is delay limitation. The outage capacity, C_0 (bits/s/Hz), is simply defined as the maximum data rate without errors. The outage probability, $p_{out}(C_0)$, can be represented as follows:

$$p_{out}(C_0) = \Pr(C < C_0) = \Pr\left(\log_2\left(1 + \left|\overline{h}_l\right|^2 \text{SNR}\right) < C_0\right).$$ (3.32)

For example, $\Pr(C < 10) = 0.1$ means the frame error rate is 0.1 with ideal code if the wireless communication system frame has 10 bits/s/Hz spectral efficiency.

3.4   Empirical Wireless Channel Models

The channel model we looked into the previous sections is to provide theoretical backgrounds but it does not include measurement data from real wireless environments. Therefore, many scientists and engineers extensively measure wireless channel parameters in real wireless environments and develop empirical wireless channel models such as Hata model [5], Walfisch-Ikegami model [5], Erceg model [6], ITU model [7], and so on. Among them, the ITU-R recommendation M.1225 is widely used as empirical channel models of cellular systems. Especially, Worldwide Interoperability for Microwave Access (WiMAX) by the Institute of Electrical and Electronics Engineers (IEEE) and Long Term Evolution (LTE) by the 3rd Generation Partnership Project (3GPP) use these models. The wireless channel models are classified into three test environments such as indoor office, pedestrian, and vehicular environment. Their path loss model for the indoor office environment [7] is expressed as follows:

$$\text{PL}_{indoor}(d_{indoor}) = 37 + 30\log_{10} d_{indoor} + 18.3n^{\left(\frac{n+2}{n+1} - 0.46\right)}$$ (3.33)

where d_{indoor} and n denote the distance (m) between a transmitter and a receiver and the number of floors, respectively. The path loss model for the pedestrian test environment [5] is

$$\text{PL}_{pedestrian}(d_{pedestrian}) = 40\log_{10} d_{pedestrian} + 30\log_{10} f + 49$$ (3.34)

where $d_{pedestrian}$ and f denote the distance (km) between a base station and a mobile station and carrier frequency, respectively. The path loss model for the vehicular test environment [5] is

$$\text{PL}_{vehicular}(d_{vehicular}) = 40(1 - 4 \times 10^{-3} h_b)\log_{10} d_{vehicular} - 18\log_{10} h_b + 21\log_{10} f + 80$$ (3.35)

where $d_{vehicular}$, h_b, and f denotes the distance (km) between a base station and a mobile station, the base station antenna height (m), and carrier frequency, respectively. In addition, they have two different delay spreads: channel A (Low delay spread) and channel B (High delay spread). Their parameters are summarized in Tables 3.1, 3.2, 3.3, and 3.4.

Summary 3.2   Multipath channels

1. The power delay profile is widely used to analyze multipath wireless channels.
2. The maximum delay spread means the total time interval we can receive the reflected signals with high energy.
3. The rms delay spread is more widely used and is inversely proportional to coherence bandwidth, B_c, as follows:

$$B_c \propto \frac{1}{\tau_{rms}}.$$

4. The coherence time, T_c, is inversely proportional to the Doppler spread as follows:

$$T_c \approx \sqrt{\frac{9}{16\pi f_m^2}} = \frac{0.423}{f_m}.$$

Table 3.1   Parameters for channel impulse response model [5]

Test environment	Channel A		Channel B	
	rms (ns)	P (%)	rms (ns)	P (%)
Indoor office	35	50	100	45
Outdoor to indoor and pedestrian	45	40	750	55
Vehicular—high antenna	370	40	4 000	55

Table 3.2   Indoor office test environment tapped-delay-line parameters [5]

Tap	Channel A		Channel B		Doppler spectrum
	Relative delay (ns)	Average power (dB)	Relative delay (ns)	Average power (dB)	
1	0	0	0	0	Flat
2	50	−3.0	100	−3.6	Flat
3	110	−10.0	200	−7.2	Flat
4	170	−18.0	300	−10.8	Flat
5	290	−26.0	500	−18.0	Flat
6	310	−32.0	700	−25.2	Flat

Table 3.3 Outdoor to indoor and pedestrian test environment tapped-delay-line parameters [5]

Tap	Channel A		Channel B		Doppler spectrum
	Relative delay (ns)	Average power (dB)	Relative delay (ns)	Average power (dB)	
1	0	0	0	0	Classic
2	110	−9.7	200	−0.9	Classic
3	190	−19.2	800	−4.9	Classic
4	410	−22.8	1200	−8.0	Classic
5	—	—	2300	−7.8	Classic
6	—	—	3700	−23.9	Classic

Table 3.4 Vehicular test environment, high antenna, tapped-delay-line parameters [5]

Tap	Channel A		Channel B		Doppler spectrum
	Relative delay (ns)	Average power (dB)	Relative delay (ns)	Average power (dB)	
1	0	0.0	0	−2.5	Classic
2	310	−1.0	300	0	Classic
3	710	−9.0	8900	−12.8	Classic
4	1090	−10.0	12900	−10.0	Classic
5	1730	−15.0	17100	−25.2	Classic
6	2510	−20.0	20000	−16.0	Classic

3.5 Problems

3.1. Consider a communication system with SNR = 10 dB and Bandwidth = 3 kHz. Calculate the channel capacity for AWGN.

3.2. Compare the channel capacity between a single antenna system and a multiple antenna system.

3.3. Define the channel capacity using the maximum mutual information.

3.4. The required channel capacity for AWGN and bandwidth are 100 kbps and 10 MHz, respectively. Find the required SNR.

3.5. Calculate the received power using the Friis transmission equation, when designing the wireless communication link with the following parameters:

Transmission power	25 W
Transmitter antenna gain	10
Receiver antenna gain	15
Carrier frequency	2 GHz
Distance between transmitter and receiver	1 km

3.6. A base station transmits a signal with the following parameters and the required received power is 1 mW. If a mobile user is located at the cell edge and path law obeys the Friis transmission model, then find the cell radius:

Transmission power	25 W
Transmitter antenna gain	1.5
Receiver antenna gain	1
Carrier frequency	1 GHz

3.7. Consider a cellular system with the following parameters and the measured power at $d_0 = 0.1$ km follows the Friis transmission model. When the minimum required power is 1 mW and the mean of the received power is log normally distributed with standard deviation 10 dB, find the cell radius which satisfies connection probability 0.8:

Transmission power	20 W
Carrier frequency	2 GHz
Transmitter antenna gain	1.6
Receiver antenna gain	1
Transmitter antenna height	20 m
Receiver antenna height	1.5 m

3.8. Consider a unit delay filter with the following:

$$h(m) = \delta(m-1) = \begin{cases} 1, & \text{when } m = 1 \\ 0, & \text{otherwise} \end{cases}.$$

Find the frequency response of the filter.

3.9. Consider a filter with the following:

$$h(m) = \delta(m) + \delta(m-L).$$

Find the frequency response of the filter when $L = 1, 3, 5$, and 10.

3.10. The multipath channel is modeled as the following FIR filter:

$$y(m) = x(m) - 1.5x(m-1) + 4.2x(m-2) - 2.1x(m-3) + 1.5x(m-4).$$

Find the z-transfer function and the phase response.

3.11. Consider the following power delay profile:

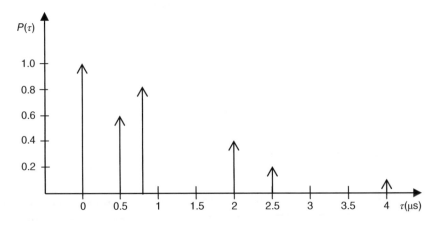

(i) Find the coherence bandwidth. (ii) Do we need an equalizer when signal bandwidth is 200 kHz? (iii) Find the coherence time when the carrier frequency is 1.8 GHz and a user mobile at a speed of 100 km/h.

3.12. Describe the advantages and disadvantages of empirical channel models.

References

[1] C. E. Shannon, "Communication in the Presence of Noise," *Proceedings of the IRE*, vol. **37**, no. 1, pp. 10–21, 1949.

[2] H. T. Friis, "A Note on a Simple Transmission Formula," *Proceedings of the IRE*, vol. **34**, pp. 254–256, 1946.

[3] T. S. Rappaport, *Wireless Communications: Principles & Practice*, Prentice Hall, Upper Saddle River, NJ, 1996.

[4] D. Tse and P. Viswanath, *Fundamentals of Wireless Communication*, Cambridge University Press, Cambridge, 2005.

[5] European Cooperative in the Field of Science and Technical Research EURO-COST 231. "Urban Transmission Loss Models for Mobile Radio in the 900- and 1,800 MHz Bands (Revision 2)," COST 231 TD(90)119 Rev. 2, The Hague, the Netherlands, September 1991.

[6] V. Erceg, L. J. Greenstein, S. Y. Tjandra, S. R. Parkoff, A. Gupta, B. Kulic, A. A. Julius, and R. Bianchi, "An Empirically Based Path Loss Model for Wireless Channels in Suburban Environments," *IEEE Journal on Selected Areas in Communications*, vol. **17**, no. 7, pp. 1205–1211, 1999.

[7] ITU-R Recommendation M.1225, "Guidelines for Evaluation of Radio Transmission Technologies for IMT 2000," February 1997. http://www.itu.int/rec/R-REC-M.1225/en (accessed April 20, 2015).

4

Optimum Receiver

We defined a signal and random process in Chapter 2 and looked into wireless channels in Chapter 3. Now, we need to know how to detect a transmitted signal correctly in a receiver. The main purpose of the receiver is to recover the transmitted signal from the distorted received signal. Decision theory is helpful for finding the original signal. Decision theories are widely used in not only wireless communication systems but also other applications. Almost everything people face in their day-to-day life is related to decisions. For example, every day, we have to decide on the color of outfit that we have to wear, what to eat, and which route to drive. For example, if we need to find a shortcut to a particular place, then we have to decide based on the following: a priori information (or a prior probability) such as weather forecast or traffic report, an occurrence (or a probability of occurrence) such as traffic jam, and a posterior information (or a posterior probability) such as data collected regarding traffic jam. Likewise, we have to decide on the messages a transmitter sends in a wireless communication system. The decision depends on a priori information such as modulation type and carrier frequency, an occurrence such as a measured signal, and a posterior information such as channel state information. The decision theory of wireless communication systems is developed to minimize the probability of error. In this chapter, several decision theories are introduced and optimum receiver is discussed.

4.1 Decision Theory

We consider a simple system model with a discrete channel as shown in Figure 4.1.

The wireless channel models we discussed in Chapter 3 are analogue channel models.

In communication theory (especially, information or coding theory), a digital channel model is mainly used such as a discrete channel, Binary Symmetric Channel (BSC), Binary Erasure Channel (BEC), and so on. In this system model, the message source produces a

Wireless Communications Systems Design, First Edition. Haesik Kim.
© 2015 John Wiley & Sons, Ltd. Published 2015 by John Wiley & Sons, Ltd.

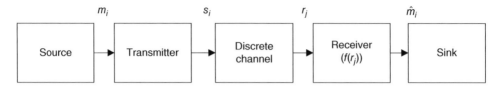

Figure 4.1   System model based on a discrete channel

discrete message, m_i, (where $i = 1, \ldots, M$) as a random variable. The probability the message, m_i, appears is a priori probability, $P(m_i)$. The transmitter produces a signal, s_i, which are the values corresponding to the message, m_i. The signal, s_i, becomes the input of the discrete channel. The output, r_j, can be expressed as a conditional probability, $P(r_j|s_i)$, which means the probability that the output, r_j, is received when the transmitter sends the signal, s_i. The receiver estimates the transmitted messages, $\widehat{m}_i$, as the output of the receiver using the decision rule, $f(\ )$, namely $f(r_j) = \widehat{m}_i$. The decision rule is a function mapping the received signal to the most probable value. The probability of error is defined as follows:

Definition 4.1   Probability of error

The probability of error is the probability that a transmitted message is not identical to an estimated message in a receiver. It is defined as follows:

$$P_\varepsilon = P\left(\widehat{m}_i \neq m_i\right).$$

The receiver is designed to minimize the probability of error, P_ε. The receiver minimizing the probability of error is called an *optimum receiver*. Let us consider a message source which can produce a random value. A transmitter sends it to a receiver over a discrete channel. In the receiver side, it would be very difficult to decide correctly if we make a decision from scratch. However, if we have a priori information, for example, the transmitted information is "0" or "1" and the probability the transmitter sends "0" is 60%, the receiver can decide correctly with 60% accuracy even if we regard all received signals as "0." The decision theory uses a priori information, a posterior information, and likelihood. The Bayesian decision rule [1] is used when there is a priori information and likelihood. More specifically, the Bayesian formula is defined as follows:

Definition 4.2   Bayesian formula

$$P(s_i|r_j) = \frac{P(r_j|s_i)P(s_i)}{P(r_j)},$$

where $P(r_j) = \sum_i P(r_j|s_i)P(s_i)$. We can interpret the Bayesian formula as follows:

$$(\text{Posteriori}) = \frac{(\text{Likelihood})\,(\text{Prior})}{(\text{Occurence})}.$$

The Bayesian decision rule with equal cost is

$$f_{\text{Bayesian}} = \begin{cases} s_0 & \text{if } P(s_0 \mid r_j) > P(s_1 \mid r_j) \\ s_1 & \text{otherwise} \end{cases}. \tag{4.1}$$

Equivalently,

$$f_{\text{Bayesian}} = \begin{cases} s_0 & \text{if } P(r_j \mid s_0)P(s_0) > P(r_j \mid s_1)P(s_1) \\ s_1 & \text{otherwise} \end{cases}. \tag{4.2}$$

The occurrence term, $P(r_j)$, is not used in the Bayesian decision rule. The probability of error is

$$P(\text{error} \mid r_j) = \min\left[P(s_0 \mid r_j),\ P(s_1 \mid r_j)\right]. \tag{4.3}$$

If the likelihood is identical, the Bayesian decision rule depends on a priori information. Likewise, if the priori information is identical, it depends on the likelihood.

Example 4.1   Bayesian decision rule
Consider a simple communication system with the following system model as shown in Figure 4.2: and a priori probability, $P(m_i)$, and conditional probability, $P(r_j \mid s_i)$, have the following probabilities:

m_i	$P(m_i)$	$P(r_0 \mid s_i)$	$P(r_1 \mid s_i)$	$P(r_2 \mid s_i)$
0	0.7	0.7	0.2	0.1
1	0.3	0.2	0.3	0.5

Design the decision rule with the minimum error probability.

Solution
Firstly, the decision rule, $f_a(\ )$, is designed by a priori probability, $P(m_i)$. Namely, the probability the transmitter sends "0" is higher than the probability the transmitter sends "1." Therefore, the estimated output, $\hat{m}_i$, is all "0" as follows:

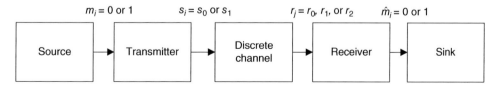

Figure 4.2   System model of Example 4.1

m_i	$P(r_0|s_i)$	$P(r_1|s_i)$	$P(r_2|s_i)$
0	0.7	0.2	0.1
1	0.2	0.3	0.5
$f_a()$	0	0	0

Secondly, the decision rule, $f_b()$, is designed by a conditional probability, $P(r_j|s_i)$. Namely, the estimated output, $\hat{m}_i$, is decided by a higher conditional probability as follows:

m_i	$P(r_0|s_i)$	$P(r_1|s_i)$	$P(r_2|s_i)$
0	0.7	0.2	0.1
1	0.2	0.3	0.5
$f_b()$	0	1	1

Lastly, the Bayesian decision rule, $f_{\text{Bayesian}}()$, is designed by both a priori probability, $P(m_i)$, and a conditional probability, $P(r_j|s_i)$. Namely, the estimated output, m_i, is basically decided by the following decision rule:

$$f_{\text{Bayesian}} = \begin{cases} s_0 & \text{if } P(s_0|r_j) > P(s_1|r_j) \\ s_1 & \text{otherwise} \end{cases}$$

or

$$f_{\text{Bayesian}} = \begin{cases} s_0 & \text{if } P(r_j|s_0)P(s_0) > P(r_j|s_1)P(s_1) \\ s_1 & \text{otherwise} \end{cases}.$$

Therefore, we can decide as follows (indicated in bold-face):

m_i	$P(s_i|r_0)$	$P(s_i|r_1)$	$P(s_i|r_2)$
0	**0.49**	**0.14**	0.07
1	0.06	0.09	**0.15**
$f_{\text{Bayesian}}()$	0	0	1

The joint probabilities have the following relationship:

$$P(m_i, r_j) = P(m_i)P(r_j \mid m_i) = P(m_i)P(r_j \mid s_i)$$

and its joint probabilities can be obtained as follows:

m_i	$P(m_i, r_0)$	$P(m_i, r_1)$	$P(m_i, r_2)$
0	0.49	0.14	0.07
1	0.06	0.09	0.15

The minimum error probability (P_e) means the maximum probability of correct decision (P_c). We can calculate $P_c^{f_a}$, $P_c^{f_b}$, and $P_c^{f_{\text{Bayesian}}}$ as follows:

$$P_c^{f_a} = P(m_i = 0, r_0) + P(m_i = 0, r_1) + P(m_i = 0, r_2)$$
$$= 0.49 + 0.14 + 0.07 = 0.7$$

$$P_c^{f_b} = P(m_i = 0, r_0) + P(m_i = 1, r_1) + P(m_i = 1, r_2)$$
$$= 0.49 + 0.09 + 0.15 = 0.73$$

$$P_c^{f_{\text{Bayesian}}} = P(m_i = 0, r_0) + P(m_i = 0, r_1) + P(m_i = 1, r_2)$$
$$= 0.49 + 0.14 + 0.15 = 0.78$$

Therefore, the Bayesian decision rule, $f_{\text{Bayesian}}(\ )$, has the minimum error probability among three decision rules.   ∎

An optimal decision means there is no better decision which brings a better result. We define the optimal decision as maximizing $P(m_i \mid r_j)$ as follows:

$$f_{\text{MAP}}(r_j) = \arg \max_{m_i} P(m_i \mid r_j). \qquad (4.4)$$

Using the Bayesian formula,

$$f_{\text{MAP}}(r_j) = \arg \max_{m_i} \frac{P(m_i)P(r_j \mid m_i)}{P(r_j)} \qquad (4.5)$$

where arg max function is defined as follows:

$$\arg \max_{x} P(x) \triangleq \{x \mid \forall y : f(y) \le f(x)\}. \qquad (4.6)$$

Therefore, the optimum receiver is designed by maximizing a posteriori probability. We call this *Maximum a Posteriori* (MAP) decision rule. When a priori probability, $P(m_i)$, is identical, the optimal decision can be changed as follows:

$$f_{\text{ML}}(r_j) = \arg \max_{m_i} P(r_j \mid m_i). \qquad (4.7)$$

The optimum receiver relies on the likelihood. Therefore, we call this *Maximum Likelihood* (ML) decision rule.

4.2   Optimum Receiver for AWGN

In this section, we consider a simple system model with the AWGN channel as shown in Figure 4.3.

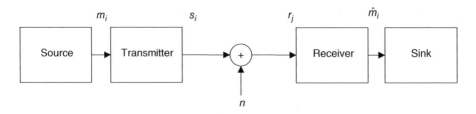

Figure 4.3   System model with AWGN

We consider two types of the message source outputs (m_i="0" or "1"), two types of the corresponding signals ($s_i = s_1$ or s_2), and $s_1 > s_2$. The probability density function of the Gaussian noise is expressed as follows:

$$p(n) = \frac{1}{\sigma\sqrt{2\pi}} \exp\left(-\frac{n^2}{2\sigma^2}\right). \tag{4.8}$$

The noise, n, is independent of the signal, s. The likelihood, $P(r_j | s_i)$ depends on the probability density function of the Gaussian noise. We can express the likelihood as follows:

$$p(r_j | s_i) = \frac{1}{\sigma\sqrt{2\pi}} \exp\left(-\frac{(r_j - s_i)^2}{2\sigma^2}\right) \tag{4.9}$$

where $i = 1$ or 2. The MAP decision rule as the optimal decision is expressed as follows:

$$p(r_j | s_1)p(m_i = 0) \ge p(r_j | s_2)p(m_i = 1) \tag{4.10}$$

$$\frac{1}{\sigma\sqrt{2\pi}} \exp\left(-\frac{(r_j - s_1)^2}{2\sigma^2}\right) p(m_i = 0) \ge \frac{1}{\sigma\sqrt{2\pi}} \exp\left(-\frac{(r_j - s_2)^2}{2\sigma^2}\right) p(m_i = 1). \tag{4.11}$$

We decide $\widehat{m_i} = 0$ if the above inequalities are satisfied and $\widehat{m_i} = 1$ otherwise. We derive a simple decision rule from the above equation as follows:

$$\exp\left(-\frac{(r_j - s_1)^2}{2\sigma^2}\right) p(m_i = 0) \ge \exp\left(-\frac{(r_j - s_2)^2}{2\sigma^2}\right) p(m_i = 1) \tag{4.12}$$

$$\left(-\frac{(r_j - s_1)^2}{2\sigma^2}\right) + \ln p(m_i = 0) \ge \left(-\frac{(r_j - s_2)^2}{2\sigma^2}\right) + \ln p(m_i = 1) \tag{4.13}$$

$$-(r_j - s_1)^2 + 2\sigma^2 \ln p(m_i = 0) \ge -(r_j - s_2)^2 + 2\sigma^2 \ln p(m_i = 1) \tag{4.14}$$

$$2r_j s_1 - s_1^2 + 2\sigma^2 \ln p(m_i = 0) \ge 2r_j s_2 - s_2^2 + 2\sigma^2 \ln p(m_i = 1) \tag{4.15}$$

$$2r_j s_1 - 2r_j s_2 \geq s_1^{\,2} - s_2^{\,2} + 2\sigma^2 \ln \frac{p(m_i = 1)}{p(m_i = 0)} \tag{4.16}$$

$$r_j \geq r_0 = \frac{s_1 + s_2}{2} + \frac{\sigma^2}{s_1 - s_2} \ln \frac{p(m_i = 1)}{p(m_i = 0)}. \tag{4.17}$$

Thus, we obtained (4.17) as a simple decision rule. When the priori probabilities $\left(p(m_i = 1) \text{ and } p(m_i = 0)\right)$ are identical, (4.17) depends on the likelihood and becomes simpler as follows:

$$r_j \geq \frac{s_1 + s_2}{2}. \tag{4.18}$$

When we have $\sigma^2 = 1, s_1 = -1,$ and $s_2 = 1,$ the likelihood, $P(r_j \mid s_i),$ can be illustrated in Figure 4.4. The ML decision rule is that $\widehat{m_i} = 1$ if r_j is greater than 0 and $\widehat{m_i} = 0$ otherwise. Now, we find the probability of error for the MAP decision rule. When s_1 is transmitted, we decide $r_j > r_0$. Likewise, when s_2 is transmitted, we decide $r_j \leq r_0$. Thus, the probability of error is

$$P_\varepsilon = p(r_j > r_0 \mid s_1) p(m_i = 0) + p(r_j \leq r_0 \mid s_2) p(m_i = 1). \tag{4.19}$$

The likelihood terms falling on the incorrect place can be expressed as follows:

$$p(r_j > r_0 \mid s_1) = \int_{r_0}^{\infty} \frac{1}{\sigma \sqrt{2\pi}} \exp\left(-\frac{(r_j - s_1)^2}{2\sigma^2}\right) dr. \tag{4.20}$$

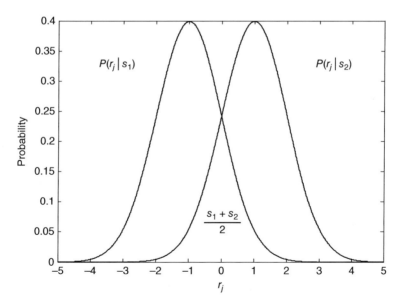

Figure 4.4   Likelihood for AWGN

Let

$$u = \frac{r_j - s_1}{\sigma}.$$ (4.21)

Then,

$$p(r_j > r_0 \mid s_1) = \int_{(r_0 - s_1)/\sigma}^{\infty} \frac{1}{\sqrt{2\pi}} \exp\left(-\frac{u^2}{2}\right) du = Q\left(\frac{r_0 - s_1}{\sigma}\right)$$ (4.22)

where $Q(x)$ is called the *complementary error function*. It is defined as follows:

$$Q(x) = \int_{x}^{\infty} \frac{1}{\sqrt{2\pi}} \exp\left(-\frac{u^2}{2}\right) du.$$ (4.23)

Likewise,

$$p(r_j \le r_0 \mid s_2) = \int_{-\infty}^{r_0} \frac{1}{\sigma\sqrt{2\pi}} \exp\left(-\frac{(r_j - s_2)^2}{2\sigma^2}\right) dr$$ (4.24)

$$p(r_j \le r_0 \mid s_2) = \int_{-\infty}^{(r_0 - s_2)/\sigma} \frac{1}{\sqrt{2\pi}} \exp\left(-\frac{u^2}{2}\right) du = 1 - Q\left(\frac{r_0 - s_2}{\sigma}\right)$$ (4.25)

$$p(r_j \le r_0 \mid s_2) = Q\left(\frac{s_2 - r_0}{\sigma}\right).$$ (4.26)

Therefore, we obtain the following equation from (4.19), (4.22), and (4.26):

$$P_\varepsilon = Q\left(\frac{r_0 - s_1}{\sigma}\right) p(m_i = 0) + Q\left(\frac{s_2 - r_0}{\sigma}\right) p(m_i = 1).$$ (4.27)

When the priori probabilities $\left(p(m_i = 1)\text{ and }p(m_i = 0)\right)$ are identical, we get

$$r_0 = \frac{s_1 + s_2}{2}$$ (4.28)

and manipulate (4.28) as follows:

$$\frac{r_0 - s_1}{\sigma} = \frac{s_2 - s_1}{2\sigma}$$ (4.29)

$$\frac{s_2 - r_0}{\sigma} = \frac{s_2 - s_1}{2\sigma}.$$ (4.30)

Therefore, the probability of error for the ML decision is

$$P_\varepsilon = Q\left(\frac{s_2 - s_1}{2\sigma}\right).$$ (4.31)

Example 4.2   Probability of error

Consider a simple communication system with the following system parameters:

	MAP decision	ML decision
$s_1 = -1$, $s_2 = 1$, and $\sigma^2 = 1$	$p(m_i = 0) = \dfrac{1}{3}$	$p(m_i = 0) = p(m_i = 1)$
	$p(m_i = 1) = \dfrac{2}{3}$	

Calculate the probability of error for the MAP and ML decision.

Solution

From (4.17),

$$r_0 = \frac{s_1 + s_2}{2} + \frac{\sigma^2}{s_1 - s_2} \ln \frac{p(m_i = 1)}{p(m_i = 0)} = -\frac{1}{2} \ln 2 = -0.3466.$$

From (4.27), the probability of error for the MAP decision is

$$P_{\varepsilon}^{MAP} = Q\left(\frac{r_0 - s_1}{\sigma}\right) p(m_i = 0) + Q\left(\frac{s_2 - r_0}{\sigma}\right) p(m_i = 1)$$

$$P_{\varepsilon}^{MAP} = Q(-0.3466 + 1)\frac{1}{3} + Q(1 + 0.3466)\frac{2}{3} \approx 0.145.$$

From (4.31), the probability of error for the ML decision is

$$P_{\varepsilon}^{ML} = Q\left(\frac{s_2 - s_1}{2\sigma}\right) = Q(1) = 0.1587. \qquad \blacksquare$$

Now, we consider a continuous time waveform channel as shown in Figure 4.5 and find an optimal receiver.

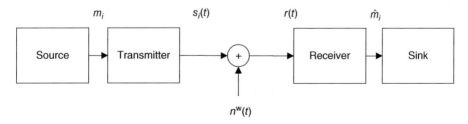

Figure 4.5   System model with waveform channel

In the waveform channel, the Gaussian noise, $n^w(t)$, is generated by a random process. The autocorrelation function is expressed as follows:

$$R_n(\tau) = \frac{N_0}{2}\delta(\tau) \ (\mathrm{W}) \tag{4.32}$$

and the power spectral density is

$$S_n(f) = \frac{N_0}{2} \ (\mathrm{W}/\mathrm{Hz}). \tag{4.33}$$

The received waveform, $r(t)$, is expressed as follows:

$$r(t) = s_i(t) + n^w(t) \tag{4.34}$$

where the signal, $s_i(t)$, can be assembled by a linear combination of N orthonormal basis function, $\phi_j(t)$, as follows:

$$s_i(t) = \sum_{j=1}^{N} s_{ij}\phi_j(t), \quad 0 \le t \le T \text{ and } i = 1,2,\dots,M \tag{4.35}$$

$$s_{ij} = \int_0^T s_i(t)\phi_j^*(t)dt, \quad i = 1,2,\dots,M \text{ and } j = 1,2,\dots,N. \tag{4.36}$$

The orthonormal basis functions, $\phi_j(t)$, are orthonormal as follows:

$$\int_0^T \phi_i(t)\phi_j(t)dt = \delta_{ij} = \begin{cases} 1, & i = j \\ 0, & i \ne j \end{cases}. \tag{4.37}$$

The coefficients, s_{ij}, are the projection of the waveform, $s_i(t)$, on the orthonormal basis and the signal set, $s_{ij}\phi_j(t)$, can be represented in an N-dimensional space geometrically. We call this *signal space* [2].

Example 4.3   Geometric representation of a signal
Consider a signal with the following parameters:

$N = 2$	$\phi_1(t) = \sqrt{\dfrac{2}{T}}\cos(2\pi f_0 t)$ and $\phi_2(t) = \sqrt{\dfrac{2}{T}}\sin(2\pi f_0 t)$
	$s_{11} = 0, s_{12} = \sqrt{E_s}$
$M = 3$	$s_{21} = \sqrt{E_s}, s_{22} = 0$
	$s_{31} = \sqrt{E_s}, s_{32} = -\sqrt{E_s}$
$s_i(t)$	$\sum_{j=1}^{2} s_{ij}\phi_j(t), i = 1,2,3$

Represent the signal, $s_i(t)$, geometrically.

Solution

The signal, $s_i(t)$ can be expressed as follows:

$$s_i(t) = s_{12}\phi_2(t) + s_{21}\phi_1(t) + s_{31}\phi_1(t) + s_{32}\phi_2(t)$$

$$= \sqrt{\frac{2E_s}{T}}\sin(2\pi f_0 t) + \sqrt{\frac{2E_s}{T}}\cos(2\pi f_0 t) + \sqrt{\frac{2E_s}{T}}\cos(2\pi f_0 t) - \sqrt{\frac{2E_s}{T}}\sin(2\pi f_0 t)$$

$$= \sqrt{\frac{8E_s}{T}}\cos(2\pi f_0 t).$$

Its geometric representation is illustrated in Figure 4.6. We can observe that signal elements are orthogonal each other. In the figure, the signal, $s_i(t)$, is represented as a bold arrow. ∎

Now, we discuss several operations in the signal space. The length of a vector, s_i, is defined as follows:

$$\|s_i\|^2 = \langle s_i, s_i \rangle = s_i^T s_i = \sum_{j=1}^{N} s_{ij}^2, \quad i = 1, 2, \dots, M. \tag{4.38}$$

If two vectors, s_i and s_k, are orthogonal, $\langle s_i, s_k \rangle = 0$. For n-tuples signal space, $\mathbb{R}^n$, the inner product of two vectors, s_i and s_k, is defined as follows:

$$\langle s_i, s_k \rangle = s_i^T s_k = \sum_{j=1}^{N} s_{ij} s_{kj}, \quad i, k = 1, 2, \dots, M. \tag{4.39}$$

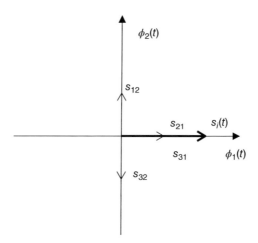

Figure 4.6   Geometric representation of the signal, $s_i(t)$

The energy of the waveform, $s(t)$, is expressed by the length of a vector as follows:

$$E_i = \int_0^T s_i(t)^2\, dt \tag{4.40}$$

$$E_i = \int_0^T \left[\sum_{j=1}^N s_{ij}\phi_j(t)\right]\left[\sum_{l=1}^N s_{il}\phi_l(t)\right] dt \tag{4.41}$$

$$E_i = \sum_{j=1}^N \sum_{l=1}^N s_{ij}s_{il} \int_0^T \phi_j(t)\phi_l(t)\, dt \tag{4.42}$$

$$E_i = \sum_{j=1}^N \sum_{l=1}^N s_{ij}s_{il}\delta_{jl} = \sum_{j=1}^N s_{ij}s_{ij} = \sum_{j=1}^N s_{ij}^2 = \|\mathbf{s}_i\|^2 . \tag{4.43}$$

The distance from 0 to a vector, $\mathbf{s}_i$, is $d_i = \|\mathbf{s}_i\| = \sqrt{\langle \mathbf{s}_i, \mathbf{s}_i \rangle}$. The distance between two vectors, $\mathbf{s}_i$ and $\mathbf{s}_k$, is

$$d_{ik} = \|\mathbf{s}_i - \mathbf{s}_k\| = \sqrt{\langle \mathbf{s}_i, \mathbf{s}_k \rangle} \tag{4.44}$$

$$d_{ik} = \sqrt{\sum_{j=1}^N (s_{ij} - s_{kj})^2} . \tag{4.45}$$

The angle, θ_{ik}, between two vectors, $\mathbf{s}_i$ and $\mathbf{s}_k$, is expressed as follows:

$$\cos\theta_{ik} = \frac{\langle \mathbf{s}_i, \mathbf{s}_k \rangle}{\|\mathbf{s}_i \mathbf{s}_k\|} . \tag{4.46}$$

We consider the vector, $\mathbf{s}_i$, as the sum of two vectors as follows:

$$\mathbf{s}_i = \mathbf{s}_{i\|\mathbf{s}_k} + \mathbf{s}_{i\perp\mathbf{s}_k} \tag{4.47}$$

where $\mathbf{s}_{i\|\mathbf{s}_k}$ is collinear with $\mathbf{s}_k$ and $\mathbf{s}_{i\perp\mathbf{s}_k}$ is orthogonal to $\mathbf{s}_k$ as shown in Figure 4.7.

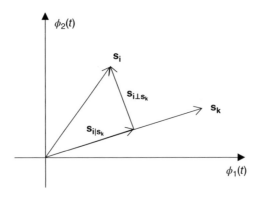

Figure 4.7   One dimensional projection

The vector, $s_{i|s_k}$, is called the projection of s_i on s_k. It is expressed as follows:

$$s_{i|s_k} = \frac{\langle s_i, s_k \rangle}{\|s_k\|^2} s_k.$$

(4.48)

We can now define *Gram–Schmidt orthogonalization process* [3] using (4.48) as follows:

Definition 4.3   Gram–Schmidt orthogonalization process

For *n*-tuples signal space, $\mathbb{R}^n$, *Gram–Schmidt orthogonalization process* is to find an orthogonal set, $S' = \{g_1, g_2, \ldots, g_n\}$, from a linearly independent set, $S = \{s_1, s_2, \ldots, s_n\}$. It is defined as follows:

$$g_i = s_i - \sum_{j=1}^{n-1} \frac{\langle s_i, g_j \rangle}{\|g_j\|^2} g_j$$

where $g_1 = s_1$, $j < i$ and $2 \le i \le n$.

The normalized vector, e_i, is expressed as follows:

$$e_i = \frac{g_i}{\|g_i\|}.$$

(4.49)

We call this vectors *Gram–Schmidt orthonormalization*.

Example 4.4   Gram–Schmidt process
Consider a signal set, **S**, with the following parameters:

$N=2$	$\phi_1(t) = \sqrt{\frac{2}{T}} \cos(2\pi f_0 t)$ and $\phi_2(t) = \sqrt{\frac{2}{T}} \sin(2\pi f_0 t)$
$M=2$	$s_{11} = 0, s_{12} = \sqrt{E_s}$
	$s_{21} = 2\sqrt{E_s}, s_{22} = -\sqrt{E_s}$
S	$\{s_1, s_2\} = \left\{ \left(0, \sqrt{E_s}\right), \left(2\sqrt{E_s}, -\sqrt{E_s}\right) \right\}$

Find the orthogonal set, $S' = \{g_1, g_2\}$ using Gram–Schmidt orthogonalization process.

Solution
From Definition 4.3, the first vector, g_1, of the orthogonal set, S', can be obtained as follows:

$$g_1 = s_1 = \left(0, \sqrt{E_s}\right).$$

The second vector, g_2, can be obtained as follows:

$$g_2 = s_2 - \sum_{j=1}^{1} \frac{s_2, g_j}{g_j^2} g_j = s_2 - \frac{s_2, g_1}{g_1^2} g_1$$

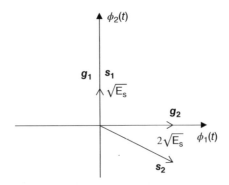

Figure 4.8   Geometric representation of the signals in S and S'

$$\mathbf{g}_2 = \left(2\sqrt{E_s}, -\sqrt{E_s}\right) - \frac{\left(2\sqrt{E_s}, -\sqrt{E_s}\right) \cdot \left(0, \sqrt{E_s}\right)}{\left(0, \sqrt{E_s}\right)^2}\left(0, \sqrt{E_s}\right)$$

$$\mathbf{g}_2 = \left(2\sqrt{E_s}, -\sqrt{E_s}\right) + \left(0, \sqrt{E_s}\right) = \left(2\sqrt{E_s}, 0\right).$$

Therefore, we can obtain the orthogonal set S' as follows:

$$S' = \{\mathbf{g}_1, \mathbf{g}_2\} = \left\{\left(0, \sqrt{E_s}\right), \left(2\sqrt{E_s}, 0\right)\right\}$$

The signals in S and S' can be represented geometrically as shown in Figure 4.8.   ∎

We now discuss the received signal. The received signal is expressed as follows:

$$r(t) = s_i(t) + n^w(t), \quad 0 \le t \le T \quad \text{and} \quad i = 1, 2, \ldots, M. \tag{4.50}$$

The vector representation of (4.50) is

$$\mathbf{R} = \mathbf{S}_i + \mathbf{N}^w \tag{4.51}$$

$$\begin{bmatrix} r_1 \\ r_2 \\ \vdots \\ r_N \end{bmatrix} = \begin{bmatrix} s_{i1} \\ s_{i2} \\ \vdots \\ s_{iN} \end{bmatrix} + \begin{bmatrix} N_1^w \\ N_2^w \\ \vdots \\ N_N^w \end{bmatrix} \tag{4.52}$$

and each element can be expressed as follows:

$$r_j = s_{ij} + N_j^w \tag{4.53}$$

$$\int_0^T r(t)\phi_j^*(t)\,dt = \int_0^T s_i(t)\phi_j^*(t)\,dt + \int_0^T n(t)\phi_j^*(t)\,dt \tag{4.54}$$

where $j = 1, 2, \ldots, N$. (4.54) means the received signal is assembled by the orthonormal basis functions, $\phi_j(t)$. The mean and variance of the received signal is expressed as follows:

$$\mu_{r_j} = \mathrm{E}\left[r_j\right] = \mathrm{E}\left[s_{ij} + N_j^w\right] = s_{ij} + \mathrm{E}\left[N_j^w\right] = s_{ij} \tag{4.55}$$

$$\sigma_{r_j}^2 = \mathrm{var}\left[r_j\right] = \mathrm{E}\left[\left(r_j - s_{ij}\right)^2\right] = \mathrm{E}\left[N_j^w\right] = \frac{N_0}{2}. \tag{4.56}$$

As we observed wireless channels in Chapter 3, the transmitted signals are experienced in different channels so that the received signal, $r(t)$, can be composed of two received signals, $r_1(t)$ and $r_2(t)$, as shown in Figure 4.9.

We consider the received signal and its projection on the two-dimensional signal space as shown in Figure 4.10.

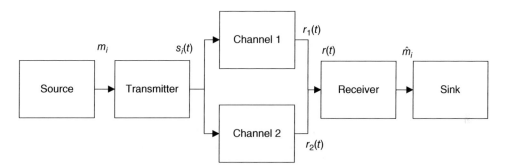

Figure 4.9   System model with two channels

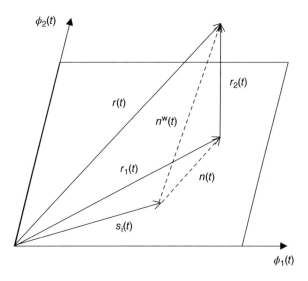

Figure 4.10   Received signal and its projection on the two-dimensional signal space

The received signal, $r(t)$ $(=r_1(t)+r_2(t))$, is composed of the transmitted signal, $s_i(t)$, and the noise, $n^w(t)$. The projection of the received signal, $r(t)$, on the two-dimensional signal space is represented as $r_1(t)$ which is composed of the transmitted signal, $s_i(t)$, and the noise, $n(t)$. Therefore, the element, $r_2(t)$, of the received signal, $r(t)$, is irrelevant to estimate the transmitted signal, $s_i(t)$ and an optimum receiver can be designed through investigating $r_1(t)$. We call this *theorem of irrelevance*.

4.3 Matched Filter Receiver

The optimum receiver is composed of mainly two parts. The first part is to decide the receive signal vector

$$\mathbf{r} = (r_1, r_2, ..., r_N) \tag{4.57}$$

where

$$r_j = \int_0^T r(t)\phi_j(t)\,dt, \quad j=1,2,...,N. \tag{4.58}$$

The second part is to determine $\hat{m}$ to minimize the probability of error. Namely, the decision rule

$$\|\mathbf{r}-\mathbf{s}_i\|^2 = \sum_{j=1}^N (r_j - s_{ij})^2 = \sum_{j=1}^N r_j^2 - 2\sum_{j=1}^N r_j s_{ij} + \sum_{j=1}^N s_{ij}^2 \tag{4.59}$$

is to minimize the probability of error. In (4.59), the term, $\sum_{j=1}^N r_j^2$, is independent of index i $(i=1,2,...,M)$. Therefore, the decision rule is to maximize the following equation:

$$\sum_{j=1}^N r_j s_{ij} - \frac{E_i}{2}, \quad \text{where } E_i = \sum_{j=1}^N s_{ij}^2. \tag{4.60}$$

An optimum correlation receiver composed of a signal detector and a signal estimator is illustrated in Figures 4.11 and 4.12.

When the orthonormal basis function, $\phi_j(t)$, is zero outside a finite time interval, $0 \le t \le T$, we can replace a multiplier and an integrator by a matched filter and a sampler in the signal detector of an optimum correlation receiver. A matched filter and a sampler can be helpful for designing an optimum receiver easily because the accurate multiplier and integrator are not easy to implement. Consider a linear filter with an impulse response, $h_j(t)=\phi_j(T-t)$. When $r(t)$ is the input of the linear filter, the output of the linear filter can be described by

$$u_j(t) = \int_{-\infty}^{\infty} r(\tau)h_j(t-\tau)\,d\tau \tag{4.61}$$

$$u_j(t) = \int_{-\infty}^{\infty} r(\tau)\phi_j(T-t+\tau)\,d\tau. \tag{4.62}$$

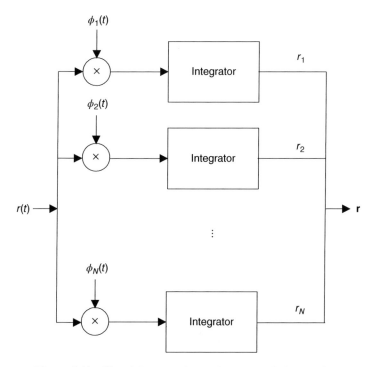

Figure 4.11   Signal detector of an optimum correlation receiver

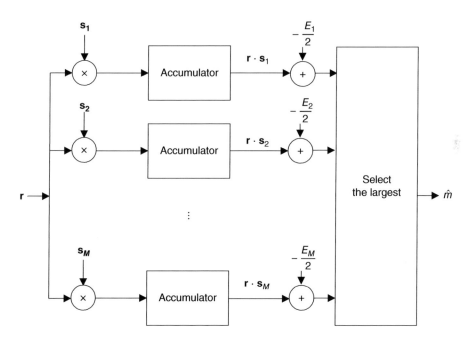

Figure 4.12   Signal estimator of an optimum correlation receiver

When the output of the linear filter is sampled at $t=T$,

$$u_j(T) = \int_{-\infty}^{\infty} r(\tau)\phi_j(\tau)\,d\tau = r_j. \qquad (4.63)$$

The impulse response of the linear filter is a delayed time-reversed version of the orthonormal basis function, $\phi_j(t)$, as shown in Figure 4.13.

We call this linear filter *matched* to $\phi_j(t)$. An optimum receiver with this linear filter is called *a matched filter receiver*. The signal detector of a matched filter receiver is illustrated in Figure 4.14.

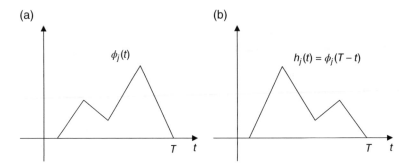

Figure 4.13   (a) Orthonormal bases function, $\phi_j(t)$, and (b) the impulse response, $h_j(t)$

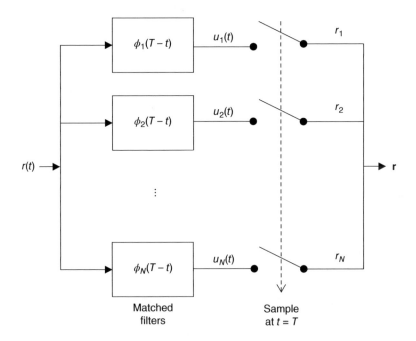

Figure 4.14   Signal detector of a matched filter receiver

4.4 Coherent and Noncoherent Detection

The matched filter receiver can be used for the coherent detection of modulation schemes. Consider a coherent M-ary Phase Shift Keying (MPSK) system. The transmitted signal, $s_i(t)$, is expressed as follows:

$$s_i(t) = \sqrt{\frac{2E_s}{T}} \cos\left(2\pi f_0 t - \frac{2\pi i}{M}\right) \qquad (4.64)$$

where $0 \le t \le T$, $i = 1,2,\ldots,M$, and E_s is the signal energy for symbol duration T. If we assume the two-dimensional signal space, the orthonormal basis function, $\phi_j(t)$ is

$$\phi_1(t) = \sqrt{\frac{2}{T}} \cos(2\pi f_0 t), \quad \phi_2(t) = \sqrt{\frac{2}{T}} \sin(2\pi f_0 t). \qquad (4.65)$$

Therefore, the transmitted signal, $s_i(t)$, is expressed as follows:

$$s_i(t) = s_{i1}\phi_1(t) + s_{i2}\phi_2(t) \qquad (4.66)$$

$$s_i(t) = \sqrt{E_s} \cos\left(\frac{2\pi i}{M}\right)\phi_1(t) + \sqrt{E_s} \sin\left(\frac{2\pi i}{M}\right)\phi_2(t). \qquad (4.67)$$

When considering Quadrature Phase Shift Keying (QPSK), $M=4$, the transmitted signal, $s_i(t)$, has four signals which is expressed by a combination of two orthonormal basis function $\phi_1(t)$ and $\phi_2(t)$. The transmitted signal, $s_i(t)$, is expressed as follows:

$$s_i(t) = \sqrt{E_s} \cos\left(\frac{\pi i}{2}\right)\phi_1(t) + \sqrt{E_s} \sin\left(\frac{\pi i}{2}\right)\phi_2(t) \qquad (4.68)$$

The transmitted signal, $s_i(t)$, can be represented in the two-dimensional signal space as shown in Figure 4.15.

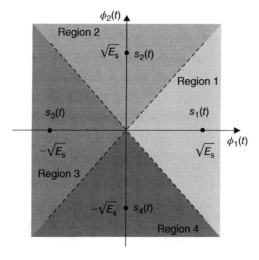

Figure 4.15 Two-dimensional signal space and decision regions for QPSK modulation

The two-dimensional signal space is divided into four regions. If the received signal, $r(t)$, falls in region 1, we decide the transmitter sent $s_1(t)$. If the received signal, $r(t)$, falls in region 2, we decide the transmitter sent $s_2(t)$. We decide $s_3(t)$ and $s_4(t)$ in the same way. The signal estimator in Figure 4.12 needs M correlators for the demodulation of MPSK. The received signal, $r(t)$, of MPSK system is expressed as follows:

$$r(t) = \sqrt{\frac{2E_s}{T}} \left(\cos\left(\frac{2\pi i}{M}\right) \cos(2\pi f_0 t) \sin\left(\frac{2\pi i}{M}\right) \sin(2\pi f_0 t) \right) + n(t) \qquad (4.69)$$

where $0 \le t \le T$, $i = 1, 2, \ldots, M$ and $n(t)$ is white Gaussian noise. The demodulator of MPSK system can be designed in a similar way to the matched filter receiver as shown in Figure 4.16.

The arctan calculation part is different from an optimum correlation receiver of Figure 4.12. In Figure 4.16, X is regarded as the in-phase part of the received signal, $r(t)$, and Y is regarded as the quadrature-phase part of the received signal, $r(t)$. The result, $\hat{\theta}_i$, of arctan calculation is compared with a prior value, θ_j. For QPSK modulation, it has four values $(0, \pi/2, \pi,$ and $3\pi/2$ in radian) as shown in Figure 4.15. We choose the smallest phase difference as the transmitted signal, $\hat{s}_i(t)$.

Unlike the coherent detection, the noncoherent detection cannot use the matched filter receiver because the receiver does not know a reference (such as the carrier phase and the carrier frequency) of the transmitted signal. The Differential Phase Shift Keying (DPSK) system is basically classified as noncoherent detection scheme but sometime it is categorized as differentially coherent detection scheme. Consider a DPSK system. The transmitted signal, $s_i(t)$, is expressed as follows:

$$s_i(t) = \sqrt{\frac{2E_s}{T}} \cos(2\pi f_0 t + \theta_i(t)) \qquad (4.70)$$

where $0 \le t \le T$ and $i = 1, 2, \ldots, M$. The received signal, $r(t)$, is expressed as follows:

$$r(t) = \sqrt{\frac{2E_s}{T}} \cos\left(2\pi f_0 t + \theta_i(t) + \theta_d\right) + n(t) \qquad (4.71)$$

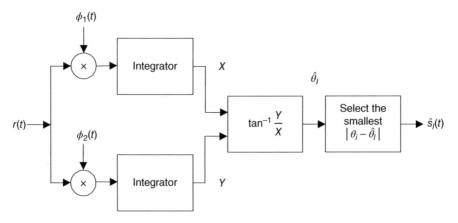

Figure 4.16   Demodulator of MPSK system

where $0 \leq t \leq T, i = 1, 2, \ldots, M, \theta_d$ is constant with a finite interval, $0 \leq \theta_d \leq 2\pi$, and $n(t)$ is white Gaussian noise. If the unknown phase, θ_d, varies relatively slower than two symbol durations $(2T)$, the phase difference between two consecutive received signal is independent of the unknown phase, θ_d as follows:

$$\left(\theta_{i+1}(T_{i+1}) + \theta_d\right) - \left(\theta_i(T_i) + \theta_d\right) = \theta_{i+1}(T_{i+1}) - \theta_i(T_i). \tag{4.72}$$

Therefore, we can use the carrier phase of the previous received signal as a reference. The modulator of binary DPSK is illustrated in Figure 4.17.

The binary message, m_i, and the delayed differentially encoded bits, d_{i-1}, generates the differentially encoded bits, d_i, by modulo 2 addition ($\oplus$). The differentially encoded bits, d_i, decide the phase shift of the transmitted signal. Table 4.1 describes an example of encoding process.

The example of the differentially encoded bits, d_i, and corresponding phase shift, $\theta_i(t)$, is illustrated in Figure 4.18.

When we detect the differentially encoded signal, we do not need to estimate a phase of carrier. Instead, we use the phase difference between the present signal phase and the previous signal phase as shown in Figure 4.19. In the DPSK demodulator, we need the orthonormal basis function, $\phi_j(t)$, as the reference carrier frequency but do not need a prior value, θ_i as the reference phase.

The DPSK modulation is simple and easy to implement because we do not need to synchronize the carrier. However, the disadvantage of the DPSK modulation is that one error of the received signal can propagate to other received signals detection because their decision is highly related.

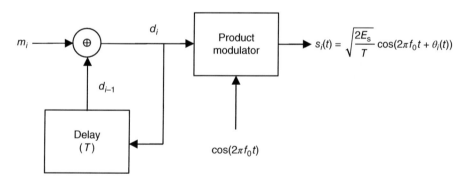

Figure 4.17 Modulator of DPSK system

Table 4.1 Example of binary DPSK encoding

i	0	1	2	3	4	5	6	7
m_i		1	0	1	0	0	1	1
d_{i-1}		1	0	0	1	1	1	0
d_i	1	0	0	1	1	1	0	1

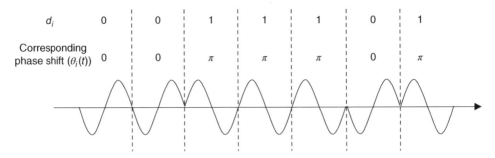

Figure 4.18 Example of phase shift and waveform of DPSK modulator

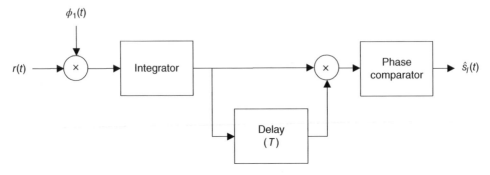

Figure 4.19 Demodulator of DPSK system

Summary 4.1 Optimum receiver

1. The receiver minimizing the probability of error is called an optimum receiver.
2. The optimum receiver is designed by maximizing a posteriori probability and the Maximum a Posteriori (MAP) decision rule is expressed as follows:

$$f_{\mathrm{MAP}}(r_j) = \arg\max_{m_i} \frac{P(m_i)P(r_j \mid m_i)}{P(r_j)}$$

where $P(m_i)$, $P(r_j|m_i)$, and $P(r_j)$ are a priori probability, a conditional probability, and an occurrence, respectively.
3. The optimum receiver relies on the likelihood and the Maximum likelihood (ML) decision rule is expressed as follows:

$$f_{\mathrm{ML}}(r_j) = \arg\max_{m_i} P(r_j \mid m_i).$$

4. A signal can be assembled by a linear combination of an N orthonormal basis function. The signal set can be represented in an N-dimensional space geometrically. We call this signal space.
5. The coherent detection needs expensive and complex carrier recovery circuit but has good performance of detection.
6. The noncoherent detection does not require expensive and complex carrier recovery circuit but has poor performance of detection. For noncoherent detection, differential technique is widely used.

4.5 Problems

4.1. Consider a simple communication system as shown in Figure 4.2 and a priori probability, $P(m_i)$, and a conditional probability, $P(r_j|s_i)$, have the following probabilities:

m_i	$P(m_i)$	$P(r_0|s_i)$	$P(r_1|s_i)$	$P(r_2|s_i)$
0	0.4	0.2	0.3	0.5
1	0.6	0.7	0.1	0.2

Design the decision rule with the minimum error probability.

4.2. Consider a simple communication system with the following system parameters:

	MAP decision	ML decision
$s_1=-1$, $s_2=1$, and $\sigma^2=1$	$p(m_i=0)=\dfrac{1}{4}$ $p(m_i=1)=\dfrac{3}{4}$	$p(m_i=0)=p(m_i=1)$

Calculate the probability of error for MAP and ML decision.

4.3. Consider a signal with the following parameters:

$N=2$	$\phi_1(t)=\sqrt{\dfrac{2}{T}}\cos(2\pi f_0 t)$ and $\phi_2(t)=\sqrt{\dfrac{2}{T}}\sin(2\pi f_0 t)$
	$s_{11}=0, s_{12}=2\sqrt{E_s}$
$M=3$	$s_{21}=-\sqrt{E_s}, s_{22}=0$
	$s_{31}=3\sqrt{E_s}, s_{32}=\sqrt{E_s}$
$s_i(t)$	$\sum_{j=1}^{2}s_{ij}\phi_j(t), i=1,2,3$

Represent the signal, $s_i(t)$, geometrically.

4.4. Consider a signal set, S, with the following parameters:

$N=2$	$\phi_1(t)=\sqrt{\dfrac{2}{T}}\cos(2\pi f_0 t)$ and $\phi_2(t)=\sqrt{\dfrac{2}{T}}\sin(2\pi f_0 t)$
	$s_{11}=-\sqrt{E_s}, s_{12}=\sqrt{E_s}$
$M=2$	$s_{21}=\sqrt{E_s}, s_{22}=-3\sqrt{E_s}$
S	$\{s_1,s_2\}=\left\{\left(-\sqrt{E_s},\sqrt{E_s}\right),\left(\sqrt{E_s},-3\sqrt{E_s}\right)\right\}$

Find the orthogonal set, $S'=\{g_1,g_2\}$ using Gram–Schmidt orthogonalization process.

4.5. A user needs to buy a new mobile phone but it is not easy to find good mobile phone $p(m_{good}) = 0.1$ and $p(m_{bad}) = 0.9$. Thus, the user decides to observe a gadget review website. The range of the grade is from 0 to 100. The mobile user's estimation of the possible losses is given in the following table and the class conditional probability densities are known to be approximated by normal distribution as $p(\text{grade} \mid m_{good}) \sim N(80,3)$ and $p(\text{grade} \mid m_{bad}) \sim N(45,10)$. Find the grade to minimize the risk.

	$\alpha(\text{decision}, m_{good})$	$\alpha(\text{decision}, m_{bad})$
Purchase	0	30
No purchase	5	0

4.6. Consider BPSK signals with $s_1 = 1$ and $s_2 = -1$. When the prior probabilities are $p(s_1) = p$ and $p(s_2) = 1 - p$, find the metrics for MAP and ML detector in AWGN.

4.7. When we transmit the signals in successive symbol intervals, each signal is interdependent and the receiver is designed by observation of the received sequence. Design the maximum likelihood sequence detector.

4.8. Consider three signals: $s_1(t) = \cos(2\pi ft)$, $s_2(t) = \sin(2\pi ft + \pi/2)$, and $s_1(t) = \cos(2\pi ft + \pi/2)$. Check whether or not they are linearly independent.

4.9. The digital message (0 or 1) can be encoded as a variation of the amplitude, phase, and frequency of a sinusoidal signal. For example, binary amplitude shift keying (BASK), binary phase shift keying (BPSK), and binary frequency shift keying (BFSK). Draw their waveform and express in signal space.

4.10. Draw the decision regions of the minimum distance receiver for 16QAM.

4.11. Compare the bit error probabilities for the following binary systems: coherent detection of BPSK, coherent detection of differentially encoded BPSK, noncoherent detection of orthogonal BFSK.

4.12. Describe the relationship between E_b/N_0 and SNR.

References

[1] E. T. Jaynes, "Bayesian Methods: General Background," In J. H. Justice (ed.), *Maximum-Entropy and Bayesian Methods in Applied Statistics*, Cambridge University Press, Cambridge, pp. 1–15, 1986.
[2] J. M. Wozencraft and I. M. Jacobs, *Principles of Communication Engineering*, John Wiley & Sons, Inc., New York, 1965.
[3] G. Arfken "Gram-Schmidt Orthogonalization," In *Mathematical Methods for Physicists*, Academic Press, Orlando, FL, 3rd edition, pp. 516–520, 1985.

5

Wireless Channel Impairment Mitigation Techniques

We realized wireless channels cause many types of wireless channel impairments such as noise, path loss, shadowing, and fading, and wireless communication systems should be designed to overcome these wireless channel impairments. There are many techniques to mitigate wireless channel impairments. For example, for the purpose of managing shadowing effects, the location of a base station is carefully selected to avoid shadowing effects. A power control scheme is also used to make a good radio link between a mobile station and a base station. For the purpose of mitigating delay spreads, Global System for Mobile Communications (GSM) system uses adaptive channel equalization techniques and Code Division Multiple Access (CDMA) system uses a rake receiver. For the purpose of mitigating the Doppler spreads, the signal bandwidth of GSM is designed to be much greater than the Doppler spread. GSM also uses a frequency correction feedback. CDMA system uses pilot channels for channel estimation. In Long-Term Evolution (LTE), a symbol time is designed to be less than channel coherence time. In Wireless Local Area Network (WLAN), the Doppler spread is not a big problem due to a low mobility. Thus, the wireless channel impairment mitigation techniques should be adopted according to system requirements and channel environments. In this chapter, fundamentals of wireless channel impairment mitigation techniques are investigated.

5.1 Diversity Techniques

Diversity techniques mitigate multipath fading effects and improve the reliability of a signal by utilizing multiple received signals with different characteristics. These are very important techniques in modern telecommunications to compensate for multipath fading effects. Diversity techniques are based on the fact that individual channels experience different levels

Wireless Communications Systems Design, First Edition. Haesik Kim.
© 2015 John Wiley & Sons, Ltd. Published 2015 by John Wiley & Sons, Ltd.

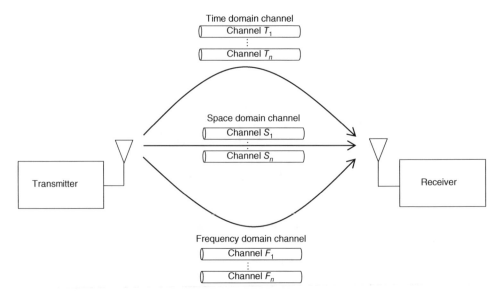

Figure 5.1   Diversity channels

of fading effects. Assume we transmit and receive same signals several times. If one received signal experiences deep fading at a specific part, another received signal may not experience deep fading and another received signal may experience a good channel at a specific part. From the probability point of view, this approach makes sense. If the probability that one received signal experiences deep fading at a specific part in one channel is p, the probability that the other received signals experience deep fading at a specific part in N channels is p^N. Therefore, the more channels we can use, the more diversity gains we can obtain. There are several types of diversity techniques in space, time, and frequency domain as shown in Figure 5.1. Their combination is possible as well. For example, space-frequency diversity, space-time diversity, space-time-frequency diversity. We will take a look into three fundamental diversity techniques in this section.

Space diversity uses multiple antennas and is classified into macroscopic diversity and microscopic diversity. Macroscopic diversity mitigates large-scale fading caused by log normal fading and shadowing. To achieve macroscopic diversity, antennas are spaced far enough and we select an antenna which is not shadowed. Thus, we can increase the signal to noise ratio. Microscopic diversity mitigates small-scale fading caused by multipath. To achieve microscopic diversity, a multiple antenna technique is used as well and an antenna is selected to have a signal with small fading effect. One example of space diversity configuration is illustrated in Figure 5.2.

Two different wireless channels are derived from the receiver with two antennas. The distance between two antennas is important to achieve the diversity gain. Basically, we expect that deep fading does not occur on both wireless channel 1 and wireless channel 2 at the same time. It is shown in Figure 5.3 that microscopic diversity improves the fading channel characteristic. When a transmitted signal arrives at a receiver via two different wireless channels, each signal experiences several deep fading. However, if we use a space diversity technique based on averaging two signals, we can obtain a better signal characteristic as shown in the average of

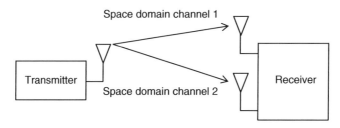

Figure 5.2 Space diversity

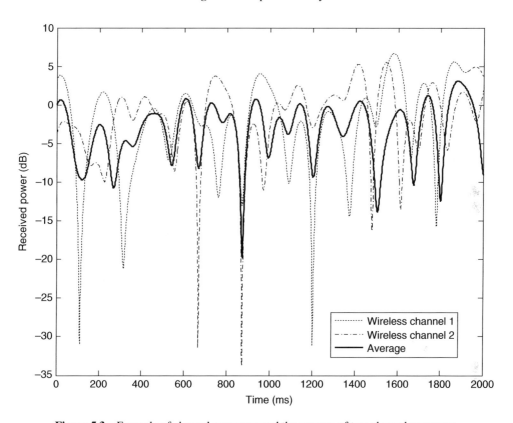

Figure 5.3 Example of channel responses and the average of two channel responses

two channel responses. Multiple Input Multiple Output (MIMO) techniques are based on space diversity. The MIMO will be discussed in Section 5.3 and Chapter 8 in detail.

Time diversity uses different time slots. Basically, consecutive signals are highly correlated in wireless channels. Thus, a time diversity technique transmits same signal sequences in different time slots. The time sequence difference should be larger than the channel coherence time. An interleaving technique is one of time diversity techniques. Figure 5.4 illustrates one example of an interleaving technique. Consider 4 codewords (aaaa, bbbb, cccc, and dddd) transmission and a deep fading channel in time slot 3 (T_3). When there is no interleaving, the codeword, cccc, in T_3 is erased due to a deep fading effect. However, if we shuffle the

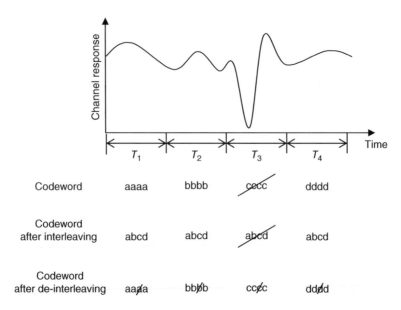

Figure 5.4   Time diversity

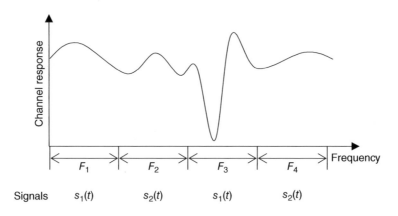

Figure 5.5   Frequency diversity

codewords and transmit them, we lose a part of the codeword, cccc, as shown in Figure 5.4. After de-interleaving, we can obtain partially erased codewords (aaa, bbb, ccc, and ddd) and recover the original codewords.

Frequency diversity uses different frequency slots. It transmits a signal through different frequencies or spreads it over a wide frequency spectrum. Frequency diversity is based on the fact that the fading effect is differently appeared in different frequencies separated by more than the channel coherence bandwidth. When the transmission bandwidth is greater than the channel coherence bandwidth (namely, it is a broadband system), the frequency diversity technique is useful. Figure 5.5 illustrates one example of frequency diversity. Consider we transmit two signals, $s_1(t)$ and $s_2(t)$, and there is a deep fading effect in frequency slot 3 (F_3).

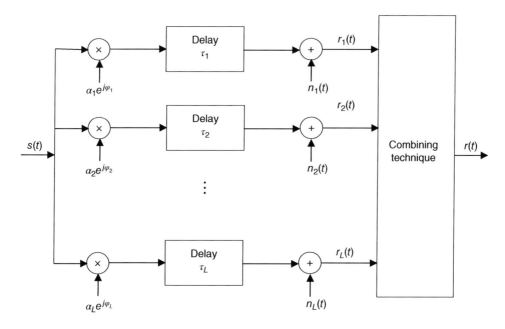

Figure 5.6   System model of diversity techniques with L channels

When we send $s_1(t)$ through frequency slot 1 and 3 (F_1 and F_3), the signal, $s_1(t)$, in frequency slot 3 (F_3) experiences a deep fading effect but the other signal, $s_1(t)$, in frequency slot 1 (F_1) does not experience a deep fading effect. If we use a combining technique or select the signal, $s_1(t)$, in frequency slot 1 (F_1) in the receiver, we can obtain the original signal.

We allocate each frequency to be larger than the channel coherence bandwidth. Frequency diversity gain can be obtained in a broadband multicarrier system. The symbol of the broadband multicarrier system is divided into narrowband subcarriers. The symbol has a long transmission period. Frequency selective fading may occur in the long transmission period but each subcarrier which is separated enough experiences flat fading.

There are several combining techniques for diversity: Maximal Ratio Combining (MRC), Equal Gain Combining (EGC), and Selection Combining (SC). The system model of diversity techniques is illustrated in Figure 5.6 to derive combining techniques. The signal, $s(t)$, is transmitted through L different channels. The each received signal, $r_l(t)$, through different channels is represented by

$$r_l(t) = g_l s(t) + n_l(t) \tag{5.1}$$

where $g_l = \alpha_l e^{j\varphi_l}$ ($l = 1, 2, \ldots, L$) represents the channel gain (α_l) and phase rotation (φ_l) and $n_l(t)$ is the Gaussian noise.

The MRC technique is a method combining all signals using weighting factors (amplitude and phase) to achieve a high SNR as shown in Figure 5.7.

In the MRC technique, the received signal, $r(t)$, is weighted by w_l and is represented by

$$r(t) = \sum_{l=1}^{L} w_l r_l(t) \tag{5.2}$$

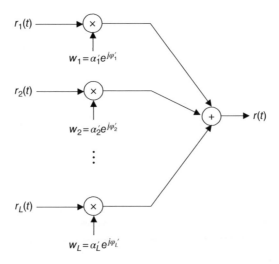

Figure 5.7   Maximal ratio combining

$$r(t) = \sum_{l=1}^{L} w_l \left(g_l s(t) + n_l(t) \right) \tag{5.3}$$

$$r(t) = \sum_{l=1}^{L} w_l g_l s(t) + \sum_{l=1}^{L} w_l n_l(t). \tag{5.4}$$

The SNR, γ, of the received signal, $r(t)$, is represented by

$$\gamma = \frac{E_s \left| \sum\limits_{l=1}^{L} w_l g_l \right|^2}{N_0 \sum\limits_{l=1}^{L} |w_l|^2}. \tag{5.5}$$

We use the Schwartz's inequality and find w_l to maximize γ as follows:

$$\left| \sum_{l=1}^{L} w_l g_l \right|^2 \leq \left(\sum_{l=1}^{L} |w_l|^2 \right) \left(\sum_{l=1}^{L} |g_l|^2 \right). \tag{5.6}$$

The equality holds if $w_l = K g_l^*$ for all l, where K is an arbitrary complex constant. Thus, we obtain the upper bound of γ as follows:

$$\gamma \leq \frac{\left(\sum\limits_{l=1}^{L} |w_l|^2 \right) \left(\sum\limits_{l=1}^{L} |g_l|^2 \right)}{\sum\limits_{l=1}^{L} |w_l|^2} \frac{E_s}{N_0} \tag{5.7}$$

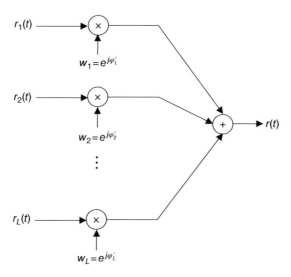

Figure 5.8   Equal gain combining

$$\gamma \leq \left(\sum_{l=1}^{L}|g_l|^2\right)\frac{E_s}{N_0} \tag{5.8}$$

$$\gamma_{\max} = \left(\sum_{l=1}^{L}|g_l|^2\right)\frac{E_s}{N_0} = \left(\sum_{l=1}^{L}\alpha_l^2\right)\frac{E_s}{N_0} \tag{5.9}$$

and the maximum SNR, $\gamma_{\max}$, can be found when we decide $w_l = g_l^*$.

The EGC technique is a method combining all signals using phase estimation and unitary weight to achieve a high SNR as shown in Figure 5.8.

In the EGC technique, the received signal, $r(t)$, is weighted by w_l and is represented by

$$r(t) = \sum_{l=1}^{L}w_l g_l s(t) + \sum_{l=1}^{L}w_l n_l(t) \tag{5.10}$$

$$r(t) = \sum_{l=1}^{L}\alpha_l e^{j(\varphi_l + \varphi_l')}s(t) + \sum_{l=1}^{L}e^{j\varphi_l'}n_l(t). \tag{5.11}$$

If we set $\varphi_l' = -\varphi_l$, the received signal, $r(t)$, is represented by

$$r(t) = \sum_{l=1}^{L}\alpha_l s(t) + \sum_{l=1}^{L}e^{-j\varphi_l}n_l(t). \tag{5.12}$$

Therefore, the SNR, γ, of the received signal, $r(t)$, is represented by

$$\gamma = \frac{E_s\left|\sum_{l=1}^{L}\alpha_l\right|^2}{N_0 L}. \tag{5.13}$$

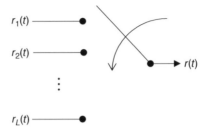

Figure 5.9   Selection combining

The SC technique is the simplest method and selects one signal with the strongest SNR as shown in Figure 5.9.

The SNR, γ, of the received signal, $r(t)$, is represented by

$$\gamma = \max\left(\gamma_1, \gamma_2, ..., \gamma_L\right). \tag{5.14}$$

Among three combining techniques, the SC technique has the lowest complex and the MRC technique has the best performance. However, in case of some specific channel condition such as deep fading, the SC technique may have a better performance because it can avoid the worst channel.

Summary 5.1   Diversity techniques

1. Diversity techniques are based on the fact that individual channels experience different levels of fading.
2. Space diversity uses multiple antennas. Macroscopic diversity mitigates large-scale fading caused by log normal fading and shadowing. Microscopic diversity mitigates small-scale fading caused by multipath fading.
3. Time diversity uses different time slots. Time diversity technique transmits same signal sequence at different time slots. An interleaving technique is one of time diversity techniques.
4. Frequency diversity uses different frequency slots and transmits a signal through different frequencies. Frequency diversity is based on the fact that the fading effect is differently appeared in different frequencies separated by more than the channel coherence bandwidth.
5. There are several combining techniques for diversity such as MRC, EGC, and SC.

5.2   Error Control Coding

The landmark paper "A Mathematical Theory of Communication" [1.4] describes one important concept about error control coding theorem. In the paper, C. Shannon predicts that it is possible to transmit information without errors over the unreliable channel and an error control coding technique exists to achieve this. In modern wireless communication systems,

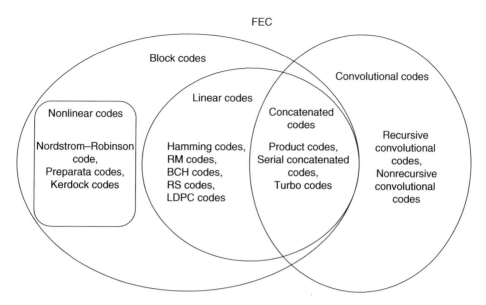

Figure 5.10   Classification of forward error correction

the error control coding technique became an essential part. The error control coding technique can be classified into *Forward Error Correction* (FEC) and *Automatic Repeat reQuest* (ARQ). The FEC does not have a feedback channel. The transmitter sends codewords including information and redundancy. In the receiver side, an error correction coding technique recovers from the corrupted codewords. On the other hand, the ARQ has a feedback channel. The ARQ technique uses an error detection technique and noiseless feedback. When errors occur, the receiver does not correct but detect errors and then requests that the transmitter sends codewords again. There are many forward error correction schemes which can be roughly classified into block codes, convolution codes, and concatenated codes as shown in Figure 5.10.

Block codes are based on algebra and algebraic geometry and developed by mainly mathematicians. The construction and lower/upper bounds of the block codes are well analyzed mathematically. The block codes are split into linear codes and nonlinear codes. Basically, nonlinear codes such as Nordstrom-Robinson code, Preparata codes, and Kerdock codes have a better performance than linear codes. However, it is not easy to implement due to their nonlinear characteristic. Therefore, linear block codes are widely used in practical systems. In the 1970s and 1980s, many beautiful linear block codes such as Hamming codes, Reed-Muller (RM) codes, Bose-Chaudhuri-Hocquenghem (BCH) codes, and Reed-Solomon (RS) codes were developed and practically used in many digital devices. *Low-Density Parity Check* (LDPC) [1] codes are one of the important linear block codes in modern wireless communication systems. The LDPC codes were not recognized for about 30 years due to a high complexity encoder and decoder. Mackey and Neal reinvented the LDPC codes [2] as practical codes to closely approach Shannon limit using an iterative belief propagation technique. *Convolutional codes* are based on a finite state machine and probability theory. One decoding scheme for the convolutional codes is the sequential decoding algorithm by Wozencraft [3]. This decoding scheme is complex so that the convolutional codes were not used widely. In the 1960s, the optimum algorithm of the convolutional codes by Viterbi was invented. This decoding scheme became a de facto standard of many wireless communication systems. *Concatenated codes* are to combine two different codes. This combination brought an amazing result. Serial concatenated codes composed of a convolutional code as an inner code and an RS code as an outer code were widely used in deep space communications in the 1970s. In 1993, *turbo codes* [4] which are parallel concatenation of Recursive Systematic Convolutional (RSC) codes were invented by C. Berrou, B. Glavieux, and P. Thitimajshima. The performance of the turbo code closely approaches Shannon limit. It opens a new era of communication systems and affects many wireless communication techniques. Besides, there are different types of error correction coding techniques. The above coding schemes assume a binary symmetric channel because this channel model is suitable for digital communication systems. However, a binary erasure channel is more accurate in a wired channel. For this channel, foundation codes such as Luby transform (LT) codes [5] and Raptor codes [6] are invented and investigated actively. Polar codes [7] are a class of capacity achieving codes for symmetric Binary Discrete Memoryless Channels (B-DMC). Although LDPC codes and turbo codes are capacity achieving codes, we do not know clearly why they achieve Shannon limit and how to construct capacity achieving codes. However, Polar codes do not have any theoretical gaps and have a well-defined rule for code construction.

In this section, we deal with two important error correction coding techniques: linear block codes and convolutional codes. Firstly, we look into Hamming codes as a linear block code. The linear block codes are one important family of error correction coding techniques. It is well developed mathematically and is conceptually useful for understanding an error correction coding technique. In addition, it has a wide range of practical applications and especially LDPC codes are important codes in the modern wireless communication systems. We design LDPC codes in Chapter 6.

5.2.1   Linear Block Codes

Let the message/information $m = (m_0, m_1, \ldots, m_{k-1})$ be a binary k-tuple. The encoder for the (n, k) linear block code over the Galois field (GF) of two elements, GF(2), generates the codeword $c = (c_0, c_1, \ldots, c_{n-1})$ with a binary n-tuple $(n > k)$. Figure 5.11 illustrates an encoder for the (n, k) linear block code.

Definition 5.1   Linear block code

A linear block code over GF(2) with codeword length n and 2^k codewords is called an (n, k) linear block code if and only if its 2^k codewords form a k-dimensional subspace of the set of all binary n-tuples and the modulo-2 sum of two codewords among 2^k codewords is also one codeword among 2^k codewords.

We define an (n, k) linear block code as follows:

A generator matrix is used to construct a codeword and has a $k \times n$ matrix for an (n, k) linear block code. Let us consider a $k \times n$ generator matrix G as follows:

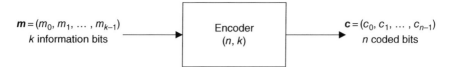

Figure 5.11   Encoder

$$G = \begin{bmatrix} g_0 \\ g_1 \\ \vdots \\ g_{k-1} \end{bmatrix} = \begin{bmatrix} g_{00} & g_{01} & \cdots & g_{0n-1} \\ g_{10} & g_{11} & & g_{1n-1} \\ \vdots & & \ddots & \vdots \\ g_{k-10} & g_{k-11} & \cdots & g_{k-1n-1} \end{bmatrix}. \quad (5.15)$$

Every codeword can be produced by an inner product of the message m and the generator matrix G. The corresponding codeword is presented as follows:

$$c = mG = \begin{bmatrix} m_0, m_1, \ldots, m_{k-1} \end{bmatrix} \begin{bmatrix} g_0 \\ g_1 \\ \vdots \\ g_{k-1} \end{bmatrix} \quad (5.16)$$

$$c = m_0 g_0 + m_1 g_1 + \cdots + m_{k-1} g_{k-1}. \quad (5.17)$$

The rows of the generator matrix G are linearly independent because we assume G has rank k and the codeword c is a k-dimensional subspace of the set of all binary n-tuples.

Example 5.1   Generator matrix
The $(7, 4)$ linear block code has the following generator matrix:

$$G = \begin{bmatrix} 1 & 0 & 0 & 0 & 1 & 1 & 0 \\ 0 & 1 & 0 & 0 & 1 & 0 & 1 \\ 0 & 0 & 1 & 0 & 0 & 1 & 1 \\ 0 & 0 & 0 & 1 & 1 & 1 & 1 \end{bmatrix}.$$

When we have the message [1 0 1 1], find the corresponding codeword.

Solution
From (5.16), the corresponding codeword is

$$c = mG = \begin{bmatrix} c_0, c_1, c_2, c_3, c_4, c_5, c_6 \end{bmatrix} = \begin{bmatrix} 1011 \end{bmatrix} \begin{bmatrix} 1 & 0 & 0 & 0 & 1 & 1 & 0 \\ 0 & 1 & 0 & 0 & 1 & 0 & 1 \\ 0 & 0 & 1 & 0 & 0 & 1 & 1 \\ 0 & 0 & 0 & 1 & 1 & 1 & 1 \end{bmatrix}.$$

In GF(2), the addition (+) and multiplication (·) are defined as following:

+	0	1
0	0	1
1	1	0

·	0	1
0	0	0
1	0	1

and logically the addition is XOR operation and the multiplication is AND operation. Thus, we obtain each coded bit as follows:

$$
\begin{aligned}
c_0 &= 1\cdot1+0\cdot0+1\cdot0+1\cdot0=1+0+0+0=1 \\
c_1 &= 1\cdot0+0\cdot1+1\cdot0+1\cdot0=0+0+0+0=0 \\
c_2 &= 1\cdot0+0\cdot0+1\cdot1+1\cdot0=0+0+1+0=1 \\
c_3 &= 1\cdot0+0\cdot0+1\cdot0+1\cdot1=0+0+0+1=1 \\
c_4 &= 1\cdot1+0\cdot1+1\cdot0+1\cdot1=1+0+0+1=0 \\
c_5 &= 1\cdot1+0\cdot0+1\cdot1+1\cdot1=1+0+1+1=1 \\
c_6 &= 1\cdot0+0\cdot1+1\cdot1+1\cdot1=1+1+0+0=0
\end{aligned}
$$

Therefore, the corresponding codeword is

$$
c = \left[c_0,c_1,c_2,c_3,c_4,c_5,c_6\right] = \left[1011010\right].
$$

■

We observe the corresponding codeword and generator matrix of Example 5.1. The codeword [1011010] is composed of the message segment ([1 0 1 1] which are the first four bits (k bits).) and the parity segment ([0 1 0] which are the next three bits (n-k bits).). The generator matrix is composed of the $k \times k$ identity matrix (I_k) segment and the P matrix segment as follows:

$$
G = \left[I_k \mid P\right] = \begin{bmatrix} 1 & 0 & 0 & 0 & 1 & 1 & 0 \\ 0 & 1 & 0 & 0 & 1 & 0 & 1 \\ 0 & 0 & 1 & 0 & 0 & 1 & 1 \\ 0 & 0 & 0 & 1 & 1 & 1 & 1 \end{bmatrix}. \tag{5.18}
$$

We call this linear systematic block code and its codeword structure is shown in Figure 5.12. The code rate, R, is defined as the ratio of the message length (k bits) to the codeword length (n bits).

There is another useful matrix called a $(n-k) \times n$ parity check matrix H. The parity check matrix is

$$
H = \left[-P^T I_{n-k}\right] = \left[P^T I_{n-k}\right] = \begin{bmatrix} 1101100 \\ 1011010 \\ 0111001 \end{bmatrix} \tag{5.19}
$$

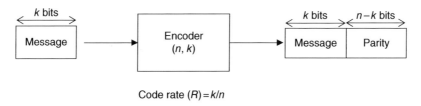

Code rate $(R) = k/n$

Figure 5.12 Linear systematic encoder

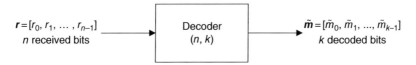

Figure 5.13 Decoder

where the negative of a number in GF (2) is simply the number. This equation is useful for decoding the received codeword. The transpose of the parity check matrix $\boldsymbol{H}$ is

$$H^T = \begin{bmatrix} P \\ I_{n-k} \end{bmatrix} = \begin{bmatrix} 110 \\ 101 \\ 011 \\ 111 \\ 100 \\ 010 \\ 001 \end{bmatrix}. \tag{5.20}$$

Now, we can have one important equation: $c \cdot H^T = 0$. Let $r = [r_0, r_1, \ldots, r_{n-1}]$ be the received vector through a noisy channel. The received vector is composed of the transmitted codeword $c = [c_0, c_1, \ldots, c_{n-1}]$ and the error vector $e = [e_0, e_1, \ldots, e_{n-1}]$ as follows:

$$r = c + e. \tag{5.21}$$

The decoder recovers the message from the received bits including the channel noise. Figure 5.13 illustrates a decoder for the (n, k) linear block code.

The syndrome decoding scheme is much simpler than many other decoding schemes. The first step is to calculate the syndrome of the received vector. The second step is to find an error location and error pattern using syndrome look-up table. The last step is to estimate the transmitted codeword from error patterns. The syndrome, S, of r is represented as

$$S = rH^T \tag{5.22}$$

$$S = (c + e) H^T = eH^T = [s_0, s_1, \ldots, s_{n-k-1}]. \tag{5.23}$$

The syndrome depends on the error pattern of the codeword. This is useful for detecting errors. When $S=0$, there are no errors and the receiver estimates r as the transmitted codeword. When $S \neq 0$, there are errors and the receiver detects an error location and pattern.

Example 5.2 Syndrome decoding

Assume that the message $m = [1011]$ is encoded by the (7,4) liner block code, the codeword $c = [1011010]$ from Example 5.1 is transmitted, and the decoder receives $r = [1010010]$. Correct an error using the syndrome decoding scheme.

Solution

From (5.22), the syndrome of r is

$$S = rH^T = [1010010] \begin{bmatrix} 110 \\ 101 \\ 011 \\ 111 \\ 100 \\ 010 \\ 001 \end{bmatrix} = [111].$$

We can generate syndrome look-up table as in Table 5.1.

From the below look-up table, the error pattern of the syndrome [111] is $e = [0001000]$. Therefore, the transmitted codeword is estimated as follows:

$$\tilde{c} = r + e = [1010010] + [0001000] = [1011010].$$

The message $m = [1011]$ is transmitted over a noisy channel and the message [1 0 1 0] of the received vector $r = [1010010]$ is not correctly delivered to the receiver. However, the syndrome decoding scheme detects and corrects an error.   ■

Table 5.1   Syndrome look-up table for the (7, 4) linear block code

Syndrome	Error pattern
000	0000000
001	0000001
010	0000010
100	0000100
111	0001000
011	0010000
101	0100000
110	1000000

We realized an error correction coding technique can correct an error. Now, we need to discuss the performance of an error correction code technique. Before doing this, we define several metrics such as Hamming distance, Hamming weight, minimum distance, minimum weight, and distance distribution. We firstly define Hamming weight and Hamming distance as follows:

Definition 5.2   Hamming weight

The Hamming weight $w_H(c)$ of a codeword c is the number of nonzero components of c.

Definition 5.3   Hamming distance

The Hamming distance $d_H(c_1, c_2)$ between a codeword c_1 and a codeword c_2 is the number of elements in which they are different.

For a linear block code, the Hamming distance between any two codewords is the Hamming weight of the difference between any two codewords. It can be described as follows:

$$d_H(c_1, c_2) = w_H(c_1 - c_2) = w_H(c_3).$$
(5.24)

Example 5.3   Hamming weight and distance

Find the Hamming weight and distance of two codewords $c_1 = [1011010]$ and $c_2 = [1010001]$.

Solution

From Definition 5.2, the Hamming weights are

$w_H(c_1) = d_H([1\ 0\ 1\ 1\ 0\ 1\ 0]) = 4$
$w_H(c_2) = d_H([1\ 0\ 1\ 0\ 0\ 0\ 1]) = 3.$

From Definition 5.3, the Hamming distance is
$d_H(c_1, c_2) = d_H([1\ 0\ 1\ 1\ 0\ 1\ 0], [1\ 0\ 1\ 0\ 0\ 0\ 1]) = 3.$ ∎

The minimum distance is one of the most important metrics when evaluating the performance of an error correction code. It guarantees error detection and correction capability. We define it as follows:

Definition 5.4   Minimum distance

Among the Hamming distances between all pairs of codewords in a block code C, the smallest Hamming distance is the minimum distance d_{min}. It is defined as follows:

$$d_{min} = \min\{d_H(c_1, c_2): c_1, c_2 \in C, c_1 \neq c_2\}.$$

For a block code C with the minimum distance d_{min}, we can calculate the guaranteed error detection and correction capability. The guaranteed error detection capability e is

$$e = d_{min} - 1. \tag{5.25}$$

This equation means a block code C can detect errors in the n bits of the codeword if $d_{min} - 1$ or fewer errors occur. The guaranteed error correction capability t is

$$t = \left\lfloor \frac{d_{min} - 1}{2} \right\rfloor. \tag{5.26}$$

This equation means a block code C can correct errors in the n bits of the codeword if $\lfloor (d_{min} - 1)/2 \rfloor$ or fewer errors occur.

Example 5.4   Error detection and correction capability
Find the error detection and correction capability of the (7, 4) Hamming code.

Solution
For a positive integer $m > 3$, a (n, k) Hamming code has the following parameters:
   The codeword length $n = 2^m - 1$
   The number of information $k = 2^m - m - 1$
   The minimum distance $d_{min} = 3$.
   From (5.25) and (5.26), the guaranteed error detection capability is

$$e = d_{min} - 1 = 3 - 1 = 2$$

and the guaranteed error correction capability is

$$t = \left\lfloor \frac{d_{min} - 1}{2} \right\rfloor = \left\lfloor \frac{3 - 1}{2} \right\rfloor = 1. \qquad \blacksquare$$

For a linear block code, the minimum distance is identical to the minimum weight of its nonzero codewords. The minimum weight is defined as follows:

Definition 5.5   Minimum weight

The minimum weight (w_{min}) of a code C is defined as follows:

$$w_{min} = \min\{w_H(c) : c \in C, c \neq 0\}.$$

It is easier to find the minimum weight than the minimum distance. However, the minimum distance does not describe the weight of other codewords. Thus, the weight distribution should be investigated. It is defined as follows:

Definition 5.6 Weight distribution

The weight distribution of an (n, k) linear block code as weight enumerator polynomial is

$$W(z) = \sum_{i=0}^{n} w_i z^i$$

where w_i is the number of codewords with Hamming weight i.

Example 5.5 Weight distribution

Find and plot the weight distribution of the (7, 4) Hamming code.

Solution

From the generator matrix of Example 5.1, we can generate all codewords and find the weight enumerator polynomial as follows:

$$W(z) = 1 + 7z^3 + 7z^4 + z^7$$

and their plot is illustrated in Figure 5.14. ∎

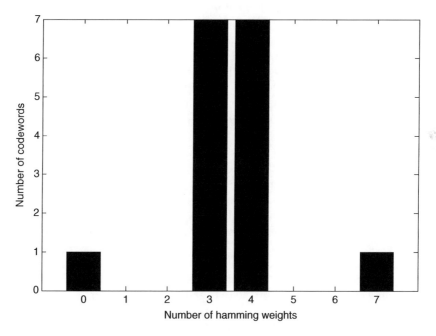

Figure 5.14 Weight distribution of the (7, 4) Hamming code

The weight distribution is calculated to evaluate the performance of an error correction code. However, when handling a nonlinear error correction code, the distance distribution is more significant than the weight distribution because the minimum distance can be obtained from the distance distribution while the weight distribution apparently cannot provide any relevant information. The distance distribution of an error correction code C of length n is the set $\{D_i(C) \mid 0 \leq i \leq n\}$, where

$$D_i(C) = \frac{1}{|C|} \sum_{c_2 \in C} \left| \{ c_1 \in C \mid d_H(c_1, c_2) = i \} \right|. \tag{5.27}$$

In error correction coding theory, the code design problem for a (n, k, d_{min}) block code is to optimize each parameter for the other two given parameters in the direction of decreasing n and increasing k and d_{min}.

5.2.2   Convolutional Codes

The convolutional codes [8] were firstly introduced by Elias in 1955. They are different from the block codes. The encoder of convolutional codes contains a shift register and each output bit depends on the previous input bits. An (n, k, L) convolutional code is designed with k input sequence, n output sequence, and L memory (shift register). The code rate R is k/n and the constraint length K is determined by $L+1$. Basically, the performance of convolutional codes is related to the code rate and the constraint length. Longer constraint length and smaller code rate bring higher coding gain. However, a trade-off exists between them. Longer constraint length means that information of each input bit is contained in a longer codeword and the encoder needs a large memory. Smaller code rate means that the codeword includes many redundancies. Thus, the decoder complexity increases and the bandwidth efficiency decreases. Basically, the convolutional encoder is described by vector or polynomial. In the vector representation, an element of the generator is "1" if an element of the constraint length is connected to a modulo-2 adder. It is "0" otherwise. In the polynomial representation, each polynomial means the connection between an element of the constraint length and a modulo-2 adder.

Assume a $(n, k, 2)$ convolutional code described by generator vectors ($g_1 = [1\ 0\ 1]$ and $g_2 = [1\ 1\ 1]$) and the input sequence $[1\ 0\ 1\ 1\ 1\ 0\ 0\ 0]$. The Figure 5.15 illustrates the convolutional encoder.

The convolutional encoder includes two shift registers. Thus, the encoder is based on a finite state machine and the states are represented by the content of the memory. Each input bit is stored in the first shift register and then it moves to the second shift register. The output of the encoder is calculated by modulo-2 addition among the input bit and the stored bits in two shift registers. The input, output, and memory state of the convolutional encoder are summarized in Table 5.2.

In Table 5.2, the initial state (time=0) of the memory is "00." When the input bit is "1," the memory state changes from "00" to "10" due to the shift register and the output is "11" which is calculated by generator vectors ($g_1 = [1\ 0\ 1]$ and $g_2 = [1\ 1\ 1]$). That is to say ["input at time=1" "M1 at time=0" "M2 at time=0"] is equal to ["1" "0" "0"] when the input bit is injected to the encoder at time=1. We calculate the output by the input bit at time=1 and the memory state at time=0. Thus, the encoder output is calculated by modulo-2 addition: (1 XOR

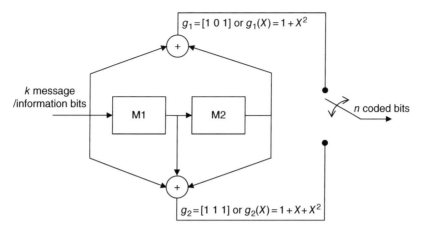

Figure 5.15   Example of the convolutional encoder

Table 5.2   Memory states (M1 M2) of the convolutional encoder

Time	Input	Memory state (M1 M2)	Output
0		00	
1	1	10	11
2	0	01	01
3	1	10	00
4	1	11	10
5	1	11	01
6	0	01	10
7	0	00	11
8	0	00	00

$0 = "1"$ and 1 XOR 0 XOR $0 = "1") = (1\ 1)$ at time $= 1$. The remaining memory state and output can be filled in the same way. We can express this table as a state diagram. The state diagram is a graph which represents the relationship among the memory status, input, and output. It has 2^L nodes as the memory state. The nodes are connected by a branch which is labeled by the input/output. The state diagram for Table 5.2 is illustrated in Figure 5.16. In this figure, the dotted branch means that the input bit of the encoder is "0" and the solid branch means that the input bit of the encoder is "1."

There is another representation which is used to describe the encoding process of convolutional codes. The trellis diagram expresses the encoding process using the memory state and time. It is very helpful of us to understand the encoding and decoding process of convolutional codes. It is also possible to describe block codes on the trellis diagram. The trellis diagram of the convolutional encoder is illustrated in Figure 5.17.

The vertical nodes mean the memory state and they are connected by a branch which is labeled by the input/output. The horizontal nodes mean the time corresponding to the memory state. The dotted line means that the input is "0" and the solid line means that

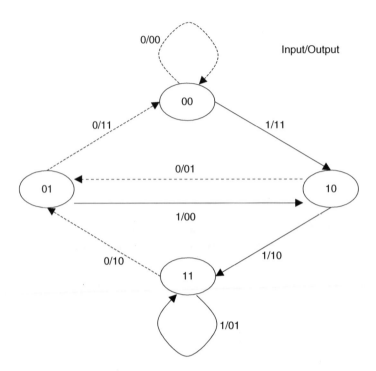

Figure 5.16 State diagram of the convolutional encoder

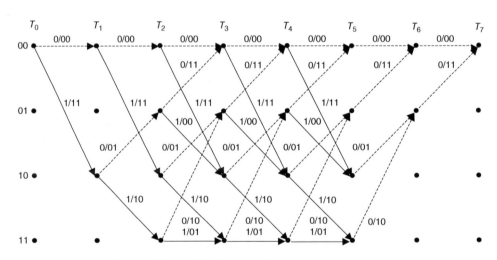

Figure 5.17 Trellis diagram of the convolutional encoder

the input is "1." In the trellis diagram, the input and output provide a unique path. Assume that the convolutional encoder has the trellis diagram of Figure 5.17, the input sequence of the convolutional encoder is [1 1 0 1 1], and the initial state (T_0) of the memory is "00." When the first input bit "1" is injected to the convolutional encoder, the memory state changes from

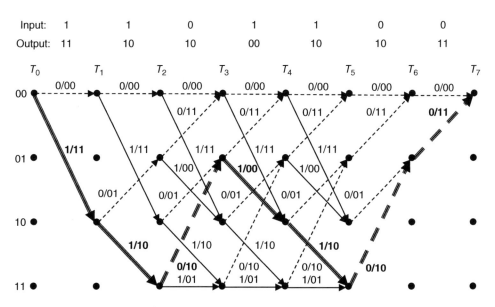

Figure 5.18   Encoding example of the convolutional encoder

"00 (T_0)" to "10 (T_1)." At this time, the output bit is "11." When the second input bit "1" is injected, the memory state changes from "10 (T_1)" to "11 (T_2)." At this time, the output bit is "10." In the same way, we can find the unique path as shown in Figure 5.18.

In a convolutional encoder, the memory should be initialized by all zero bits so that the memory can be cleared out. We add zero tail bits (which are same size as the number of shift registers) to k information bits. In this example, we added 2 tail bits [0 0] to 5 information bits. Therefore, the effective code rate is 5/14.

The Viterbi algorithm provides us with a low complexity Maximum Likelihood (ML) solution. Thus, it is widely used in many areas. The Viterbi algorithm is composed of three steps. The first step is to measure a distance between the transmitted codeword and the received codeword. The distance can be the Hamming distance for a hard decision input or the Euclidean distance for a soft decision input. In the Viterbi algorithm, this distance is called a branch metric, $\mathrm{BM}_i^{l \to m}$, which is related to the likelihood $p(r|c)$ (where r and c are the received codeword and the transmitted codeword, respectively). For a hard decision input, the branch metric is calculated as follows:

$$\mathrm{BM}_i^{l \to m} = d_{\mathrm{H}}(r_i, c_i) \tag{5.28}$$

where l and m are the states at T_{i-1} and T_i, respectively.

The second step is to calculate a path metric, PM_i^m, and select one path. A node at T_i is connected by several branches from T_{i-1} to T_i. We should select the most likely branch path according to path metric calculations as follows:

$$\mathrm{PM}_i^m = \mathrm{Min}(\mathrm{PM}_{i-1}^l + \mathrm{BM}_i^{l \to m}, \mathrm{PM}_{i-1}^{l'} + \mathrm{BM}_i^{l' \to m}). \tag{5.29}$$

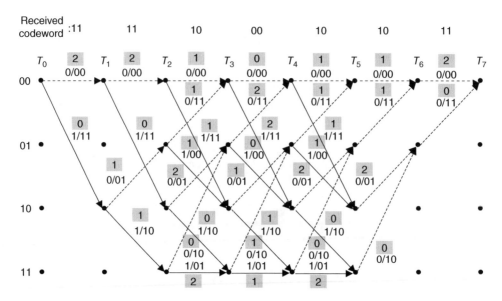

Figure 5.19 Branch metric calculations

The third step is to perform a traceback process. Basically, we assume that the encoding process ends at all zero state due to zero tail bits. We select a path and find a message/information by tracing the path from right to left on the trellis diagram. When the encoder and the trellis diagram are given in Figures 5.15 and 5.18, respectively, we assume an error occurs to the fourth bit of the transmitted codeword and the received codeword is [11 **11** 10 00 10 10 11]. As the first step of the Viterbi algorithm, we calculate branch metrics using the Hamming distance. The branch metric $BM_1^{00 \rightarrow 00} = 2$ (transition from T_0 to T_1 and from state 00 to state 00) is calculated by the Hamming distance between the received coded bits [11] and the branch value from state 00 to state 00 [00]. The $BM_1^{00 \rightarrow 10} = 0$ (transition from T_0 to T_1 and from state 00 to state 10) is calculated by the Hamming distance between the received coded bits [11] and the branch value from state 00 to state 10 [11]. In the same way, we calculate all branch metrics as shown in Figure 5.19.

The second step is the path metric calculation. Each node at T_1 and T_2 is connected to only one node in the previous state. Thus, we compute the following path metrics:

$$PM_1^{00} = BM_1^{00 \rightarrow 00} = 2 \tag{5.30}$$

$$PM_1^{10} = BM_1^{00 \rightarrow 10} = 0 \tag{5.31}$$

$$PM_2^{00} = PM_1^{00} + BM_2^{00 \rightarrow 00} = 2 + 2 = 4 \tag{5.32}$$

$$PM_2^{01} = PM_1^{10} + BM_2^{10 \rightarrow 01} = 0 + 1 = 1 \tag{5.33}$$

$$PM_2^{10} = PM_1^{00} + BM_2^{00 \rightarrow 10} = 2 + 0 = 2. \tag{5.34}$$

$$PM_2^{11} = PM_1^{10} + BM_2^{10 \rightarrow 11} = 0 + 1 = 1. \tag{5.35}$$

Each node at T_3 is connected to two nodes in the previous state. Thus, we need to select one path. The path metrics are calculated as follows:

$$PM_3^{00} = Min(PM_2^{00} + BM_3^{00\to00}, PM_2^{01} + BM_3^{01\to00}) \tag{5.36}$$
$$= Min(4+1, 1+1) = 2$$

$$PM_3^{01} = Min(PM_2^{10} + BM_3^{10\to01}, PM_2^{11} + BM_3^{11\to01}) \tag{5.37}$$
$$= Min(2+2, 1+0) = 1$$

$$PM_3^{10} = Min(PM_2^{00} + BM_3^{00\to10}, PM_2^{01} + BM_3^{01\to10}) \tag{5.38}$$
$$= Min(4+1, 1+1) = 2$$

$$PM_3^{11} = Min(PM_2^{10} + BM_3^{10\to11}, PM_2^{11} + BM_3^{11\to11}) \tag{5.39}$$
$$= Min(2+0, 1+2) = 2.$$

In the same way, we compute all path metrics as shown in Figure 5.20. Now, it is ready to select one survival path according to the smallest path metric. We carry out the trackback as the third step. From T_7 to T_0, we select and follow the smallest path metric $(1\to1\to1\to1\to1\to1\to0)$. We finally find the output of the traceback process $(0\to0\to1\to1\to0\to1\to1)$ as shown in Figure 5.20. The output is reversed. Therefore, the output of the Viterbi algorithm is [1 1 0 1 1 0 0]. Now, we can compare this output with the message of Figure 5.18 and observe the error is corrected.

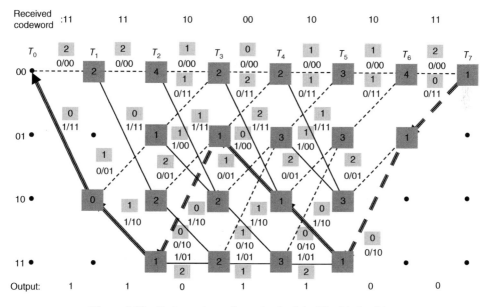

Figure 5.20   Path metrics and traceback of the Viterbi algorithm

The Bit Error Rate (BER) is defined as the number of bit errors divided by the total transmitted bits. This metric is widely used to evaluate the performance of error correction codes. The coding gain is E_b/N_0 difference between the uncoded system and the coded system at a specific BER. The BER performance of several error correction codes (Hamming code, Golay code, and Convolutional code) in AWGN environment is illustrated in Figure 5.21.

In this figure, the solid line (-) is the uncoded system performance. The circle line (-O-) is the (7, 4, 3) Hamming code performance. The square line (-□-) is the (24, 12, 8) Golay code performance. The diamond line (-◊-) is the convolutional code with the constraint length=7, the code rate=1/2, and the hard decision input. The asterisk line (-*-) is the convolutional code with the constraint length=7, the code rate=1/2, and the soft decision input. The first two error correction codes (the (7, 4, 3) Hamming code and the (24, 12, 8) Golay code) are linear block codes. The performance of the error correction codes depends on the minimum distance and the codeword length. Since the (24, 12, 8) Golay code has a higher minimum distance and longer codeword length than the (7, 4, 3) Hamming code, the Golay code has better performance than the Hamming code. The coding gains of the (7, 4, 3) Hamming code and the (24, 12, 8) Golay code at BER $= 10^{-6}$ are about 0.8 and 2.2 dB, respectively. The next two error correction

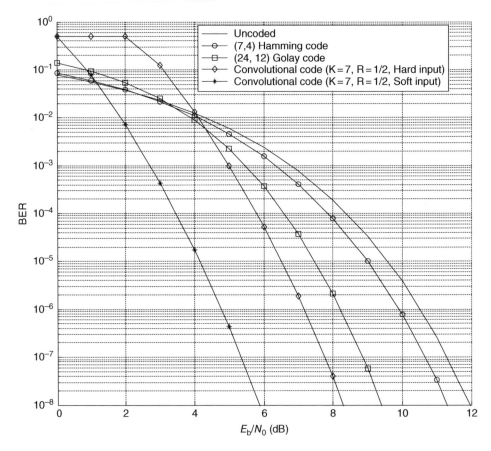

Figure 5.21   BER of error correction codes

codes are convolutional codes. They are different from the input type. The hard decision input is expressed as either "0" and "1" or "−1" and "1." On the other hand, the soft decision input is expressed as the 2^b-1 level where b is the bit resolution. If we consider the three bits soft decision input, we have eight levels from 0 (000) to 7 (111). We can allocate four levels (000–011) to zero and four levels (100–111) to one. The input value 0 (000) is the most confident zero, the input value 1 (001) is the second most confident zero, and the input value 3 (011) is the least confident zero. The input value 7 (111) is the most confident one, the input value 6 (110) is the second most confident one, and the input value 4 (100) is the least confident one. The coding gains of the convolutional codes with the hard decision and the soft decision at BER $= 10^{-6}$ are about 3.8 and 5.9 dB, respectively.

Summary 5.2   Error control coding

1. C. Shannon predicts that it is possible to transmit information without errors over the unreliable channel and an error control coding technique exists to achieve this.
2. Forward error correction (FEC) does not have a feedback channel. In the receiver, the corrupted errors are recovered by an error correction coding technique.
3. Automatic repeat request (ARQ) has a feedback channel and uses error detection technique and noiseless feedback.
4. The block codes are based on algebra and algebraic geometry and developed by mainly mathematicians. The construction and lower/upper bounds of the block codes are well analyzed mathematically.
5. The convolutional codes are based on a finite state machine and probability theory. The Viterbi algorithm is widely used in many communication systems.
6. The weight distribution is calculated to evaluate the performance of an error correction code.
7. The bit error rate (BER) is defined as the number of bit errors divided by the total transmitted bits.

5.3   MIMO

MIMO techniques use the multiple antennas at a transmitter and receiver and improve the performances of wireless communication systems. The MIMO techniques are very effective to mitigate the degradation of fading channels and enhance the link quality between a transmitter and a receiver. Especially, it improves Signal to Noise Ratio (SNR), Signal to Interference plus Noise Ratio (SINR), spectral efficiency, and error probability. The MIMO techniques are classified into *spatial diversity techniques*, *spatial multiplexing techniques*, and *beamforming techniques*. Each technique targets to improve different aspects of wireless communication system performances. Spatial diversity techniques target to decrease the error probability. A transmitter sends multiple copies of the same data sequence and a receiver combines them as shown in Figure 5.22.

As we reviewed diversity techniques in Section 5.1, the multiple same data sequences experience statistically independent channels and the receiver can obtain spatial diversity gain. In the diversity techniques, there are two types of diversities: transmit diversity and receive diversity. For transmit diversity, the transmitter has multiple antennas and pre-processing

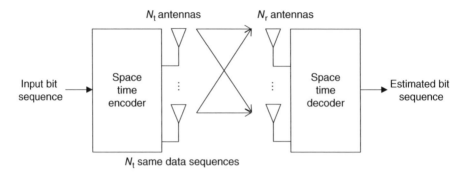

Figure 5.22   MIMO as spatial diversity technique

blocks for combining the multiple same data sequences. We typically assume the receiver has channel knowledge. Several well-known spatial diversity techniques are Space-Time Block Codes (STBCs) [9, 10] and Space-Time Trellis Codes (STTCs) for improving the reliability of the data transmission. The STBC provides us with diversity gain only. However, the STTC uses convolutional codes and provides us with both code gain and diversity gain [11]. For receive diversity, the receiver has multiple antennas and combining techniques such as MRC, EGC, and SC.

Spatial multiplexing techniques substantially increase spectral efficiency. A transmitter sends N_t data sequences simultaneously in the different antennas and same frequency band and a receiver detects them using an interference cancellation algorithm as shown in Figure 5.23.

This technique improves spectral efficiency because multiple data sequences are transmitted in parallel. Thus, the spectral efficiency is improved by increasing the number of the transmit antennas (N_t). This technique requires channel knowledge at the receiver as well. It can be combined with beamforming techniques when channel knowledge is available at both the transmitter and the receiver. In addition, we can expand it into multi-user MIMO or Space-Division Multiple Access (SDMA). Multi-user MIMO techniques assign each data sequence to each user as shown in Figure 5.24.

It is especially useful for uplink systems due to the limited number of antennas at mobile stations. Sometime, this is called collaborative MIMO or collaborative spatial multiplexing. One of the well-known spatial multiplexing techniques is the Bell laboratories layered space time (BLAST) technique [12].

Beamforming techniques are signal processing techniques for directional transmission as shown in Figure 5.25.

The beamformer with N_t antenna elements combines Radio Frequency (RF) signals of each antenna element to have a certain direction by adjusting phases and weighting of RF signals. Thus, this technique brings antenna gain and suppresses interferences in multiuser environments. The beamforming technique was investigated in the radar technology since the 1960s. However, this technique was paid attention in the 1990s again since cellular systems gave rise to a new wireless communication. In the modern wireless communications, the cell size is getting smaller and the number of cells is getting bigger. The interferences among cells became serious problem. Thus, the interference mitigation technique among

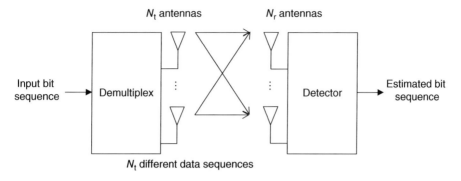

Figure 5.23   MIMO as spatial multiplexing technique

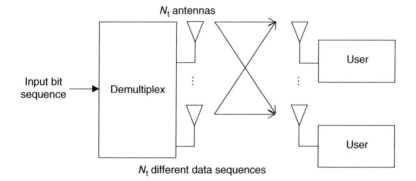

Figure 5.24   Multiuser MIMO

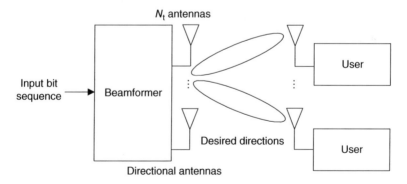

Figure 5.25   MIMO as beamforming technique

cells is an essential part of wireless communication systems. The beamforming technique is very effective to mitigate interferences.

In this section, we will discuss the fundamentals of MIMO techniques. Firstly, we will look into spatial diversity techniques. We consider a MISO system with two transmit

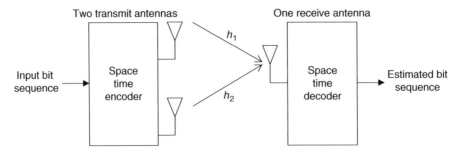

Figure 5.26   Alamouti scheme with 2×1 antennas

Table 5.3   Signal mapping of Alamouti scheme with 2×1 antennas

	Time t	Time $t+1$
Antenna 1	$s[t]$	$-s^*[t+1]$
Antenna 2	$s[t+1]$	$s^*[t]$

antennas and one receive antenna as shown in Figure 5.26 and carry out two dimensional (space and time) signal mapping as shown in Table 5.3. This technique is known as Alamouti scheme [9].

The transmit symbols make a pair. At the time index t, $s[t]$ and $s[t+1]$ are simultaneously transmitted via antenna 1 and antenna 2, respectively. At the time index $t+1$, $-s^*[t+1]$ and $s^*[t]$ are simultaneously transmitted via antenna 1 and antenna 2, respectively. The operation "*" is the complex conjugate. Each symbol experiences different channel responses (h_1 and h_2). We assume the channel responses are not changed during the transmission. The received symbols:

$$y[t] = h_1 s[t] + h_2 s[t+1] + n[t] \tag{5.40}$$

$$y[t+1] = -h_1 s^*[t+1] + h_2 s^*[t] + n[t+1]. \tag{5.41}$$

These equations can be rewritten in the matrix form as follows:

$$\begin{bmatrix} y[t] \\ y^*[t+1] \end{bmatrix} = \begin{bmatrix} h_1 & h_2 \\ h_2^* & -h_1^* \end{bmatrix} \begin{bmatrix} s[t] \\ s[t+1] \end{bmatrix} + \begin{bmatrix} n[t] \\ n^*[t+1] \end{bmatrix} \tag{5.42}$$

$$\boldsymbol{y} = \boldsymbol{H}\boldsymbol{s} + \boldsymbol{n}. \tag{5.43}$$

The combining technique in the receiver is performed. Based on the orthogonal properties of $\boldsymbol{H}$ matrix, we have the following equation:

$$\boldsymbol{H}\boldsymbol{H}^H = \left(|h_1|^2 + |h_2|^2 \right) \boldsymbol{I}_2 \tag{5.44}$$

where $()^H$ and I_2 are Hermitian matrix and 2×2 identity matrix, respectively. Thus, we have the following combined symbols:

$$y' = H^H y = H^H Hs + H^H n \qquad (5.45)$$

$$y' = \left(|h_1|^2 + |h_2|^2 \right) s + n'. \qquad (5.46)$$

The combined symbols are sent to ML detector and we estimate the transmitted symbol as shown in Figure 5.27.

As another example of spatial diversity techniques, we consider a SIMO system with N_r antennas as shown in Figure 5.28.

We transmit $s[t]$ and receive $y_i[t]$ via receive antenna i at the time index t. Therefore, the received symbol $y_i[t]$ is expressed as follows:

$$y_i[t] = h_i s[t] + n_i[t] \qquad (5.47)$$

where h_i is the channel response and follows complex normal distribution. The receiver collects $y_i[t]$ using combining techniques in Section 5.1 and obtains more reliable received symbols. When dealing with spatial diversity techniques, it is important to maintain uncorrelated antennas. Under the uncorrelated condition, we can obtain diversity gain which means SNR or SINR increases. If antennas are strongly correlated, we cannot obtain diversity gain.

Secondly, we look into the BLAST technique [12] as one of key spatial multiplexing techniques. There are two types of BLAST techniques. Diagonal BLAST (D-BLAST) can achieve near the Shannon limit but it has significant complexity. On the other hand, Vertical BLAST

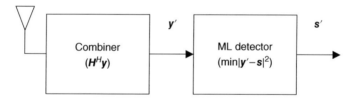

Figure 5.27 ML detection

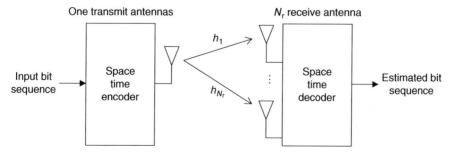

Figure 5.28 Spatial diversity technique with $1 \times N_r$ antennas

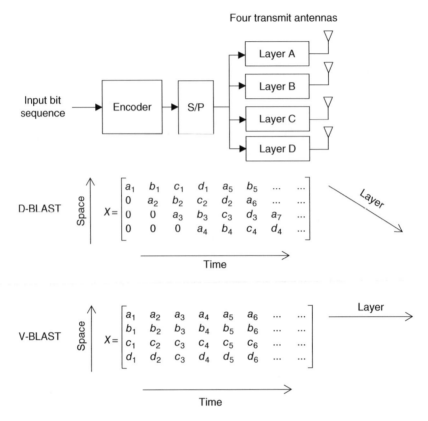

Figure 5.29   D-BLAST and V-BLAST transmitter and data sequences

(V-BLAST) has a lower capacity but the complexity is low. Figure 5.29 illustrates the transmitter architecture with four antennas and data sequence mapping of D-BLAST and V-BLAST.

In the BLAST system, N_t data sequences are transmitted simultaneously in the same frequency band and different antennas and these various data sequences can be separated at the receiver. The transmitter does not need channel state information and the total transmit power is maintained regardless of the number of transmit antennas. The receiver is based on interference mitigation techniques such as Zero Forcing (ZF), Minimum Mean Squared Error (MMSE), and Successive Interference Cancellation (SIC). In the MIMO system as spatial multiplexing, the complexity is a critical issue. The higher number of antennas brings a better system performance but significantly increases system complexity. ZF detection uses the estimated channel response matrix H. Its complexity is low and system performance is also low. MMSE detection considers both H and the noise variance. Thus, its complexity is higher than ZF detection. Its performance is better than ZF detection at a low or middle SNR but their performance becomes similar at a high SNR. ML detection is the optimal solution. However, the complexity is very high because it checks all possible hypotheses. V-BLAST provides good trade-off between system performance and complexity. V-BLAST requires multiple and successive calculations and is based on SIC using QR decomposition. Assume a MIMO system with N_t transmit antennas and N_r receive antennas ($N_t = N_r = N$). The QR decomposition of the channel response matrix H is defined as follows:

$$H = QR \tag{5.48}$$

where the matrix Q is an orthogonal matrix as follows:

$$Q^H Q = I_N. \tag{5.49}$$

The matrix R is an upper triangular matrix as follows:

$$R = \begin{bmatrix} r_{11} & \cdots & r_{1N} \\ \vdots & \ddots & \vdots \\ 0 & \cdots & r_{NN} \end{bmatrix}. \tag{5.50}$$

When we transmit $s_i[t]$ at receive antenna i and receive $y_i[t]$ at receive antenna i, the received symbol $y_i[t]$ is expressed as follows:

$$\begin{bmatrix} y_1[t] \\ \vdots \\ y_N[t] \end{bmatrix} = \begin{bmatrix} h_{11} & \cdots & h_{1N} \\ \vdots & \ddots & \vdots \\ h_{N1} & \cdots & h_{NN} \end{bmatrix} \begin{bmatrix} s_1[t] \\ \vdots \\ s_N[t] \end{bmatrix} + \begin{bmatrix} n_1[t] \\ \vdots \\ n_N[t] \end{bmatrix} \tag{5.51}$$

$$y = Hs + n. \tag{5.52}$$

At the receiver, we calculate the following equation:

$$\tilde{s} = Q^H y = Q^H(Hs + n) = Q^H QRs + \tilde{n} = Rs + \tilde{n} \tag{5.53}$$

$$\begin{bmatrix} \tilde{s}_1[t] \\ \vdots \\ \tilde{s}_N[t] \end{bmatrix} = \begin{bmatrix} r_{11} & \cdots & r_{1N} \\ \vdots & \ddots & \vdots \\ 0 & \cdots & r_{NN} \end{bmatrix} \begin{bmatrix} s_1[t] \\ \vdots \\ s_N[t] \end{bmatrix} + \begin{bmatrix} n_1[t] \\ \vdots \\ n_N[t] \end{bmatrix}. \tag{5.54}$$

Due to the upper triangular structure, the last element $\tilde{s}_N[t]$ of the matrix $\tilde{s}$ in (5.54) is not affected by interferences and can be directly separated. The element $\tilde{s}_{N-1}[t]$ can be obtained by subtracting the N row from the $N-1$ row. Likewise, we can find the other elements.

Thirdly, beamforming techniques use array gain and control the direction of signals by adjusting the magnitude and phase at each antenna array. The array gain means a power gain of multiple antennas with respect to single antenna. When we allocate each antenna in a line and by equal spaces and assume each antenna is strongly correlated, the plane wave departs from each antenna in a time interval as shown in Figure 5.30. In this figure, the plane wave departs from omnidirectional antennas in the time interval $d \sin\theta$.

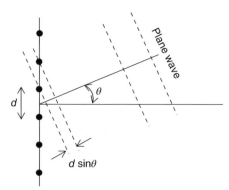

Omnidirectional antennas

Figure 5.30   Simple beamformer

Thus, we can calculate the delay of departure among antennas as follows:

$$\tau = \frac{d \sin \theta}{c} \tag{5.55}$$

where c is the speed of light. Each signal $s_i(t)$ at each omnidirectional antenna can be approximately expressed as follows:

$$s_1(t) = s(t) \tag{5.56}$$

$$s_2(t) = s(t - \tau) \approx s(t)e^{-j\theta} \tag{5.57}$$

$$\vdots$$

$$s_{N_t}(t) = s(t - N_t\tau) \approx s(t)e^{-jN_t\theta} \tag{5.58}$$

$$\mathbf{s} = \mathbf{a}s(t) = \begin{bmatrix} 1 \\ e^{-j\theta} \\ e^{-j2\theta} \\ \vdots \\ e^{-jN_t\theta} \end{bmatrix} s(t) \tag{5.59}$$

where $\mathbf{a}$ as the antenna array steering vector controls the direction of the signals. In addition, we can add the weighting vector $\mathbf{w}$ to the Equation 5.59 as follows:

$$\mathbf{s} = \mathbf{w}^H \mathbf{a}s(t) = \begin{bmatrix} w_1 & w_2 & w_3 & \cdots & w_{N_t} \end{bmatrix} \begin{bmatrix} 1 \\ e^{-j\theta} \\ e^{-j2\theta} \\ \vdots \\ e^{-jN_t\theta} \end{bmatrix} s(t). \tag{5.60}$$

Thus, the signal power is strengthened in the desired direction and weakened in the undesired direction. The beamforming performance depends on finding the suitable weighting vector w, antenna array arrangement, distance d between antenna arrays, and signal correlation.

Summary 5.3   MIMO

1. MIMO techniques are classified into spatial diversity techniques, spatial multiplexing techniques, and beamforming techniques.
2. Spatial diversity techniques targets to decrease the error probability. The transmitter sends multiple copies of the same data sequence and the receiver combines them.
3. Spatial multiplexing techniques substantially increase spectral efficiency. The transmitter sends N_t data sequences simultaneously in the different antennas and same frequency band and the receiver detects them using interference cancellation algorithm.
4. Beamforming techniques are signal processing techniques for directional transmission. The beamformer with N_t antenna elements combines RF signals of each antenna element to have a certain direction through adjusting phases and weighting of RF signals. Thus, this technique brings antenna gain and suppresses interferences in multiuser environments.

5.4   Equalization

Due to a time dispersive channel by multipath fading, Inter-Symbol Interference (ISI) occurs and an equalizer plays an important role in ISI compensation. It literally equalizes the time dispersive channel in the receiver. Figure 5.31 illustrates the channel model for equalization.

In this figure, $F(f)$ is the combined frequency response of the transmit filter, channel, and receive filter as follows:

$$F(f) = H_t(f) \cdot H_c(f) \cdot H_r(f) \tag{5.61}$$

and $G(f)$ includes $H_{eq}(f)$ in $F(f)$ as follows:

$$G(f) = F(f) \cdot H_{eq}(f). \tag{5.62}$$

The frequency response of the equalizer should be designed to satisfy the following equation:

$$G(f) = F(f) \cdot H_{eq}(f) = 1 \tag{5.63}$$

$$H_{eq}(f) = \frac{1}{F(f)}. \tag{5.64}$$

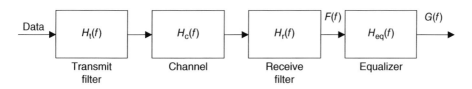

Figure 5.31   Channel model for equalization

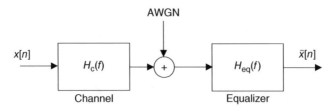

Figure 5.32   Channel model for zero forcing equalizer

In time domain, the combined impulse response $g(t)$ can be obtained by inverse Fourier transforms of $G(f)$. It is expressed as follows:

$$g(t) = IDFT(G(t))   \tag{5.65}$$

The equalization is to find the inverse of the combined frequency response $F(f)$. The zero forcing equalizer is one of simple equalizers using the inverse of the channel frequency response as shown in Figure 5.32.

In this figure, the zero forcing equalizer is designed as follows:

$$H_{eq}(f) = \frac{1}{H_c(f)}.   \tag{5.66}$$

Thus, the combined frequency response provides us with a flat frequency response $H_{eq}(f){\cdot}H_c(f) = 1$. This technique basically ignores AWGN. The complexity is low but the performance is not good. However, the equalization in the Orthogonal Frequency Division Multiplexing (OFDM) system is not much important because multipath fading is compensated in the OFDM system itself. Each subcarrier of the OFDM symbol experiences a flat fading channel. Thus, the OFDM system does not need a strong equalization and the one-tap zero forcing equalizer in frequency domain is enough to compensate subcarrier distortions.

In order to estimate how well an equalizer works, the Mean Squared Error (MSE) is used as the metric. It is defined as the mean squared error between the received signal and the desired signal as follows:

Definition 5.7   Mean squared error

$$\text{MSE} = \sum_{n=0}^{N-1}\left(\tilde{x}[n] - x[n]\right)^2.$$

For the unknown parameter $\hat{\theta}$, MSE is defined as follows:

$$\text{MSE}(\hat{\theta}) = E\left[(\hat{\theta} - \theta)^2\right].$$

Figure 5.33   Frame structure including preambles and data symbols

Now, we should estimate the channel response $H_c(f)$. We call this channel estimation. There are two types of channel estimations. The first type is to use training symbols (preambles) or pilot symbols and the other type is a blind channel estimation. The method using preambles in the OFDM system is to reserve several OFDM symbols. The receiver knows which symbol the transmitter sends without interpolation. Figure 5.33 illustrates the frame structure including training symbols and data symbols.

If we have a long preamble, we can estimate a wireless channel accurately. However, a long preamble means a high redundancy and the spectral efficiency is low. Thus, it is one important design issue in wireless communication systems design. In addition, the preambles have been utilized to compensate the impairments during synchronization. It affects to Peak-to-Average Power Ratio (PAPR) of the OFDM system. Thus, it is highly desirable to design the preambles capable of reducing the PAPR.

There are three types of pilot structures in the OFDM system. The first is a block-type pilot structure. It allocates a pilot signal to all subcarriers in a specific period as shown in Figure 5.34. The pilot allocation is one of the important parameters for the OFDM system design. In Ref. [13], the pilot interval is described in frequency and time domain.

This structure is suitable for a frequency selective channel. The pilot interval in time domain should be satisfied by the following equation:

$$T_p \leq \frac{1}{2 f_d T_{\text{OFDM}}}$$ (5.67)

where f_d is the Doppler spread and T_{OFDM} is the OFDM symbol duration. Another pilot structure is a comb-type pilot structure. It allocates a pilot signal to all time slots in a specific period as shown in Figure 5.35.

This structure is suitable for a time selective channel. The pilot interval in frequency domain should be satisfied by the following equation:

$$f_p \leq \frac{1}{2 \tau_{\text{max}} \Delta f}.$$ (5.68)

where τ_{max} is the maximum delay spread and Δf is the subcarrier spacing in frequency domain. The last pilot structure is a lattice-type pilot structure which is combination of both the block-type pilot structure and the comb-type pilot structure. It allocates a pilot signal to a part of subcarriers and time slots as maintaining specific interval as shown in Figure 5.36.

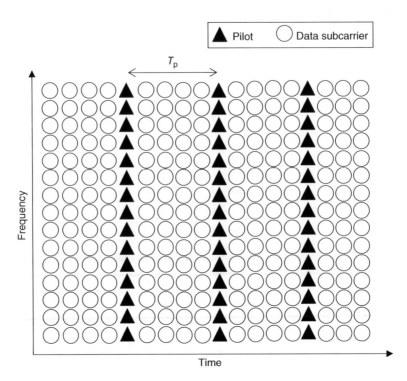

Figure 5.34 Block-type pilot structure in the OFDM system

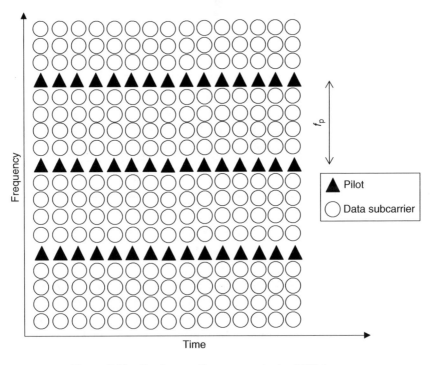

Figure 5.35 Comb-type pilot structure in the OFDM system

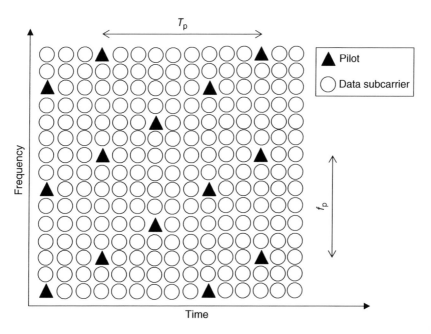

Figure 5.36 Lattice-type pilot structure in the OFDM system

This structure is suitable for a frequency and time selective channel. The pilot interval in time domain and frequency domain should be satisfied by the following both equations:

$$T_p \leq \frac{1}{2f_d T_{\text{OFDM}}} \quad \text{and} \quad f_p \leq \frac{1}{2\tau_{\text{max}}\Delta f}.$$ (5.69)

As we can observe Figures 5.34, 5.35, and 5.36, pilot signals do not cover a whole OFDM symbol. Thus, interpolation should be carried out at data subcarrier positions.

The Least Square (LS) estimation and the MMSE estimation are important channel estimations based on the training symbols or pilot symbols. When we have the received signal Y as follows:

$$Y = HX + N_{\text{awgn}}$$ (5.70)

where H and N_{awgn} are the channel response and AWGN, respectively, the LS estimation ignores AWGN and calculate the following simple equation:

$$\hat{H}_{\text{LS}} = X^{-1}Y$$ (5.71)

where X and Y are the transmitted training symbol (or pilot symbol after interpolation) matrix and the received symbol matrix, respectively. In the receiver, we have both matrices and find $\hat{H}_{\text{LS}}$. In the OFDM system, the LS estimation for each subcarrier is expressed as follows:

$$\hat{H}_{\text{LS}}[k] = \frac{Y[k]}{X[k]}$$ (5.72)

$$\hat{H}_{\text{LS}} = \left[\frac{y_0}{x_0}, \frac{y_1}{x_1}, \dots, \frac{y_{N-1}}{x_{N-1}} \right]$$ (5.73)

where N is the total number of subcarriers. The MSE of the LS estimation is derived as follows:

$$\text{MSE}_{\text{LS}} = E\left[(H - \hat{H}_{\text{LS}})^{\text{H}} (H - \hat{H}_{\text{LS}}) \right] \tag{5.74}$$

$$\text{MSE}_{\text{LS}} = E\left[(H - X^{-1}Y)^{\text{H}} (H - X^{-1}Y) \right] \tag{5.75}$$

$$\text{MSE}_{\text{LS}} = E\left[\left(H - X^{-1}\left(HX + N_{\text{awgn}}\right)\right)^{\text{H}} \left(H - X^{-1}\left(HX + N_{\text{awgn}}\right)\right) \right] \tag{5.76}$$

$$\text{MSE}_{\text{LS}} = E\left[\left(X^{-1}N_{\text{awgn}}\right)^{\text{H}} \left(X^{-1}N_{\text{awgn}}\right) \right] = \frac{1}{\text{SNR}}. \tag{5.77}$$

The MMSE estimation based on Bayesian estimation is to minimize MSE between the actual channel response H and the LS estimated channel response H_{LS}. The idea of the MMSE estimation is illustrated in Figure 5.37.

The MMSE estimation has the property of orthogonality as follows:

$$E\left[eH_{\text{LS}}^{\text{H}} \right] = 0. \tag{5.78}$$

Using this orthogonality, we drive the MMSE weighting W_{MMSE} as follows:

$$E\left[(H - H_{\text{MMSE}})H_{\text{LS}}^{\text{H}} \right] = 0 \tag{5.79}$$

$$E\left[(H - W_{\text{MMSE}}H_{\text{LS}})H_{\text{LS}}^{\text{H}} \right] = 0 \tag{5.80}$$

$$E\left[HH_{\text{LS}}^{\text{H}} - W_{\text{MMSE}}H_{\text{LS}}H_{\text{LS}}^{\text{H}} \right] = 0 \tag{5.81}$$

$$R_{HH_{\text{LS}}} - W_{\text{MMSE}}R_{H_{\text{LS}}H_{\text{LS}}} = 0 \tag{5.82}$$

$$W_{\text{MMSE}} = R_{HH_{\text{LS}}} R_{H_{\text{LS}}H_{\text{LS}}}^{-1} \tag{5.83}$$

where $R_{HH_{\text{LS}}}$ and $R_{H_{\text{LS}}H_{\text{LS}}}$ are covariance matrix between the actual channel and the estimated channel and covariance matrix of the estimated channels, respectively. We can obtain the MMSE estimation as follows:

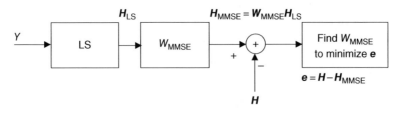

Figure 5.37   MMSE approach

$$H_{\text{MMSE}} = R_{HH_{LS}} R_{H_{LS}H_{LS}}{}^{-1} H_{LS} \qquad (5.84)$$

$$H_{\text{MMSE}} = R_{HH} \left(R_{HH} + \sigma_n^2 (XX^{\text{H}})^{-1} \right)^{-1} H_{LS} \qquad (5.85)$$

where σ_n^2 is the variance of the AWGN. The complexity of (5.85) is very high. Thus, we assume the same modulation scheme on each subcarrier and replace the term $(XX^{\text{H}})^{-1}$ with $E\left[(XX^{\text{H}})^{-1}\right]$. We simplify (5.85) as follows:

$$H_{\text{MMSE}} = R_{HH} \left(R_{HH} + \frac{\beta}{\text{SNR}} I \right)^{-1} H_{LS} \qquad (5.86)$$

where I is the identity matrix and SNR and β are defined as follows:

$$\text{SNR} = \frac{E[x_k]^2}{\sigma_n^2} \qquad (5.87)$$

$$\beta = E[x_k]^2 E\left[\frac{1}{x_k}\right]^2 . \qquad (5.88)$$

The parameter β depends on the modulation scheme. $\beta = 17/9$ when the modulation is 16 QAM. The detailed derivation is described in Ref. [14]. The MMSE estimation has the optimal solution in MSE point of view but its computational complexity is very high to obtain correlation function and carry out matrix operation. Figure 5.38 illustrates simple MSE comparison of the LS estimation and the MMSE estimation.

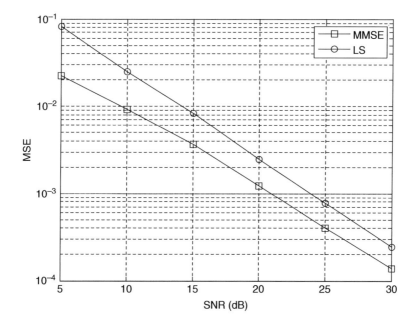

Figure 5.38 Comparison of the LS estimation and MMSE estimation

Summary 5.4   Equalization

1. The ISI occurs due to a time dispersive channel by multipath and an equalizer plays an important role in ISI compensation.
2. The equalization is to find the inverse of the combined impulse response.
3. The zero forcing equalizer is designed using the inverse of the channel frequency response. It ignores AWGN.
4. There are two types of channel estimations. The first type is to use training symbols (preambles) or pilot symbols and the other type is a blind channel estimation.
5. There are three types of pilot structures in OFDM system: block type, comb type, and lattice type.
6. The pilot signals do not cover a whole OFDM symbol. Thus, interpolation should be carried out at data subcarrier positions.

5.5   OFDM

C. Robert in Bell Labs proposed the OFDM patent in 1966 [15] but this technique did not attract a lot of public attention at that time. After two decades, L. Cimini suggested its use in mobile communications in 1985 [16]. People began to pay attention to this technique because of a number of advantages. Firstly, the OFDM equalizer is much simpler to implement than those in Code Division Multiple Access (CDMA). Secondly, the OFDM system is almost completely resistant to multipath fading due to very long symbols. Lastly, the OFDM system is ideally suited to MIMO techniques due to easy matching of transmit signals to the uncorrelated wireless channel. On the other hands, the disadvantage of the OFDM system is firstly it is sensitive to frequency errors and phase noises due to close subcarrier spacing. Secondly, it is sensitive to the Doppler shift which creates interferences between subcarriers. Thirdly, it creates a high peak to average power ratio. Lastly, it is more complex than other communication systems when handling interferences at the cell edge.

The OFDM technique is based on Frequency Division Multiplexing (FDM) which transmits multiple signals in multiple frequencies simultaneously. Figure 5.39 illustrates FDM symbols with three carriers with different carrier frequencies and each subcarrier is separated by a guard band.

At the receiver, individual subcarriers are detected and demodulated. One disadvantage of the FDM is a long guard band between the carriers. This long guard band makes spectral efficiency of the FDM system worse. On the other hand, the OFDM uses the similar concept but increases the spectral efficiency by reducing the guard band between the subcarriers. This can be achieved by orthogonality characteristic of the OFDM system.

Figure 5.40 illustrates OFDM symbols with three subcarriers. These subcarriers are overlapped. A part of the subcarrier C passes through the frequencies of the subcarrier $C-1$ and the subcarrier $C+1$. The side lobes radiated by the adjacent subcarriers ($C-1$ and $C+1$) cause an interference to the subcarrier C. However, this overlapping is acceptable due to the orthogonality. The OFDM system uses multiple subcarriers. Thus, it needs multiple local oscillators to generate them and multiple modulators to transmit them. However, a practical OFDM system uses Fast Fourier Transform (FFT) to generate this parallel data sequences.

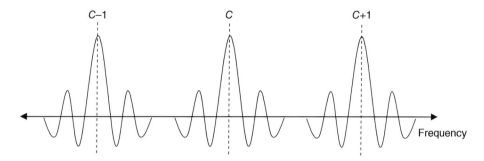

Figure 5.39 FDM with three carriers

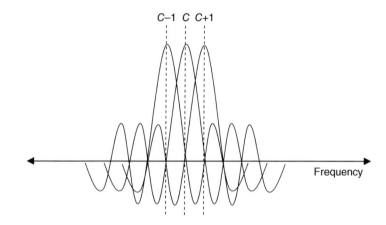

Figure 5.40 OFDM with three subcarriers

This is a big benefit because a local oscillator is expensive and not easy to implement. In the transmitter of the OFDM system, the data sequences are passed to Inverse FFT (IFFT) and these data sequences are converted into parallel data sequences which are combined by multiple subcarriers with maintaining the orthogonality between subcarriers. In the receiver, the parallel data sequences are converted into the serial data sequences by FFT. Although the OFDM system overcomes interferences in frequency domain by orthogonality, the interference problem still exists in time domain. One of the major problems in wireless communication systems is Inter-ISI. This is caused by multipath as we discussed in Chapter 3 and one important reason is a distorted original signal. In the OFDM system, a Cyclic Prefix (CP) or Zero Padding (ZP) is used to mitigate the effects of multipath propagation. This can be represented as a guard period which is located just in front of the data and is able to mitigate delay spreads.

We consider an OFDM system with N parallel data sequences as shown in Figure 5.41.

The baseband modulated symbol of the OFDM system can be represented as follows:

$$x(t) = \frac{1}{\sqrt{T_s}} \sum_{k=0}^{N-1} X_k e^{j2\pi f_k \frac{t}{T_s}}, \quad nT_s \leq t \leq (n+1)T_s \tag{5.89}$$

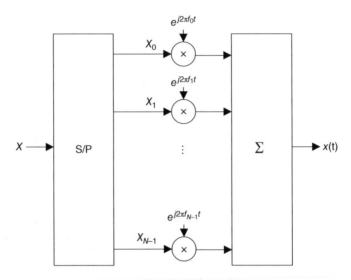

Figure 5.41   OFDM transmitter with N parallel data sequence

where X_k is the baseband modulated symbol such as BPSK, QPSK, or QAM and N is the total number of subcarriers. In this OFDM symbol, we can observe one subcarrier $\left(c_{k_1}(t)\right)$ is orthogonal to another subcarrier $\left(c_{k_2}(t)\right)$ as follows:

$$\frac{1}{T_s}\int_0^{T_s}c_{k_1}(t)c_{k_2}^*(t)dt = \frac{1}{T_s}\int_0^{T_s}e^{j2\pi f_{k_1}\frac{t}{T_s}}e^{-j2\pi f_{k_2}\frac{t}{T_s}}dt \tag{5.90}$$

$$=\frac{1}{T_s}\int_0^{T_s}e^{j2\pi\left(f_{k_1}-f_{k_2}\right)\frac{t}{T_s}}dt = \begin{cases}1, k_1 = k_2 \\ 0, k_1 \neq k_2\end{cases}. \tag{5.91}$$

The subcarrier spacing is expressed as follows:

$$\Delta f = f_k - f_{k-1} = \frac{1}{T_s}. \tag{5.92}$$

Thus, (5.89) is expressed as follows:

$$x(t) = \frac{1}{\sqrt{T_s}}\sum_{k=0}^{N-1}X_k e^{j2\pi k\frac{t}{T_s}}, \quad nT_s \leq t \leq (n+1)T_s \tag{5.93}$$

In addition, we can regard this signal as a discrete OFDM symbol when sampling the signal in every T_s/N. Thus, the OFDM symbol is expressed as follows:

$$x(n) = x\left(\frac{nT_s}{N}\right) = \frac{1}{\sqrt{T_s}}\sum_{k=0}^{N-1}X_k e^{j2\pi k\frac{1}{T_s}\frac{nT_s}{N}} = \frac{1}{\sqrt{T_s}}\sum_{k=0}^{N-1}X_k e^{j\frac{2\pi kn}{N}}. \tag{5.94}$$

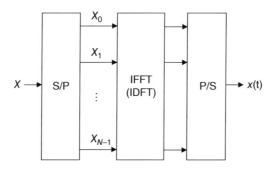

Figure 5.42   OFDM Transmitter using IFFT/IDFT

We represent the OFDM transmitter using IFFT (Inverse Discrete Fourier Transform, IDFT) as shown in Figure 5.42.

When we insert a cyclic prefix as a guard interval, we have the following OFDM symbol:

$$x(t) = \frac{1}{\sqrt{T_s}} \sum_{k=0}^{N-1} X_k e^{j2\pi k \frac{t}{T_s}}, \quad nT_s - T_g \leq t \leq (n+1)T_s \tag{5.95}$$

where T_g is a cyclic prefix length. This baseband signal is up-converted to a carrier frequency f_c and we obtain the following the transmitted OFDM signal:

$$s(t) = Re\left\{\sqrt{2}x(t)e^{j2\pi f_c t}\right\}. \tag{5.96}$$

The complex baseband signal $x(t)$ is represented in terms of real and imaginary part as follows:

$$x(t) = x_I(t) + jx_Q(t). \tag{5.97}$$

(5.96) is rewritten as follows:

$$s(t) = \sqrt{2}\left(x_I(t)\cos 2\pi f_c t - x_Q(t)\sin 2\pi f_c t\right). \tag{5.98}$$

The up-conversion from the baseband signal $x(t)$ to the passband signal $s(t)$ is illustrated in Figure 5.43.

In the receiver, we detect the following received signal $r(t)$:

$$r(t) = s(t) + n(t) \tag{5.99}$$

and then perform down-conversion from the passband signal to the baseband signal as shown in Figure 5.44.

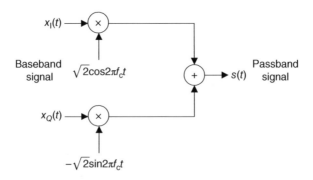

Figure 5.43   Up-conversion from the baseband signal to the passband signal

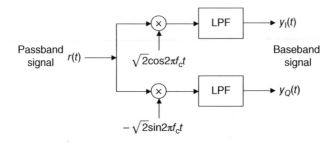

Figure 5.44   Down-conversion from the passband signal to the baseband signal

We perform synchronization process using the baseband signal $y(t)$. The OFDM signal is very sensitive to synchronization errors such as ISI and inter-carrier interference. Thus, this process is very important and should be implemented very carefully. Generally, the synchronization of the OFDM system is composed of three stages which are symbol timing synchronization, carrier frequency/phase offset synchronization, and sampling clock/sampling frequency synchronization. This synchronization process will be dealt in Chapter 10 in detail. After removing the CP, the baseband signal is extracted by FFT process as follows:

$$y_k(n) = \frac{1}{\sqrt{T_s}} \int_{nT_s}^{(n-1)T_s} y(t)e^{-j2\pi k\frac{t}{T_s}} dt = \tilde{X}_k + I_k + N_k \tag{5.100}$$

where $\tilde{X}_k$, I_k, and N_k are the estimated signal, interference, and AWGN, respectively. Equation (5.100) includes interferences. Therefore, we need to remove the undesired part and equalization should be carried out. The channel estimation and equalization will be dealt in Chapter 9 in detail. The blocks of the OFDM system are illustrated in Figure 5.45. The orthogonal frequency division multiple access (OFDMA) is a multiple access scheme based on the OFDM technique.

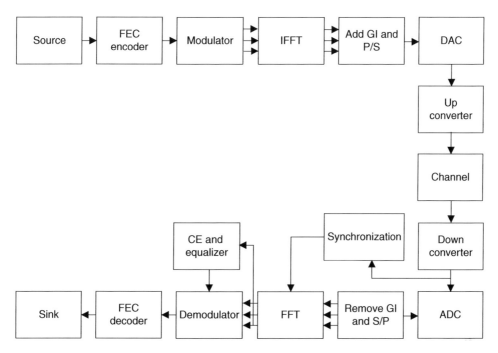

Figure 5.45   OFDM-based communication system

The subcarriers in the OFDM system are allocated to users. However, the subcarriers are shared by multiple users in the OFDMA system. The OFDMA system uses not only time domain resource but also frequency domain resource. Thus, we can achieve a higher spectral efficiency than the other multiple access schemes. In addition, its structure is well matched with MIMO system. Therefore, many broadband wireless communication systems adopted MIMO-OFDM/OFDMA system.

Summary 5.5   OFDM

1. The OFDM system has many advantages: (i) OFDM equalizers are much simpler to implement than those in CDMA. (ii) OFDM is almost completely resistant to multi-path fading due to very long symbols. (iii) OFDM is ideally suited to MIMO due to easy matching of transmit signals to the uncorrelated wireless channel.
2. The disadvantages of the OFDM system are (i) it is sensitive to frequency errors and phase noises due to close subcarrier spacing, (ii) it is sensitive to the Doppler shift which creates interferences between subcarriers, (iii) it creates a high PAPR, and (iv) it is more complex than other communication systems when handling interferences at the cell edge.

5.6 Problems

5.1. Describe which diversity techniques are used in LTE standard.

5.2. Determine the optimal weighting coefficients of a MRC receiver when the noises are uniform but the fading gains are different in each branch.

5.3. The diversity order means the number of independent paths over a wireless channel. Compare the bit error probabilities of diversity techniques for diversity orders 1, 2, 4, and 8.

5.4. The repetition code is defined as the message bit is repeated $N-1$ times. When $N=5$, find the minimum distance and the guaranteed error correction capability.

5.5. The Cyclic Redundancy Check (CRC) codes are one of the most common coding schemes in wireless communication systems. Show that all codewords of the CRC codes have the cyclic shift property.

5.6. Show that the minimum distance cannot obtain from the weight distribution when dealing with nonlinear error correction codes.

5.7. The Bose-Chaudhuri-Hocquenghem (BCH) codes have the following parameters:

$$n = 2^m - 1$$

$$k \geq n - mt$$

$$d_{min} \geq 2t + 1$$

$$m \geq 3$$

$$t \leq \frac{2^m - 1}{2}.$$

5.8. Consider a (15, 7, 2) BCH code with generator polynomial $= X^8 + X^7 + X^6 + X^4 + 1$. Design the BCH encoder and decoder.

5.9. Plot the state diagram and trellis diagram of the convolutional code with generator vectors ($g_1 = [1\ 0\ 1\ 1]$ and $g_2 = [1\ 1\ 0\ 1]$).

5.10. Consider the convolutional encoder with generator vectors ($g_1 = [1\ 0\ 0]$, $g_2 = [1\ 0\ 1]$, and $g_3 = [1\ 1\ 1]$). Describe the encoding process.

5.11. Compare the BER of convolutional codes with constraint length 3, 5, and 7.

5.12. Find the transfer function of convolutional code in Problem 5.4.

5.13. Show that the puncturing schemes reduce the free distance.

5.14. Describe the relationship between the performance and the degree of quantization of the input signal in the Viterbi decoder.

5.15. In IS-95 CDMA standard, the convolutional code has the following parameters: code rate $= 1/2$, constraint length $= 9$, and generator vectors ($g_1 = 753$ (octal) and $g_1 = 561$ (octal)). Design the convolutional encoder and Viterbi decoder for IS-95 CDMA standard.

5.16. In the single input single output (SISO), the capacity C is defined as follows:

$$C_{siso} = \log_2(1 + SNR).$$

5.17. Find the capacity C_{mimo} for MIMO system with N transmit and M receive antennas.

5.18. Explain why space time coding outperforms spatial multiplexing at low SNR and spatial multiplexing outperforms space time coding at high SNR.

5.19. Describe which MIMO techniques are used in LTE standard.

5.20. Compare the time domain equalizer with the frequency domain equalizer.

5.21. Describe the pros and cons of the linear equalizer and the nonlinear equalizer.

5.22. Compare the single-carrier transmission with the multi-carrier transmission.

5.23. Compare the spectral efficiencies among OFDMA, CDMA, and TDMA.

References

[1] R. G. Gallager, *Low Density Parity Check Codes*, MIT Press, Cambridge, 1963.
[2] D. J. C. MacKay and R. M. Neal, "Near Shannon Limit Performance of Low Density Parity Check Codes," *Electronics Letters*, vol. **32**, p. 1645, 1996.
[3] J. M. Wozencraft, "Sequential Decoding for Reliable Communication," *IRE National Convention Recode*, vol. **5**, pt. 2, pp. 11–25, 1957.
[4] B. Berrou, B. Glavieux, and P. Thitimajshima, "Near Shannon Limit Error-Correcting Coding and Decoding: Turbo-Code," *IEEE International Conference on Communications 1993 (ICC 93)*, vol. **2**, pp. 1064–1070, 1993.
[5] M. Luby, "LT Codes," *Proceedings of the IEEE Symposium on the Foundations of Computer Science*, pp. 271–280, November 2002.
[6] A. Shokrollahi, "Raptor Codes," *IEEE Transactions on Information Theory*, vol. **52**, no. 6), pp. 2551–2567, 2006.
[7] E. Arikan, "Channel Polarization: A Method for Constructing Capacity-Achieving Codes for Symmetric Binary-Input Memoryless Channels," *IEEE Transactions on Information Theory*, vol. **55**, no. 7, pp. 3051–3073, 2009.
[8] P. Elias, "Coding for Noisy Channels," *IRE Convention Recode*, vol. **3**, pt. 4, pp. 37–46, 1955.
[9] S. M. Alamouti, "A Simple Transmit Diversity Technique for Wireless Communication," *IEEE Journal on Selected Areas in Communications*, vol. **16**, no. 8, pp. 1451–1458, 1998.
[10] V. Tarokh, H. Jafarkhani, and A. R. Calderbank, "Space-Time Block Codes from Orthogonal Designs," *IEEE Transactions on Information Theory*, vol. **45**, no. 5, pp. 1456–1467, 1999.
[11] V. Tarokh, N. Seshadri, and A. R. Calderbank, "Space-Time Codes for High Data Rate Wireless Communication: Performance Criterion and Code Construction," *IEEE Transactions on Information Theory*, vol. **44**, no. 2, pp. 744–765, 1998.

[12] G. J. Foschini, "Layered space-time architecture for wireless communication in a fading environment when using multi-element antennas," *Bell Labs Technical Journal*, vol. **1**, pp. 41–59, 1996.

[13] M. K. Ozdemir and H. Arslan, "Channel Estimation for Wireless OFDM Systems," *IEEE Communications Surveys & Tutorials*, vol. **9**, no. 2, pp. 18–48, 2007.

[14] O. Edfors, M. Sandell, J. van de Beek, S. K. Wilson, and P. O Börjesson, "OFDM Channel Estimation by Singular Value Decomposition," *Proceedings of the 46th IEEE Vehicular Technology Conference (VTC spring'96)*, Atlanta, GA, pp. 923–927, April–May 1996.

[15] C. Robert, "Orthogonal Frequency Multiplex Data Transmission System," US Patent 3488445.

[16] L. J. Cimini, Jr., "Analysis and Simulation of a Digital Mobile Channel Using Orthogonal Frequency Division Multiplexing," *IEEE Transactions on Communications*, vol. **COM-33**, no. 7, pp. 665–675, 1985.

Part II

Wireless Communications Blocks Design

6

Error Correction Codes

In Part II, we design key wireless communication blocks such as error correction codes, OFDM, MIMO, channel estimation, equalization, and synchronization. The floating point design step is to check the reliability of each block in wireless communication system design flows. In this step, we select suitable algorithms and evaluate their performances. The floating point design is getting more important because the modern wireless communication system design flow is automated and many parts of the back-end design flow are already optimized. Thus, we focus on the floating point design in this book. In Part III, we will deal with the wireless communication system design flow and system integration in detail. An error correction code is an essential part of wireless communication systems. In this chapter, we design the turbo encoder/decoder, turbo product encoder/decoder, and Low-Density Parity Check (LDPC) encoder/decoder. These codes opened a new era of coding theory because their performance is close to the Shannon limit. We look into design criteria, describe the algorithms, and discuss hardware implementation issues.

6.1 Turbo Codes

6.1.1 Turbo Encoding and Decoding Algorithm

The turbo codes [Ref. 4 in Chapter 5] achieved very low error probability which is close to the Shannon limit. They became an essential part of many broadband wireless communication systems. The idea of the revolutionary error correction codes originates from *concatenated encoding*, *randomness*, and *iterative decoding*. The turbo encoder is composed of two Recursive Systematic Convolutional (RSC) encoders as component codes. Two component encoders are concatenated in parallel. The important design criteria of the turbo codes are to find suitable component codes which maximize the effective free distance [1] and to optimize the

Wireless Communications Systems Design, First Edition. Haesik Kim.
© 2015 John Wiley & Sons, Ltd. Published 2015 by John Wiley & Sons, Ltd.

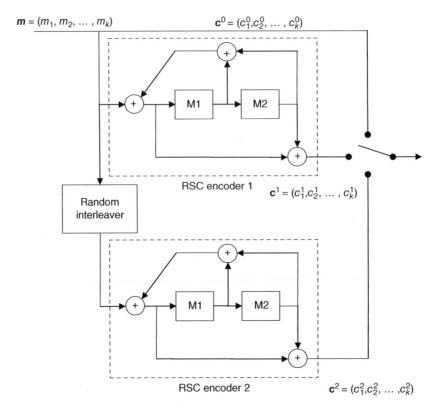

Figure 6.1   Turbo encoder

weight distribution of the codewords at a low $E_b N_0$ [2]. A random interleaver is placed between two component encoders. The information (or message) bits are encoded by one component encoder and reordered by a random interleaver. The permuted information bits are encoded by another component encoder. Then, we can puncture parity bits to adjust the code rate. The weight distribution of the turbo codes is related to how the codeword from one component encoder correlates the codeword from another component encoder. The random interleaver plays an important role as preventing codeword paring between two component encoders and makes codewords from each component encoder statistically independent. A random interleaver design is not easy because there is no standardized method. However, the important design criteria of a random interleaver are to find a low complexity implementation and produce both low-weight codewords from one component encoder and high-weight codewords from another component encoder simultaneously. One example of the turbo encoder structure with code rate 1/3 is illustrated in Figure 6.1.

As we can observe from Figure 6.1, the turbo encoder is systematic because the information sequence m is same as coded bit c^0. Each RSC encoder includes two shift registers (memories). Thus, a finite state machine represents the state as the content of the memory. The output of each RSC encoder is calculated by modulo-2 addition among the input bit and the stored bits in two shift registers. The memory states and the outputs corresponding to the inputs are summarized in Table 6.1. We can transform this table into a state diagram or a trellis

Table 6.1 Memory states (M1 M2), inputs, and outputs of RSC encoder

Time	Input (m_i)	Memory state (M1 M2)	Output $\left(c_i^0, c_i^1\right)$
0		00	
1	1	10	11
2	1	01	10
3	0	10	00
4	0	11	01
5	1	11	10
6	0	01	01
7	1	00	11
8	0	00	00

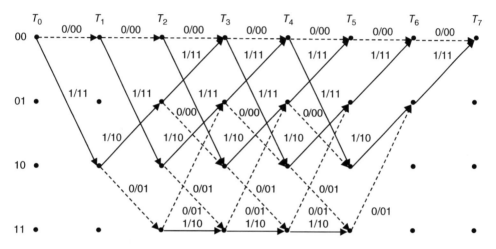

Figure 6.2 Trellis diagram of RSC encoder

diagram as we did in Chapter 5. The trellis diagram is widely used to represent turbo encoding process. Thus, we express this table as the trellis diagram as shown in Figure 6.2.

The turbo decoder based on an iterative scheme is composed of two Soft Input Soft Output (SISO) component decoders. Figure 6.3 illustrates the turbo decoder structure.

As we can observe from Figure 6.3, two component decoders give and take their output as a priori information and reduce the error probability of the original information bits. The SISO

Summary 6.1   Turbo code design

Two important design criteria of turbo codes are to:

1. Find component codes maximizing the effective free distance and optimizing the weight distribution of the codewords at a low $E_b N_0$.
2. Find a low complexity random interleaver preventing codeword paring between two component encoders.

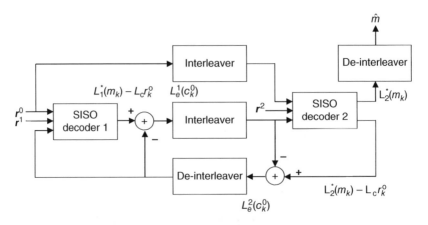

Figure 6.3 Turbo decoder

decoder is based on the Maximum a Posteriori (MAP) algorithm (also known as Bahl-Cocke-Jelinek-Raviv (BCJR) algorithm). The RSC encoders 1 and 2 of the turbo encoder correspond to the SISO decoders 1 and 2 of the turbo decoder, respectively. The turbo decoder receives three inputs (r^0, r^1, and r^2). The SISO decoder 1 uses the information sequence (r^0), the parity sequence (r^1) by RSC encoder 1, and a priori (or extrinsic) information. The SISO decoder 2 uses the interleaved information sequence (r^0), the parity sequence (r^2) by RSC encoder 2, and a priori information. The SISO decoder 1 starts decoding without a priori information at first and provides the SISO decoder 2 with the output as a priori information. The SISO decoder 2 starts decoding with this priori information, interleaved information r^0, and parity r^2. In the iteration process, the SISO decoder 1 receives a priori information from the SISO decoder 2 and decodes with a priori information, information r^0, and parity r^1 again. This iteration enables the decoder to reduce the error probability of the information bits but brings a long latency. In addition, the coding gain decreases rapidly at the high number of iteration.

Due to the high complexity, long latency, and high energy consumption of the MAP algorithm, it was not widely used until the turbo codes appear. Both the MAP algorithm and the Viterbi algorithm have similar decoding performance but the Viterbi algorithm does not produce soft outputs. Thus, the Viterbi algorithm is not suitable for the component decoder of the turbo decoder. The MAP decoder formulates the bit probabilities as soft outputs using the Log Likelihood Ratio (LLR). The soft output of the first MAP decoder is defined as follows:

$$L_1(m_k) = \log\left[\frac{P\left(m_k = +1 \middle| \left(r_k^0, r_k^1\right)\right)}{P\left(m_k = -1 \middle| \left(r_k^0, r_k^1\right)\right)}\right]. \tag{6.1}$$

The LLR calculates and compares the probabilities when the information m_k is +1 and −1 after observing two inputs $\left(r_k^0 \text{ and } r_k^1\right)$. We calculate the probability when the information m_k is +1 using the Bayes' theorem as follows:

$$P\left(m_k = +1 \middle| \left(r_k^0, r_k^1\right)\right) = \frac{P\left(m_k = +1, \left(r_k^0, r_k^1\right)\right)}{P\left(r_k^0, r_k^1\right)} \tag{6.2}$$

$$= \frac{\sum_{(s',s)\in S^p} P\left(s_{k-1} = s', s_k = s \middle| \left(r_k^0, r_k^1\right)\right)}{P\left(r_k^0, r_k^1\right)} \tag{6.3}$$

$$= \frac{\sum_{(s',s)\in S^p} P\left(s_{k-1} = s', s_k = s, \left(r_k^0, r_k^1\right)\right) \middle/ P\left(r_k^0, r_k^1\right)}{P\left(r_k^0, r_k^1\right)}. \tag{6.4}$$

Likewise, we calculate the probability when the information m_k is -1 as follows:

$$P\left(m_k = -1 \middle| \left(r_k^0, r_k^1\right)\right) = \frac{P\left(m_k = -1, \left(r_k^0, r_k^1\right)\right)}{P\left(r_k^0, r_k^1\right)} \tag{6.5}$$

$$= \frac{\sum_{(s',s)\in S^n} P\left(s_{k-1} = s', s_k = s \middle\| \left(r_k^0, r_k^1\right)\right)}{P\left(r_k^0, r_k^1\right)} \tag{6.6}$$

$$= \frac{\sum_{(s',s)\in S^n} P\left(s_{k-1} = s', s_k = s, \left(r_k^0, r_k^1\right)\right) \middle/ P\left(r_k^0, r_k^1\right)}{P\left(r_k^0, r_k^1\right)}. \tag{6.7}$$

Now, we reformulate the LLR in terms of the trellis diagram as follows:

$$L_1(m_k) = \log\left[\frac{\sum_{(s',s)\in S^p} P\left(s_{k-1} = s', s_k = s, \left(r_k^0, r_k^1\right)\right)}{\sum_{(s',s)\in S^n} P\left(s_{k-1} = s', s_k = s, \left(r_k^0, r_k^1\right)\right)}\right] \tag{6.8}$$

where s_k denotes the state at time k in the trellis diagram, s' and s are a state of the trellis diagram, and S^p and S^n are transition sets of the trellis diagram when $m_k = +1$ and $m_k = -1$, respectively. Figure 6.4 illustrates the transition set of Figure 6.2 trellis diagram.

The numerator and denominator probability term of the LLR can be rewritten as follows:

$$P\left(s_{k-1} = s', s_k = s, \left(r_k^0, r_k^1\right)\right) = P\left(s', s, \left(r_k^0, r_k^1\right)\right) \tag{6.9}$$

$$= P\left(s', s, \left(r_{k-1}^0, r_{k-1}^1\right), \left(r_k^0, r_k^1\right), \left(r_{k+1}^0, r_{k+1}^1\right)\right) \tag{6.10}$$

$$= P\left(\left(r_{k+1}^0, r_{k+1}^1\right) \middle| s', s, \left(r_{k-1}^0, r_{k-1}^1\right), \left(r_k^0, r_k^1\right)\right) P\left(s', s, \left(r_{k-1}^0, r_{k-1}^1\right), \left(r_k^0, r_k^1\right)\right) \tag{6.11}$$

$$= P\left(\left(r_{k+1}^0, r_{k+1}^1\right) \middle| s', s, \left(r_{k-1}^0, r_{k-1}^1\right), \left(r_k^0, r_k^1\right)\right) P\left(s, \left(r_k^0, r_k^1\right) \middle| s', \left(r_{k-1}^0, r_{k-1}^1\right)\right) P\left(s', \left(r_{k-1}^0, r_{k-1}^1\right)\right) \tag{6.12}$$

$$= P\left(\left(r_{k+1}^0, r_{k+1}^1\right) \middle| s\right) P\left(s, \left(r_k^0, r_k^1\right) \middle| s'\right) P\left(s', \left(r_{k-1}^0, r_{k-1}^1\right)\right) \tag{6.13}$$

$$= P\left(\left(r_{k+1}^0, r_{k+1}^1\right) \middle| s_k = s\right) P\left(s_k = s, \left(r_k^0, r_k^1\right) \middle| s_{k-1} = s'\right) P\left(s_{k-1} = s', \left(r_{k-1}^0, r_{k-1}^1\right)\right) \tag{6.14}$$

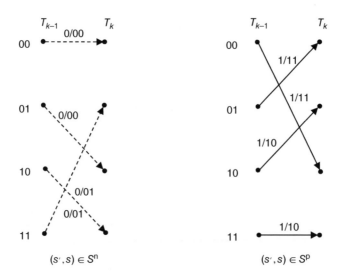

Figure 6.4   Transition sets of Figure 6.2 trellis diagram

Then, we define the forward path metric $(\alpha_{k-1}(s_{k-1} = s'))$, the backward path metric $(\beta_k(s_k = s))$, and the branch metric $(\gamma_k(s_{k-1} = s', s_k = s))$ as follows:

$$\alpha_{k-1}(s_{k-1} = s') = P\left(s_{k-1} = s', \left(r_{k-1}^0, r_{k-1}^1\right)\right) \qquad (6.15)$$

$$\beta_k(s_k = s) = P\left(\left(r_{k+1}^0, r_{k+1}^1\right)\Big| s_k = s\right) \qquad (6.16)$$

$$\gamma_k\left(s_{k-1} = s', s_k = s\right) = P\left(s_k = s, \left(r_k^0, r_k^1\right)\Big| s_{k-1} = s'\right). \qquad (6.17)$$

Thus, (6.9) is rewritten as follows:

$$P\left(s_{k-1} = s', s_k = s, \left(r_k^0, r_k^1\right)\right) \qquad (6.18)$$
$$= \beta_k(s_k = s)\gamma_k(s_{k-1} = s', s_k = s)\alpha_{k-1}(s_{k-1} = s')$$

and (6.8) is rewritten using the above three metrics as follows:

$$L_1(m_k) = \log\left[\frac{\sum_{(s',s)\in S^p} \beta_k(s_k = s)\gamma_k(s_{k-1} = s', s_k = s)\alpha_{k-1}(s_{k-1} = s')}{\sum_{(s',s)\in S^n} \beta_k(s_k = s)\gamma_k(s_{k-1} = s', s_k = s)\alpha_{k-1}(s_{k-1} = s')}\right]. \qquad (6.19)$$

The forward path metric is calculated from time index $k-1$ to time index k as follows:

$$\alpha_k(s_k = s) = P\left(s_k = s, \left(r_k^0, r_k^1\right)\right) \qquad (6.20)$$

$$= \sum_{s_{k-1}} P\left(s_{k-1} = s', \left(r_{k-1}^0, r_{k-1}^1\right)\right) P\left(s_k = s, \left(r_k^0, r_k^1\right)\Big| s_{k-1} = s'\right) \qquad (6.21)$$

$$= \sum_{s_{k-1}} \alpha_{k-1}(s_{k-1} = s')\gamma_k(s_{k-1} = s', s_k = s). \qquad (6.22)$$

The initial value of the forward path metric is given as follows:

$$\alpha_0(s_k = s) = \begin{cases} 1, & s = 0 \\ 0, & s \neq 0 \end{cases} \tag{6.23}$$

because the RSC encoder starts at state 0 (00 in the example). The backward path metric is calculated from time index $k+1$ to time index k as follows:

$$\beta_k(s_k = s') = P\left(\left(r_{k+1}^0, r_{k+1}^1\right) \middle| s_k = s'\right) \tag{6.24}$$

$$= \sum_{s_{k+1}} P\left(s_{k+1} = s, \left(r_{k+1}^0, r_{k+1}^1\right) \middle| s_k = s'\right) P\left(\left(r_{k+2}^0, r_{k+2}^1\right) \middle| s_{k+1} = s\right) \tag{6.25}$$

$$= \sum_{s_{k+1}} \gamma_{k+1}(s_k = s', s_{k+1} = s)\beta_{k+1}(s_{k+1} = s). \tag{6.26}$$

Likewise, the initial value of the backward path metric is given as follows:

$$\beta_K(s_k = s') = \begin{cases} 1, & s' = 0 \\ 0, & s' \neq 0 \end{cases} \tag{6.27}$$

because we generally terminate the trellis at state 0 (00 in the example). The branch metric is rewritten as follows:

$$\gamma_k(s_{k-1} = s', s_k = s) = P\left(s_k = s, \left(r_k^0, r_k^1\right) \middle| s_{k-1} = s'\right) \tag{6.28}$$

$$= P\left(\left(r_k^0, r_k^1\right) \middle| s_{k-1} = s', s_k = s\right) P\left(s_k = s \middle| s_{k-1} = s'\right) \tag{6.29}$$

$$= P\left(\left(r_k^0, r_k^1\right) \middle| \left(c_k^0, c_k^1\right)\right) P\left(c_k^0\right) \tag{6.30}$$

where $P\left(\left(r_k^0, r_k^1\right) \middle| \left(c_k^0, c_k^1\right)\right)$ and $P\left(c_k^0\right)$ represent a likelihood probability and a priori probability, respectively. The likelihood probability is expressed as follows:

$$P\left(\left(r_k^0, r_k^1\right) \middle| \left(c_k^0, c_k^1\right)\right) = P\left(r_k^0 \middle| c_k^0\right) P\left(r_k^1 \middle| c_k^1\right) \tag{6.31}$$

$$= \frac{\omega_0}{\sqrt{2\pi\sigma^2}} e^{-\frac{\left(r_k^0 - c_k^0\right)^2}{2\sigma^2}} \frac{\omega_1}{\sqrt{2\pi\sigma^2}} e^{-\frac{\left(r_k^1 - c_k^1\right)^2}{2\sigma^2}} \tag{6.32}$$

$$= A_k e^{\left(r_k^0 c_k^0 - r_k^1 c_k^1\right)/\sigma^2} \tag{6.33}$$

where ω_0 and ω_1 are very small values and these terms are ignored. The priori probability is expressed as follows:

$$P\left(c_k^0\right) = \frac{e^{L_a\left(c_k^0\right)/2}}{1 + e^{L_a\left(c_k^0\right)}} e^{c_k^0 L_a\left(c_k^0\right)/2} \tag{6.34}$$

$$= B_k \, e^{c_k^0 L_a \left(c_k^0 \right)/2} \tag{6.35}$$

where

$$L_a \left(c_k^0 \right) = \log \left[\frac{P \left(c_k^0 = +1 \right)}{P \left(c_k^0 = -1 \right)} \right]. \tag{6.36}$$

The branch metric is rewritten as follows:

$$\gamma_k (s_{k-1} = s', s_k = s) = A_k B_k e^{\left(c_k^0 L_a \left(c_k^0 \right) + c_k^0 L_c r_k^0 + c_k^1 L_c r_k^1 \right)/2} \tag{6.37}$$

where

$$L_c = \frac{2}{\sigma^2}. \tag{6.38}$$

Now, we reformulate LLR as follows:

$$L_1 (m_k) = \log \left[\frac{\sum_{(s',s) \in S^P} \beta_k (s_k = s) \gamma_k (s_{k-1} = s', s_k = s) \alpha_{k-1} (s_{k-1} = s')}{\sum_{(s',s) \in S^n} \beta_k (s_k = s) \gamma_k (s_{k-1} = s', s_k = s) \alpha_{k-1} (s_{k-1} = s')} \right] \tag{6.39}$$

$$= \log \left[\frac{\sum_{(s',s) \in S^P} \beta_k (s_k = s) \left(e^{\left(c_k^0 L_a \left(c_k^0 \right) + c_k^0 L_c r_k^0 + c_k^1 L_c r_k^1 \right)/2} \right) \alpha_{k-1} (s_{k-1} = s')}{\sum_{(s',s) \in S^n} \beta_k (s_k = s) \left(e^{\left(-c_k^0 L_a \left(c_k^0 \right) - c_k^0 L_c r_k^0 + c_k^1 L_c r_k^1 \right)/2} \right) \alpha_{k-1} (s_{k-1} = s')} \right] \tag{6.40}$$

$$= \log \left[\frac{\sum_{(s',s) \in S^P} e^{c_k^0 L_a \left(c_k^0 \right)/2} e^{c_k^0 L_c r_k^0 /2} \beta_k \left(s_k = s \right) e^{c_k^1 L_c r_k^1 /2} \alpha_{k-1} (s_{k-1} = s')}{\sum_{(s',s) \in S^n} e^{-c_k^0 L_a \left(c_k^0 \right)/2} e^{-c_k^0 L_c r_k^0 /2} \beta_k \left(s_k = s \right) e^{c_k^1 L_c r_k^1 /2} \alpha_{k-1} (s_{k-1} = s')} \right] \tag{6.41}$$

where $e^{\pm c_k^0 L_a \left(c_k^0 \right)/2}$ and $e^{\pm c_k^0 L_c r_k^0 /2}$ do not depend on state transition. Thus, (6.41) is written as follows:

$$L_1 (m_k) = L_a \left(c_k^0 \right) + L_c r_k^0 + \log \left[\frac{\sum_{(s',s) \in S^P} \beta_k (s_k = s) e^{c_k^1 L_c r_k^1 /2} \alpha_{k-1} (s_{k-1} = s')}{\sum_{(s',s) \in S^n} \beta_k (s_k = s) e^{c_k^1 L_c r_k^1 /2} \alpha_{k-1} (s_{k-1} = s')} \right] \tag{6.42}$$

$$= L_a \left(c_k^0 \right) + L_c r_k^0 + L_e \left(c_k^0 \right) \tag{6.43}$$

where $L_a \left(c_k^0 \right)$ is the initial priori information of the MAP algorithm, $L_c r_k^0$ is the channel value, and $L_e \left(c_k^0 \right)$ is the extrinsic information. The extrinsic information is used as a priori information of the second MAP decoder.

Summary 6.2   MAP decoder design

1. The forward path metric is calculated as follows:

$$\alpha_k(s_k = s) = \sum_{s_{k-1}} \alpha_{k-1}(s_{k-1} = s')\gamma_k(s_{k-1} = s', s_k = s)$$

$$\alpha_0(s_k = s) = \begin{cases} 1, & s = 0 \\ 0, & s \neq 0 \end{cases}.$$

2. The backward path metric is calculated as follows:

$$\beta_k(s_k = s') = \sum_{s_{k+1}} \gamma_{k+1}(s_k = s', s_{k+1} = s)\beta_{k+1}(s_{k+1} = s)$$

$$\beta_K(s_k = s') = \begin{cases} 1, & s' = 0 \\ 0, & s' \neq 0 \end{cases}.$$

3. The branch metric is calculated as follows:

$$\gamma_k(s_{k-1} = s', s_k = s) = A_k B_k e^{\left(c_k^0 L_a\left(c_k^0\right) + c_k^0 L_c r_k^0 + c_k^1 L_c r_k^1\right)/2}.$$

4. The LLR is calculated as follows:

$$L(m_k) = \log\left[\frac{\sum_{(s',s)\in S^p}\beta_k(s_k = s)\gamma_k(s_{k-1} = s', s_k = s)\alpha_{k-1}(s_{k-1} = s')}{\sum_{(s',s)\in S^n}\beta_k(s_k = s)\gamma_k(s_{k-1} = s', s_k = s)\alpha_{k-1}(s_{k-1} = s')}\right].$$

6.1.2 *Example of Turbo Encoding and Decoding*

Consider the turbo encoder with code rate 1/3 as shown in Figure 6.1 and the trellis diagram of the RSC encoder as shown in Figure 6.2. When the input sequence is [1 1 0 0 1], the initial state (T_0) of the memory is "00," and the two tailing bits are [0 1], we can express the input, output, and memory state in the trellis diagram as shown in Figure 6.5. As we can observe from Figure 6.5, there is the unique path in the trellis diagram of the RSC encoder 1.

Figure 6.5 shows that the output of the RSC encoder 1 is [**11 10 00 01 10 01 11**]. We define the random interleaver of the turbo encoder as shown in Figure 6.6.

After the information sequence is passed through the random interleaver, the input of the RSC encoder 2 becomes [0 1 1 0 1 1 0]. Likewise, we have the unique path in the trellis diagram of the RSC encoder 2 as shown in Figure 6.7.

The output of the RSC encoder 2 is [00 11 10 00 10 11 00]. Thus, the output of the turbo encoder is $c^0 = [c_1^0, c_2^0, c_3^0, c_4^0, c_5^0, c_6^0, c_7^0] = $ [1 1 0 0 1 0 1], $c^1 = [c_1^1, c_2^1, c_3^1, c_4^1, c_5^1, c_6^1, c_7^1] = $ [1 0 0 1 0 1 1], and $c^2 = [c_1^2, c_2^2, c_3^2, c_4^2, c_5^2, c_6^2, c_7^2] = $ [0 1 0 0 0 1 0].

We map the output into the symbol of BPSK modulation and transmit it sequentially. Thus, the transmitted symbol sequence is $[c_1^0, c_1^1, c_1^2, c_2^0, c_2^1, c_2^2, \dots, c_7^0, c_7^1, c_7^2] \rightarrow$ [+1 +1 −1 +1

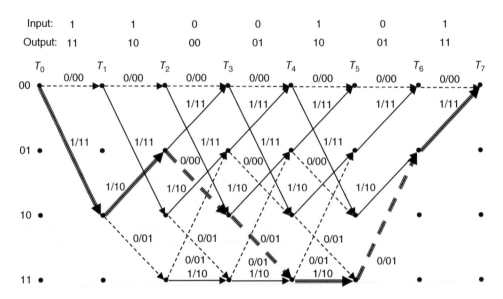

Figure 6.5 Encoding example of RSC encoder 1

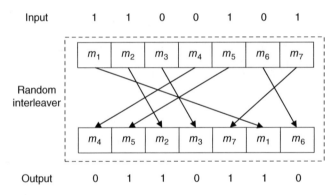

Figure 6.6 Random interleaver of the turbo encoder

$-1 +1 -1 -1 -1 -1 +1 -1 +1 -1 -1 -1 +1 +1 +1 +1 -1$]. In the channel, the transmitted symbol sequence is added by Gaussian noise as shown in Figure 6.8. In the receiver, the turbo decoder as shown in Figure 6.3 uses the received symbol sequence $[r^0 \ r^1 \ r^2]$ as the soft decision input. The first MAP decoder (SISO decoder 1) of the turbo decoder starts decoding. As we observed the above MAP algorithm, it is much more complex than the Viterbi algorithm. Thus, we simplify the MAP algorithm with a view to reducing implementation complexity. The Log-MAP algorithm and the Max-log-MAP algorithm as suboptimal algorithms provide us with a reasonable complexity. The Log-MAP algorithm is based on the following approximation:

$$\ln(e^x + e^y) \approx \max(x, y) + \ln(1 + e^{-|x-y|}). \tag{6.44}$$

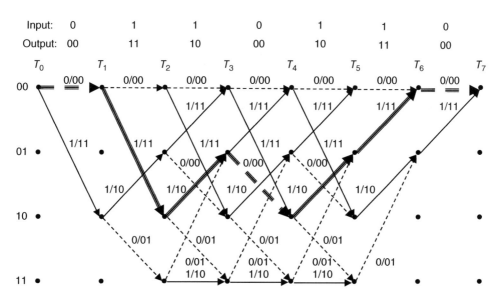

Figure 6.7   Encoding example of RSC encoder 2

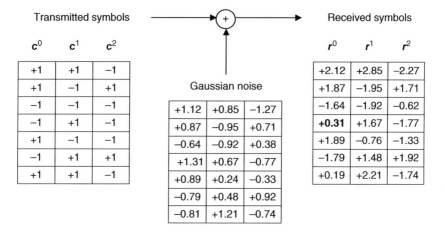

Figure 6.8   Channel noise example

The Max-log-MAP algorithm is based on the following approximation:

$$\ln(e^x + e^y) \approx \max(x, y). \tag{6.45}$$

According to Ref. [3], The Max-log-MAP algorithm has less complexity than the Log-MAP algorithm but brings about 0.35 dB performance degradation in the turbo decoding. Besides these approximations, the Soft Output Viterbi Algorithm (SOVA) which is modified by the Viterbi algorithm enables the decoder to emit soft decision outputs. Its complexity is decreased by about half of the MAP algorithm. However, the performance drop is about 0.6 dB.

In this section, we describe the Max-log-MAP algorithm. The forward path metric is simplified as follows:

$$\alpha_k{}^*(s_k = s) = \ln \alpha_k(s_k = s) \tag{6.46}$$

$$= \ln\left(\sum_{s_{k-1}} \alpha_{k-1}(s_{k-1} = s')\gamma_k(s_{k-1} = s', s_k = s)\right) \tag{6.47}$$

$$= \ln\left(\sum_{s_{k-1}} e^{(\alpha_{k-1}{}^*(s_{k-1}=s')+\gamma_k{}^*(s_{k-1}=s',s_k=s))}\right) \tag{6.48}$$

$$\approx \max_{s_{k-1}}(\alpha_{k-1}{}^*(s_{k-1} = s') + \gamma_k{}^*(s_{k-1} = s', s_k = s)). \tag{6.49}$$

and the initial value of the forward path metric is given as follows:

$$\alpha_0{}^*(s_k = s) = \begin{cases} 0, & s = 0 \\ -\inf, & s \neq 0 \end{cases}. \tag{6.50}$$

The backward path metric is simplified as follows:

$$\beta_{k-1}{}^*(s_{k-1} = s') = \ln \beta_{k-1}(s_{k-1} = s') \tag{6.51}$$

$$= \ln\left(\sum_{s_k} \gamma_k(s_{k-1} = s', s_k = s)\beta_k(s_k = s)\right). \tag{6.52}$$

$$= \ln\left(\sum_{s_k} e^{(\gamma_k{}^*(s_{k-1}=s',s_k=s)+\beta_k{}^*(s_k=s))}\right) \tag{6.53}$$

$$\approx \max_{s_k}(\gamma_k{}^*(s_{k-1} = s', s_k = s) + \beta_k{}^*(s_k = s)) \tag{6.54}$$

and the initial value of the backward path metric is given as follows:

$$\beta_K{}^*(s_k = s') = \begin{cases} 0, & s' = 0 \\ -\inf, & s' \neq 0 \end{cases}. \tag{6.55}$$

The branch metric is simplified as follows:

$$\gamma_k{}^*(s_{k-1} = s', s_k = s) = \ln \gamma_k(s_{k-1} = s', s_k = s) \tag{6.56}$$

$$= \ln\left(A_k B_k e^{\left(c_k^0 L_a\left(c_k^0\right)+c_k^0 L_c r_k^0 +c_k^1 L_c r_k^1\right)/2}\right) \tag{6.57}$$

$$\approx \left(c_k^0 L_a\left(c_k^0\right)+c_k^0 L_c r_k^0 +c_k^1 L_c r_k^1\right)\Big/2. \tag{6.58}$$

where we assume

$$L_c = \frac{2}{\sigma^2} = 2. \tag{6.59}$$

Thus, we obtain the simplified branch metric as follows:

$$\gamma_k^*\left(s_{k-1} = s', s_k = s\right) \approx c_k^0 L_a\left(c_k^0\right) + c_k^0 r_k^0 + c_k^1 r_k^1. \tag{6.60}$$

Using these simplified metrics and the trellis diagram, we firstly calculate the branch metrics of the first MAP decoder as follows:

$$\gamma_1^*(s_0 = 00, s_1 = 00) = (-1) \cdot (+2.12) + (-1) \cdot (+0.85) = -2.97$$

$$\gamma_1^*(s_0 = 00, s_1 = 10) = (+1) \cdot (+2.12) + (+1) \cdot (+0.85) = +2.97$$

$$\gamma_2^*(s_1 = 00, s_2 = 00) = (-1) \cdot (+1.87) + (-1) \cdot (-1.95) = +0.08$$

$$\gamma_2^*(s_1 = 00, s_2 = 10) = (+1) \cdot (+1.87) + (+1) \cdot (-1.95) = -0.08$$

$$\gamma_2^*(s_1 = 10, s_2 = 01) = (+1) \cdot (+1.87) + (-1) \cdot (-1.95) = +3.82$$

$$\gamma_2^*(s_1 = 10, s_2 = 11) = (-1) \cdot (+1.87) + (+1) \cdot (-1.95) = -3.82$$

$$\gamma_3^*(s_2 = 00, s_3 = 00) = (-1) \cdot (-1.64) + (-1) \cdot (-1.92) = +3.56$$

$$\gamma_3^*(s_2 = 00, s_3 = 10) = (+1) \cdot (-1.64) + (+1) \cdot (-1.92) = -3.56$$

$$\gamma_3^*(s_2 = 01, s_3 = 00) = (+1) \cdot (-1.64) + (+1) \cdot (-1.92) = -3.56$$

$$\gamma_3^*(s_2 = 01, s_3 = 10) = (-1) \cdot (-1.64) + (-1) \cdot (-1.92) = +3.56$$

$$\gamma_3^*(s_2 = 10, s_3 = 01) = (+1) \cdot (-1.64) + (-1) \cdot (-1.92) = +0.28$$

$$\gamma_3^*(s_2 = 10, s_3 = 11) = (-1) \cdot (-1.64) + (+1) \cdot (-1.92) = -0.28$$

$$\gamma_3^*(s_2 = 11, s_3 = 01) = (-1) \cdot (-1.64) + (+1) \cdot (-1.92) = -0.28$$

$$\gamma_3^*(s_2 = 11, s_3 = 11) = (+1) \cdot (-1.64) + (-1) \cdot (-1.92) = +0.28$$

$$\gamma_4^*(s_3 = 00, s_4 = 00) = (-1) \cdot (+0.31) + (-1) \cdot (+1.67) = -1.98$$

$$\gamma_4^*(s_3 = 00, s_4 = 10) = (+1) \cdot (+0.31) + (+1) \cdot (+1.67) = +1.98$$

$$\gamma_4^*(s_3 = 01, s_4 = 00) = (+1) \cdot (+0.31) + (+1) \cdot (+1.67) = +1.98$$

$$\gamma_4^*(s_3 = 01, s_4 = 10) = (-1) \cdot (+0.31) + (-1) \cdot (+1.67) = -1.98$$

$$\gamma_4^*(s_3 = 10, s_4 = 01) = (+1) \cdot (+0.31) + (-1) \cdot (+1.67) = -1.36$$

$$\gamma_4^*(s_3 = 10, s_4 = 11) = (-1) \cdot (+0.31) + (+1) \cdot (+1.67) = +1.36$$

$$\gamma_4^*(s_3 = 11, s_4 = 01) = (-1) \cdot (+0.31) + (+1) \cdot (+1.67) = +1.36$$

$$\gamma_4^*(s_3 = 11, s_4 = 11) = (+1) \cdot (+0.31) + (-1) \cdot (+1.67) = -1.36$$

$$\gamma_5^*(s_4 = 00, s_5 = 00) = (-1) \cdot (+1.89) + (-1) \cdot (-0.76) = -1.13$$

$$\gamma_5^*(s_4 = 00, s_5 = 10) = (+1) \cdot (+1.89) + (+1) \cdot (-0.76) = +1.13$$

$$\gamma_5^*(s_4 = 01, s_5 = 00) = (+1)\cdot(+1.89)+(+1)\cdot(-0.76) = +1.13$$

$$\gamma_5^*(s_4 = 01, s_5 = 10) = (-1)\cdot(+1.89)+(-1)\cdot(-0.76) = -1.13$$

$$\gamma_5^*(s_4 = 10, s_5 = 01) = (+1)\cdot(+1.89)+(-1)\cdot(-0.76) = +2.65$$

$$\gamma_5^*(s_4 = 10, s_5 = 11) = (-1)\cdot(+1.89)+(+1)\cdot(-0.76) = -2.65$$

$$\gamma_5^*(s_4 = 11, s_5 = 01) = (-1)\cdot(+1.89)+(+1)\cdot(-0.76) = -2.65$$

$$\gamma_5^*(s_4 = 11, s_5 = 11) = (+1)\cdot(+1.89)+(-1)\cdot(-0.76) = +2.65$$

$$\gamma_6^*(s_5 = 00, s_6 = 00) = (-1)\cdot(-1.79)+(-1)\cdot(+1.48) = +0.31$$

$$\gamma_6^*(s_5 = 01, s_6 = 00) = (+1)\cdot(-1.79)+(+1)\cdot(+1.48) = -0.31$$

$$\gamma_6^*(s_5 = 10, s_6 = 01) = (+1)\cdot(-1.79)+(-1)\cdot(+1.48) = -3.27$$

$$\gamma_6^*(s_5 = 11, s_6 = 01) = (-1)\cdot(-1.79)+(+1)\cdot(+1.48) = +3.27$$

$$\gamma_7^*(s_6 = 00, s_7 = 00) = (-1)\cdot(+0.19)+(-1)\cdot(+2.21) = -2.4$$

$$\gamma_7^*(s_6 = 01, s_7 = 00) = (+1)\cdot(+0.19)+(+1)\cdot(+2.21) = +2.4.$$

Figure 6.9 illustrates the branch metric calculations of the first MAP decoder on the trellis diagram.

From (6.49), the forward path metrics are calculated as follows:

$$\alpha_0^*(s_0 = 00) = 0$$

$$\alpha_1^*(s_1 = 00) = \max_{s_0}(0+(-2.97)) = -2.97$$

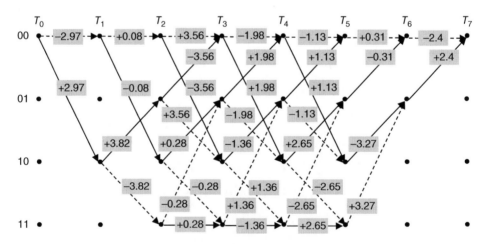

Figure 6.9   Branch metric calculations of the first MAP decoder

$$\alpha_1^*(s_1 = 10) = \max_{s_0}(0 + (+2.97)) = +2.97$$

$$\alpha_2^*(s_2 = 00) = \max_{s_1}((-2.97) + (+0.08)) = -2.89$$

$$\alpha_2^*(s_2 = 01) = \max_{s_1}((+2.97) + (+3.82)) = +6.79$$

$$\alpha_2^*(s_2 = 10) = \max_{s_1}((-2.97) + (-0.08)) = -3.05$$

$$\alpha_2^*(s_2 = 11) = \max_{s_1}((+2.97) + (-3.82)) = -0.85$$

$$\alpha_3^*(s_3 = 00) = \max_{s_2}(((-2.89) + (+3.56)),((+6.79) + (-3.56))) = +3.23$$

$$\alpha_3^*(s_3 = 01) = \max_{s_2}(((-3.05) + (+0.28)),((-0.85) + (-0.28))) = -1.13$$

$$\alpha_3^*(s_3 = 10) = \max_{s_2}(((-2.89) + (-3.56)),((+6.79) + (+3.56))) = +10.35$$

$$\alpha_3^*(s_3 = 11) = \max_{s_2}(((-3.05) + (-0.28)),((-0.85) + (+0.28))) = -0.57$$

$$\alpha_4^*(s_4 = 00) = \max_{s_3}(((+3.23) + (-1.98)),((-0.85) + (+1.98))) = +1.25$$

$$\alpha_4^*(s_4 = 01) = \max_{s_3}(((+10.35) + (-1.36)),((-0.57) + (+1.36))) = +8.99$$

$$\alpha_4^*(s_4 = 10) = \max_{s_3}(((+3.23) + (+1.98)),((-1.13) + (-1.98))) = +5.21$$

$$\alpha_4^*(s_4 = 11) = \max_{s_3}(((+10.35) + (+1.36)),((-0.57) + (-1.36))) = +11.71$$

$$\alpha_5^*(s_5 = 00) = \max_{s_4}(((+1.25) + (-1.13)),((+8.99) + (+1.13))) = +10.12$$

$$\alpha_5^*(s_5 = 01) = \max_{s_4}(((+5.21) + (+2.65)),((+11.71) + (-2.65))) = +9.06$$

$$\alpha_5^*(s_5 = 10) = \max_{s_4}(((+1.25) + (+1.13)),((+8.99) + (-1.13))) = +7.86$$

$$\alpha_5^*(s_5 = 11) = \max_{s_4}(((+5.21) + (-2.65)),((+11.71) + (+2.65))) = +14.36$$

$$\alpha_6^*(s_6 = 00) = \max_{s_5}(((+10.12) + (+0.31)),((+9.06) + (-0.31))) = +10.43$$

$$\alpha_6^*(s_6 = 01) = \max_{s_5}(((+7.86) + (-3.27)),((+14.36) + (+3.27))) = +17.63$$

$$\alpha_7^*(s_7 = 00) = \max_{s_6}(((+10.43) + (-2.4)),((+17.63) + (+2.4))) = +20.03.$$

Figure 6.10 illustrates the forward path metric calculations of the first MAP decoder on the trellis diagram.

From (6.54), the backward path metrics are calculated as follows:

$$\beta_7^*(s_7 = 00) = 0$$

$$\beta_6^*(s_6 = 00) = \max_{s_7}(0 + (-2.4)) = -2.4$$

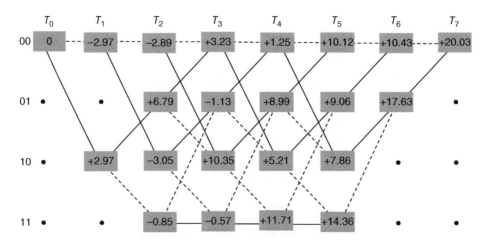

Figure 6.10 Forward path metric calculations of the first MAP decoder

$$\beta_6^*(s_6 = 01) = \max_{s_7}(0 + (+2.4)) = +2.4$$

$$\beta_5^*(s_5 = 00) = \max_{s_6}((-2.4) + (+0.31)) = -2.09$$

$$\beta_5^*(s_5 = 01) = \max_{s_6}((-2.4) + (-0.31)) = -2.71$$

$$\beta_5^*(s_5 = 10) = \max_{s_6}((+2.4) + (-3.27)) = -0.87$$

$$\beta_5^*(s_5 = 11) = \max_{s_6}((+2.4) + (+3.27)) = +5.67$$

$$\beta_4^*(s_4 = 00) = \max_{s_5}(((-2.09) + (-1.13)), ((-0.87) + (+1.13))) = +0.26$$

$$\beta_4^*(s_4 = 01) = \max_{s_5}(((-2.09) + (+1.13)), ((-0.87) + (-1.13))) = -0.96$$

$$\beta_4^*(s_4 = 10) = \max_{s_5}(((-2.71) + (+2.65)), ((+5.67) + (-2.65))) = +3.02$$

$$\beta_4^*(s_4 = 11) = \max_{s_5}(((-2.71) + (-2.65)), ((+5.67) + (+2.65))) = +8.32$$

$$\beta_3^*(s_3 = 00) = \max_{s_4}(((+0.26) + (-1.98)), ((+3.02) + (+1.98))) = +5$$

$$\beta_3^*(s_3 = 01) = \max_{s_4}(((+0.26) + (+1.98)), ((+3.02) + (-1.98))) = +2.24$$

$$\beta_3^*(s_3 = 10) = \max_{s_4}(((-0.96) + (-1.36)), ((+8.32) + (+1.36))) = +9.68$$

$$\beta_3^*(s_3 = 11) = \max_{s_4}(((-0.96) + (+1.36)), ((+8.32) + (-1.36))) = +6.96$$

$$\beta_2^*(s_2 = 00) = \max_{s_3}(((+5) + (+3.56)), ((+9.68) + (-3.56))) = +8.56$$

$$\beta_2^*(s_2 = 01) = \max_{s_3}(((+5) + (-3.56)), ((+9.68) + (+3.56))) = +13.24$$

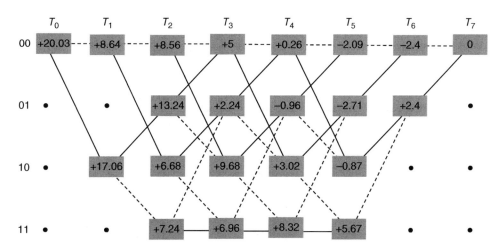

Figure 6.11 Backward path metric calculations of the first MAP decoder

$$\beta_2^*(s_2 = 10) = \max_{s_3}(((+2.24)+(+0.28)),((+6.96)+(-0.28))) = +6.68$$

$$\beta_2^*(s_2 = 11) = \max_{s_3}(((+2.24)+(-0.28)),((+6.96)+(+0.28))) = +7.24$$

$$\beta_1^*(s_1 = 00) = \max_{s_2}(((+8.56)+(+0.08)),((+6.68)+(-0.08))) = +8.64$$

$$\beta_1^*(s_1 = 10) = \max_{s_2}(((+13.24)+(+3.82)),((+7.24)+(-3.82))) = +17.06$$

$$\beta_0^*(s_0 = 00) = \max_{s_1}(((+8.64)+(-2.97)),((+17.06)+(+2.97))) = +20.03.$$

Figure 6.11 illustrates the backward path metric calculations of the first MAP decoder on the trellis diagram.

If the packet length is long, the forward and backward path metric values grow constantly and overflow in the end. Thus, we generally normalize the forward and backward path metrics at each time slot because the difference between the path metrics is important. There are two ways to normalize path metrics. The first approach is to subtract a specific value from each metric. The specific value can be selected by the average or the largest path metric of the other path metrics in each time slot. Besides, we can divide the path metrics by the sum of them in each time slot. This approach is simple but not efficient to implement hardware because the additional calculation is needed and it increases the latency. The other approach is to let overflow happen and observe them. For example, we divide the path metric memory into three parts (low, middle, and high). In a specific time slot, one path metric is in high part and the other path metric is in low part. We select the path metric in high part.

Finally, we simplify (6.39) using the Max-log-MAP algorithm and have the following LLR:

$$L_1^*(m_k) = \max_{(s',s)\in S^p} \left(\alpha_{k-1}^*(s_{k-1} = s') + \gamma_k^* \left(s_{k-1} = s', s_k = s \right) + \beta_k^*(s_k = s) \right)$$

$$- \max_{(s',s)\in S^n} \left(\alpha_{k-1}^* \left(s_{k-1} = s' \right) + \gamma_k^* \left(s_{k-1} = s', s_k = s \right) + \beta_k^*(s_k = s) \right). \tag{6.61}$$

Thus, we calculate the soft decision outputs of the first MAP decoder as follows:

$$L_1^*(m_1) = \max_{(s_0,s_1)\in S^P} ((0)+(+2.97)+(+17.06))$$

$$- \max_{(s_0,s_1)\in S^n} ((0)+(-2.97)+(+8.64)) = +14.36$$

$$L_1^*(m_2) = \max_{(s_1,s_2)\in S^P} (((-2.97)+(-0.08)+(+6.68)),((+2.97)+(+3.82)+(+13.24)))$$

$$- \max_{(s_1,s_2)\in S^n} (((-2.97)+(+0.08)+(+8.56)),((+2.97)+(-3.82)+(+7.24))) = +13.64$$

$$L_1^*(m_3) = \max_{(s_2,s_3)\in S^P} (((-2.89)+(-3.56)+(+9.68)),((+6.79)+(-3.56)+(+5)),$$
$$((-3.05)+(+0.28)+(+2.24)),((-0.85)+(+0.28)+(+6.96)))$$
$$- \max_{(s_2,s_3)\in S^n} (((-2.89)+(+3.56)+(+5)),((+6.79)+(+3.56)+(+9.68)),$$
$$((-3.05)+(-0.28)+(+6.96)),((-0.85)+(-0.28)+(+2.24))) = -11.8$$

$$L_1^*(m_4) = \max_{(s_3,s_4)\in S^P} (((+3.23)+(+1.98)+(+3.02)),((-1.13)+(+1.98)+(+0.26)),$$
$$((+10.35)+(-1.36)+(-0.96)),((-0.57)+(-1.36)+(+8.32)))$$
$$- \max_{(s_3,s_4)\in S^n} (((+3.23)+(-1.98)+(+0.26)),((-1.13)+(-1.98)+(+3.02)),$$
$$((+10.35)+(+1.36)+(+8.32)),((-0.57)+(+1.36)+(-0.96))) = -11.8$$

$$L_1^*(m_5) = \max_{(s_4,s_5)\in S^P} (((+1.25)+(+1.13)+(-0.87)),((+8.99)+(+1.13)+(-2.09)),$$
$$((+5.21)+(+2.65)+(-2.71)),((+11.71)+(+2.65)+(+5.67)))$$
$$- \max_{(s_4,s_5)\in S^n} (((+1.25)+(-1.13)+(-2.09)),((+8.99)+(-1.13)+(-0.87)),$$
$$((+5.21)+(-2.65)+(+5.67)),((+11.71)+(-2.65)+(-2.71))) = +11.8$$

$$L_1^*(m_6) = \max_{(s_5,s_6)\in S^P} (((+9.06)+(-0.31)+(-2.4)),((+7.86)+(-3.27)+(+2.4)))$$

$$- \max_{(s_5,s_6)\in S^n} (((+10.12)+(+0.31)+(-2.4)),((+14.36)+(+3.27)+(+2.4))) = -13.04$$

$$L_1^*(m_7) = \max_{(s_6,s_7)\in S^P} ((+17.63)+(+2.4)+(0)) - \max_{(s_6,s_7)\in S^n} ((+10.43)+(-2.4)+(0)) = +12$$

From (6.43), we express the extrinsic information for the second MAP decoder as follows:

$$L_e^1\left(c_k^0\right) = L_1^*(m_k) - L_a\left(c_k^0\right) - L_c r_k^0 \tag{6.62}$$

and it is calculated as shown in Table 6.2.

The extrinsic information $\left(L_e^1\left(c_k^0\right)\right)$ is used for a priori information of the second MAP decoder after interleaving. It becomes the input of the second MAP decoder together with the interleaved information r^0 and the parity r^2 from the RSC encoder 2. Thus, we have

Table 6.2 Extrinsic information for the second MAP decoder in the first iteration

k	$L_e^1\left(c_k^0\right)$	$L_1^*(m_k)$	$L_a\left(c_k^0\right)$	$L_c r_k^0$
1	+10.12	+14.36	0	2·(+2.12)
2	+9.9	+13.64	0	2·(+1.87)
3	−8.52	−11.8	0	2·(−1.64)
4	−12.42	−11.8	0	2·(+0.31)
5	+8.02	+11.8	0	2·(+1.89)
6	−9.46	−13.04	0	2·(−1.79)
7	+11.62	+12	0	2·(+0.19)

the following inputs: $L_a\left(c_k^0\right) = [-12.42 + 8.02 + 9.9 - 8.52 + 11.62 + 10.12 - 9.46]$, $r^0 = [+0.31 +1.89 +1.87 -1.64 +0.19 +2.12 -1.79]$ and $r^2 = [-2.27 +1.71 -0.62 -1.77 -1.33 +1.92 -1.74]$. In the same way as the first MAP decoding, we start decoding. Firstly, the second MAP decoders are calculated as follows:

$$\gamma_1^*(s_0 = 00, s_1 = 00) = (-1)\cdot(+12.42) + (-1)\cdot(+0.31) + (-1)\cdot(-2.27) = -10.46$$

$$\gamma_1^*(s_0 = 00, s_1 = 10) = (+1)\cdot(+12.42) + (+1)\cdot(+0.31) + (+1)\cdot(-2.27) = +10.46$$

$$\gamma_2^*(s_1 = 00, s_2 = 00) = (-1)\cdot(+8.02) + (-1)\cdot(+1.89) + (-1)\cdot(+1.71) = -11.62$$

$$\gamma_2^*(s_1 = 00, s_2 = 10) = (+1)\cdot(+8.02) + (+1)\cdot(+1.89) + (+1)\cdot(+1.71) = +11.62$$

$$\gamma_2^*(s_1 = 10, s_2 = 01) = (+1)\cdot(+8.02) + (+1)\cdot(+1.89) + (-1)\cdot(+1.71) = +8.2$$

$$\gamma_2^*(s_1 = 10, s_2 = 11) = (-1)\cdot(+8.02) + (-1)\cdot(+1.89) + (+1)\cdot(+1.71) = -8.2$$

$$\gamma_3^*(s_2 = 00, s_3 = 00) = (-1)\cdot(+9.9) + (-1)\cdot(+1.87) + (-1)\cdot(-0.62) = -11.15$$

$$\gamma_3^*(s_2 = 00, s_3 = 10) = (+1)\cdot(+9.9) + (+1)\cdot(+1.87) + (1)\cdot(-0.62) = +11.15$$

$$\gamma_3^*(s_2 = 01, s_3 = 00) = (+1)\cdot(+9.9) + (+1)\cdot(+1.87) + (+1)\cdot(-0.62) = +11.15$$

$$\gamma_3^*(s_2 = 01, s_3 = 10) = (-1)\cdot(+9.9) + (-1)\cdot(+1.87) + (-1)\cdot(-0.62) = -11.15$$

$$\gamma_3^*(s_2 = 10, s_3 = 01) = (+1)\cdot(+9.9) + (+1)\cdot(+1.87) + (-1)\cdot(-0.62) = +12.39$$

$$\gamma_3^*(s_2 = 10, s_3 = 11) = (-1)\cdot(+9.9) + (-1)\cdot(+1.87) + (+1)\cdot(-0.62) = -12.39$$

$$\gamma_3^*(s_2 = 11, s_3 = 01) = (-1)\cdot(+9.9) + (-1)\cdot(+1.87) + (+1)\cdot(-0.62) = -12.39$$

$$\gamma_3^*(s_2 = 11, s_3 = 11) = (+1)\cdot(+9.9) + (+1)\cdot(+1.87) + (-1)\cdot(-0.62) = +12.39$$

$$\gamma_4^*(s_3 = 00, s_4 = 00) = (-1)\cdot(-8.52) + (-1)\cdot(-1.64) + (-1)\cdot(-1.77) = +11.93$$

$$\gamma_4^*(s_3 = 00, s_4 = 10) = (+1)\cdot(-8.52) + (+1)\cdot(-1.64) + (+1)\cdot(-1.77) = -11.93$$

$$\gamma_4^*(s_3 = 01, s_4 = 00) = (+1)\cdot(-8.52) + (+1)\cdot(-1.64) + (+1)\cdot(-1.77) = -11.93$$

$$\gamma_4^*(s_3 = 01, s_4 = 10) = (-1) \cdot (-8.52) + (-1) \cdot (-1.64) + (-1) \cdot (-1.77) = +11.93$$

$$\gamma_4^*(s_3 = 10, s_4 = 01) = (+1) \cdot (-8.52) + (+1) \cdot (-1.64) + (-1) \cdot (-1.77) = -8.39$$

$$\gamma_4^*(s_3 = 10, s_4 = 11) = (-1) \cdot (-8.52) + (-1) \cdot (-1.64) + (+1) \cdot (-1.77) = +8.39$$

$$\gamma_4^*(s_3 = 11, s_4 = 01) = (-1) \cdot (-8.52) + (-1) \cdot (-1.64) + (+1) \cdot (-1.77) = +8.39$$

$$\gamma_4^*(s_3 = 11, s_4 = 11) = (+1) \cdot (-8.52) + (+1) \cdot (-1.64) + (-1) \cdot (-1.77) = -8.39$$

$$\gamma_5^*(s_4 = 00, s_5 = 00) = (-1) \cdot (+11.62) + (-1) \cdot (+0.19) + (-1) \cdot (-1.33) = -10.48$$

$$\gamma_5^*(s_4 = 00, s_5 = 10) = (+1) \cdot (+11.62) + (+1) \cdot (+0.19) + (+1) \cdot (-1.33) = +10.48$$

$$\gamma_5^*(s_4 = 01, s_5 = 00) = (+1) \cdot (+11.62) + (+1) \cdot (+0.19) + (+1) \cdot (-1.33) = +10.48$$

$$\gamma_5^*(s_4 = 01, s_5 = 10) = (-1) \cdot (+11.62) + (-1) \cdot (+0.19) + (-1) \cdot (-1.33) = -10.48$$

$$\gamma_5^*(s_4 = 10, s_5 = 01) = (+1) \cdot (+11.62) + (+1) \cdot (+0.19) + (-1) \cdot (-1.33) = +13.14$$

$$\gamma_5^*(s_4 = 10, s_5 = 11) = (-1) \cdot (+11.62) + (-1) \cdot (+0.19) + (+1) \cdot (-1.33) = -13.14$$

$$\gamma_5^*(s_4 = 11, s_5 = 01) = (-1) \cdot (+11.62) + (-1) \cdot (+0.19) + (+1) \cdot (-1.33) = -13.14$$

$$\gamma_5^*(s_4 = 11, s_5 = 11) = (+1) \cdot (+11.62) + (+1) \cdot (+0.19) + (-1) \cdot (-1.33) = +13.14$$

$$\gamma_6^*(s_5 = 00, s_6 = 00) = (-1) \cdot (+10.12) + (-1) \cdot (+2.12) + (-1) \cdot (+1.92) = -14.16$$

$$\gamma_6^*(s_5 = 01, s_6 = 00) = (+1) \cdot (+10.12) + (+1) \cdot (+2.12) + (1) \cdot (+1.92) = +14.16$$

$$\gamma_6^*(s_5 = 10, s_6 = 01) = (+1) \cdot (+10.12) + (+1) \cdot (+2.12) + (-1) \cdot (+1.92) = +10.32$$

$$\gamma_6^*(s_5 = 11, s_6 = 01) = (-1) \cdot (+10.12) + (-1) \cdot (+2.12) + (+1) \cdot (+1.92) = -10.32$$

$$\gamma_7^*(s_6 = 00, s_7 = 00) = (-1) \cdot (-9.46) + (-1) \cdot (-1.79) + (-1) \cdot (-1.74) = +12.99$$

$$\gamma_7^*(s_6 = 01, s_7 = 00) = (+1) \cdot (-9.46) + (+1) \cdot (-1.79) + (+1) \cdot (-1.74) = -12.99.$$

Figure 6.12 illustrates the branch metric calculations of the second MAP decoder on the trellis diagram.

From (6.49), the forward path metrics are calculated as follows:

$$\alpha_0^*(s_0 = 00) = 0$$

$$\alpha_1^*(s_1 = 00) = \max_{s_0}(0 + (-10.46)) = -10.46$$

$$\alpha_1^*(s_1 = 10) = \max_{s_0}(0 + (+10.46)) = +10.46$$

$$\alpha_2^*(s_2 = 00) = \max_{s_1}((-10.46) + (-11.62)) = -22.08$$

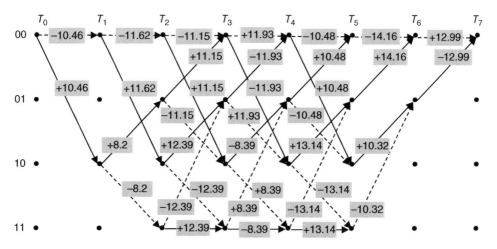

Figure 6.12   Branch metric calculations of the second MAP decoder

$$\alpha_2^*(s_2 = 01) = \max_{s_1}((+10.46) + (+8.2)) = +18.66$$

$$\alpha_2^*(s_2 = 10) = \max_{s_1}((-10.46) + (+11.62)) = +1.16$$

$$\alpha_2^*(s_2 = 11) = \max_{s_1}((+10.46) + (-8.2)) = +2.26$$

$$\alpha_3^*(s_3 = 00) = \max_{s_2}(((-22.08) + (-11.15)), ((+18.66) + (+11.15))) = +29.81$$

$$\alpha_3^*(s_3 = 01) = \max_{s_2}(((+1.16) + (+12.39)), ((+2.26) + (-12.39))) = +13.55$$

$$\alpha_3^*(s_3 = 10) = \max_{s_2}(((-22.08) + (+11.15)), ((+18.66) + (-11.15))) = +7.51$$

$$\alpha_3^*(s_3 = 11) = \max_{s_2}(((+1.16) + (-12.39)), ((+2.26) + (+12.39))) = +14.65$$

$$\alpha_4^*(s_4 = 00) = \max_{s_3}(((+29.81) + (+11.93)), ((+18.66) + (-11.93))) = +41.74$$

$$\alpha_4^*(s_4 = 01) = \max_{s_3}(((+7.51) + (-8.39)), ((+14.65) + (+8.39))) = +23.04$$

$$\alpha_4^*(s_4 = 10) = \max_{s_3}(((+29.81) + (-11.93)), ((+13.55) + (+11.93))) = +25.48$$

$$\alpha_4^*(s_4 = 11) = \max_{s_3}(((+7.51) + (+8.39)), ((+14.65) + (-8.39))) = +15.9$$

$$\alpha_5^*(s_5 = 00) = \max_{s_4}(((+41.74) + (-10.48)), ((+23.04) + (+10.48))) = +33.52$$

$$\alpha_5^*(s_5 = 01) = \max_{s_4}(((+25.48) + (+13.14)), ((+15.9) + (-13.14))) = +38.62$$

$$\alpha_5^*(s_5 = 10) = \max_{s_4}(((+41.74) + (+10.48)), ((+23.04) + (-10.48))) = +52.22$$

$$\alpha_5^*(s_5=11) = \max_{s_4}(((+25.48)+(-13.14)),((+15.9)+(+13.14))) = +29.04$$

$$\alpha_6^*(s_6=00) = \max_{s_5}(((+33.52)+(-14.16)),((+38.62)+(+14.16))) = +52.78$$

$$\alpha_6^*(s_6=01) = \max_{s_5}(((+52.22)+(+10.32)),((+29.04)+(-10.32))) = +62.54$$

$$\alpha_7^*(s_7=00) = \max_{s_6}(((+52.78)+(+12.99)),((+62.54)+(-12.99))) = +65.77.$$

Figure 6.13 illustrates the forward path metric calculations of the first MAP decoder on the trellis diagram.

From (6.54), the backward path metrics are calculated as follows:

$$\beta_7^*(s_7=00) = 0$$

$$\beta_6^*(s_6=00) = \max_{s_7}(0+(+12.99)) = +12.99$$

$$\beta_6^*(s_6=01) = \max_{s_7}(0+(-12.99)) = -12.99$$

$$\beta_5^*(s_5=00) = \max_{s_6}((+12.99)+(-14.16)) = -1.17$$

$$\beta_5^*(s_5=01) = \max_{s_6}((+12.99)+(+14.16)) = +27.15$$

$$\beta_5^*(s_5=10) = \max_{s_6}((-12.99)+(+10.32)) = -2.67$$

$$\beta_5^*(s_5=11) = \max_{s_6}((-12.99)+(-10.32)) = -23.31$$

$$\beta_4^*(s_4=00) = \max_{s_5}(((-1.17)+(-10.48)),((-2.67)+(+10.48))) = +7.81$$

$$\beta_4^*(s_4=01) = \max_{s_5}(((-1.17)+(+10.48)),((-2.67)+(-10.48))) = +9.31$$

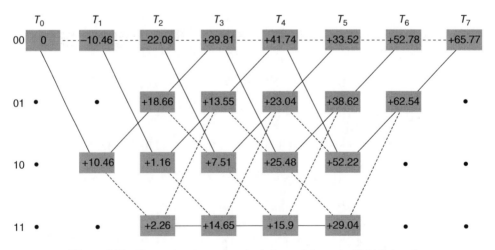

Figure 6.13   Forward path metric calculations of the second MAP decoder

$$\beta_4^*(s_4=10) = \max_{s_5}(((+27.15)+(+13.14)),((-23.31)+(-13.14))) = +40.29$$

$$\beta_4^*(s_4=11) = \max_{s_5}(((+27.15)+(-13.14)),((-23.31)+(+13.14))) = +14.01$$

$$\beta_3^*(s_3=00) = \max_{s_4}(((+7.81)+(+11.93)),((+40.29)+(-11.93))) = +28.36$$

$$\beta_3^*(s_3=01) = \max_{s_4}(((+7.81)+(-11.93)),((+40.29)+(+11.93))) = +52.22$$

$$\beta_3^*(s_3=10) = \max_{s_4}(((+9.31)+(-8.39)),((+14.01)+(+8.39))) = +22.4$$

$$\beta_3^*(s_3=11) = \max_{s_4}(((+9.31)+(+8.39)),((+14.01)+(-8.39))) = +17.7$$

$$\beta_2^*(s_2=00) = \max_{s_3}(((+28.36)+(-11.15)),((+22.4)+(+11.15))) = +33.55$$

$$\beta_2^*(s_2=01) = \max_{s_3}(((+28.36)+(+11.15)),((+22.4)+(-11.15))) = +39.51$$

$$\beta_2^*(s_2=10) = \max_{s_3}(((+52.22)+(+12.39)),((+17.7)+(-12.39))) = +64.61$$

$$\beta_2^*(s_2=11) = \max_{s_3}(((+52.22)+(-12.39)),((+17.7)+(+12.39))) = +39.83$$

$$\beta_1^*(s_1=00) = \max_{s_2}(((+33.55)+(-11.62)),((+64.61)+(+11.62))) = +76.23$$

$$\beta_1^*(s_1=10) = \max_{s_2}(((+39.51)+(+8.2)),((+39.83)+(-8.2))) = +47.71$$

$$\beta_0^*(s_0=00) = \max_{s_1}(((+76.23)+(-10.46)),((+47.71)+(+10.46))) = +65.77.$$

Figure 6.14 illustrates the backward path metric calculations of the first MAP decoder on the trellis.

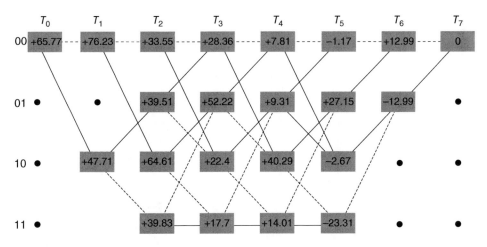

Figure 6.14   Backward path metric calculations of the second MAP decoder

Finally, we calculate the soft decision outputs of the second MAP decoder as follows:

$$L_2^*(m_1) = \max_{(s_0,s_1)\in S^p} ((0)+(+10.46)+(+47.71)) - \max_{(s_0,s_1)\in S^n} ((0)+(-10.46)+(+76.23)) = -7.6$$

$$L_2^*(m_2) = \max_{(s_1,s_2)\in S^p} (((-10.46)+(+11.62)+(+64.61)),((+10.46)+(+8.2)+(+39.51)))$$
$$- \max_{(s_1,s_2)\in S^n} (((-10.46)+(-11.62)+(+33.55)),((+10.46)+(-8.2)+(+39.83))) = +23.68$$

$$L_2^*(m_3) = \max_{(s_2,s_3)\in S^p} (((-22.08)+(+11.15)+(+22.4)),((+18.66)+(+11.15)+(+28.36)),((+1.16)$$
$$+(+12.39)+(+52.22)),((+2.26)+(+12.39)+(+17.7)))$$
$$- \max_{(s_2,s_3)\in S^n} (((-22.08)+(-11.15)+(+28.36)),((+18.66)+(-11.15)+(+22.4)),$$
$$((+1.16)+(-12.39)+(+17.7)),((+2.26)+(-12.39)+(+52.22))) = +23.68$$

$$L_2^*(m_4) = \max_{(s_3,s_4)\in S^p} (((+29.81)+(-11.93)+(+40.29)),((+13.55)+(-11.93)+(+7.81)),((+7.51)$$
$$+(-8.39)+(+9.31)),((+14.65)+(-8.39)+(+14.01)))$$
$$- \max_{(s_3,s_4)\in S^n} (((+29.81)+(+11.93)+(+7.81)),((+13.55)+(+11.93)+(+40.29)),$$
$$((+7.51)+(+8.39)+(+14.01)),((+14.65)+(+8.39)+(+9.31))) = -7.6$$

$$L_2^*(m_5) = \max_{(s_4,s_5)\in S^p} (((+41.74)+(+10.48)+(-2.67)),((+23.04)+(+10.48)+(-1.17)),((+25.48)$$
$$+(+13.14)+(+27.15)),((+15.9)+(+13.14)+(-23.31)))$$
$$- \max_{(s_4,s_5)\in S^n} (((+41.74)+(-10.48)+(-1.17)),((+23.04)+(-10.48)+(-2.67)),$$
$$((+25.48)+(-13.14)+(-23.31)),((+15.9)+(-13.14)+(+27.15))) = +35.68$$

$$L_2^*(m_6) = \max_{(s_6,s_7)\in S^p} (((+38.62)+(+14.16)+(+12.99)),((+52.22)+(+10.32)+(-12.99)))$$
$$- \max_{(s_5,s_6)\in S^n} (((+33.52)+(-14.16)+(+12.99)),((+29.04)+(-10.32)+(-12.99))) = +33.42$$

$$L_2^*(m_7) = \max_{(s_6,s_7)\in S^p} ((+62.54)+(-12.99)+(0)) - \max_{(s_6,s_7)\in S^n} ((+52.78)+(+12.99)+(0)) = -16.22$$

Thus, the output of the second MAP decoder is [−7.6 +23.68 +23.68 −7.6 +35.68 +33.42 −16.22]. After de-interleaving, we can obtain [+33.42 +23.68 −7.6 −7.6 +23.68 −16.22 +35.68] and calculate the extrinsic information for the first MAP decoder in the second iteration as shown in Table 6.3.

This extrinsic information $\left(L_e^2\left(c_k^0\right)\right)$ is used for the priori information of the first MAP decoder in the second iteration. The iteration can continue. Basically, the error correction capability increases while the iteration continues. It will converge to a stable value around six or seven iterations. In this example, we obtained [+1 +1 −1 −1 +1 −1 +1] as the output of the turbo decoder after hard decision. Now, we can observe that we received the corrupted fourth bit but it was fixed by the turbo decoder.

Table 6.3 Extrinsic information for the first MAP decoder in the second iteration

k	$L_e^2\left(c_k^0\right)$	$L_2^*(m_k)$	$L_a\left(c_k^0\right)$	$L_c r_k^0$
1	+19.06	+33.42	+10.12	2·(+2.12)
2	+10.04	+23.68	+9.9	2·(+1.87)
3	+4.2	−7.6	−8.52	2·(−1.64)
4	+4.2	−7.6	−12.42	2·(+0.31)
5	+11.88	+23.68	+8.02	2·(+1.89)
6	−3.18	−16.22	−9.46	2·(−1.79)
7	+23.68	+35.68	+11.62	2·(+0.19)

Summary 6.3   Max-log-MAP algorithm

1. The forward path metric is calculated as follows:

$$\alpha_k^*(s_k = s) \approx \max_{s_{k-1}}\left(\alpha_{k-1}^*(s_{k-1} = s') + \gamma_k^*(s_{k-1} = s', s_k = s)\right)$$

$$\alpha_0^*(s_k = s) = \begin{cases} 0, & s = 0 \\ -\text{inf}, & s \neq 0 \end{cases}.$$

2. The backward path metric is calculated as follows:

$$\beta_{k-1}^*(s_{k-1} = s') \approx \max_{s_k}\left(\gamma_k^*(s_{k-1} = s', s_k = s) + \beta_k^*(s_k = s)\right)$$

$$\beta_K^*(s_k = s') = \begin{cases} 0, & s' = 0 \\ -\text{inf}, & s' \neq 0 \end{cases}.$$

3. The branch metric is calculated as follows:

$$\gamma_k^*(s_{k-1} = s', s_k = s) \approx \frac{\left(c_k^0 L_a\left(c_k^0\right) + c_k^0 L_c r_k^0 + c_k^1 L_c r_k^1\right)}{2}$$

$$L_c = \frac{2}{\sigma^2} = 2.$$

4. The LLR is calculated as follows:

$$L^*(m_k) = \max_{(s',s)\in S^P}\left(\alpha_{k-1}^*(s_{k-1} = s') + \gamma_k^*(s_{k-1} = s', s_k = s) + \beta_k^*(s_k = s)\right)$$

$$- \max_{(s',s)\in S^n}\left(\alpha_{k-1}^*(s_{k-1} = s') + \gamma_k^*(s_{k-1} = s', s_k = s) + \beta_k^*(s_k = s)\right)$$

6.1.3   Hardware Implementation of Turbo Encoding and Decoding

In the previous sections, we derived the turbo encoder and decoder algorithm and understood their encoding and decoding process. Now, we look into hardware implementation issues of the turbo encoder and decoder. There are several key design requirements such as power,

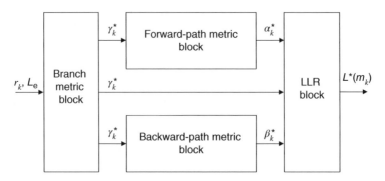

Figure 6.15   MAP decoder architecture

complexity, throughput, and latency. When dealing with a mobile application, the low power is highly required and power consumption becomes a key design parameter. When dealing with a high-speed communication system, the throughput and the latency become key design parameters. The complexity is directly related to design cost and system performance. It is always a key parameter we should carefully consider. Thus, we firstly look into a simple hardware implementation of the turbo encoder and decoder in this section. Then, we estimate the design parameters and identify the target design goal.

The hardware architecture of the turbo encoder is same as the conceptual diagram in Figure 6.1. As we can observe Figure 6.1, the turbo encoder is composed of shift registers and modulo 2 adders. It can be easily implemented. The random interleaver of the turbo encoder can be implemented by a look-up table. However, the disadvantage of a look-up table is high latency. Thus, several standards such as 3GPP [4] or 3GPP2 [5] use a simple digital logic to implement it. A hardware architecture of the MAP decoder as a SISO decoder includes four main blocks as shown in Figure 6.15.

As we investigated in Sections 6.1.1 and 6.1.2, the MAP decoder receives the soft decision input and firstly calculates branch metrics indicating the Euclidean distances. As we discussed in Chapter 5, the soft decision input can be expressed as the 2^b-1 level where b is the bit resolution. This bit level significantly affects the complexity of the MAP decoder. It becomes a basis of each metric calculation. Basically, we should carefully choose the bit level because a higher bit level brings a higher performance and complexity. The optimal quantization width (T) depends on a noise level as shown in Figure 6.16. It can be calculated as follows:

$$T = \sqrt{\frac{N_0}{2^b}} \tag{6.63}$$

where N_0 is a noise power spectral density. The four or three bits soft decision input is widely used because it is good trade-off between performance and complexity. The branch metric block can be simply implemented by addition and multiplication.

Secondly, we calculate the forward and backward path metric selecting one path in the trellis diagram. As we can observe the trellis diagram in Figure 6.2, the trellis diagram at each stage has symmetry characteristic. We call this symmetry a butterfly structure. Two states are paired in a butterfly structure as shown in Figure 6.17. The butterfly structure is useful for simplifying the forward and backward path metric calculation.

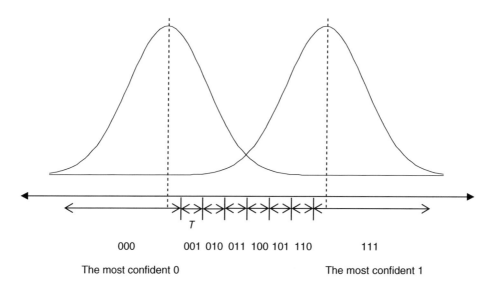

Figure 6.16   Quantization width

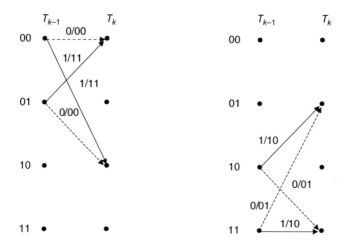

Figure 6.17   Butterfly structure of Figure 6.2 trellis diagram

These are based on the Add-Compare-Select (ACS) block as shown in Figure 6.18. The ACS block in the figure is designed for the Log-MAP algorithm. In the figure, the dashed lines and blocks represent the calculation of the term "$\ln(1+e^{-|x-y|})$."

The term of log calculation (or correction function) is not easy to implement in a digital logic. It has a high complexity. Figure 6.19 illustrates the correction function. Thus, it can be implemented by a Look-Up Table. If we remove the dashed lines and blocks, it is the ACS block of the Max-log-MAP algorithm. In the forward and backward path metric calculation block, the normalization block should be employed to prevent an overflow as mentioned in Section 6.1.2. Subtractive normalization can be implemented by subtraction and average.

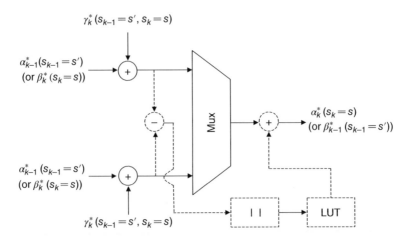

Figure 6.18   Add-Compare-Select block

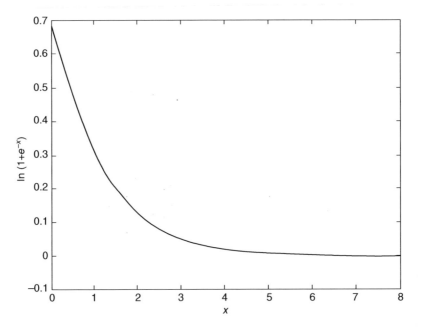

Figure 6.19   Correction function

Thirdly, we design the LLR block using addition, subtraction, and compare. In order to calculate the LLR, we need all branch metrics and forward and backward path metrics. It requires very large memory to hold them. Thus, the sliding window technique [6] is widely used to minimize the memory size. The sliding widow technique defines a window size (L) which is independent of the initial condition. In the window, the forward and backward path metric can be calculated. The calculation result in the window is almost same as the calculation result in the whole frame. Most importantly, it reduces the latency because it enables us to calculate

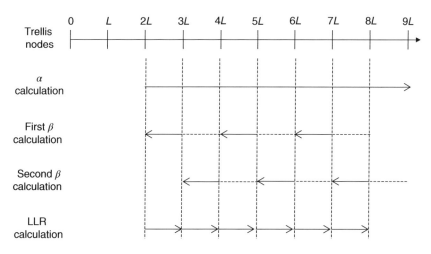

Figure 6.20   Timing of the sliding window technique

them in parallel. Figure 6.20 illustrates the timing of the forward and backward calculation using the sliding window technique. In the figure, the dashed line of the backward calculation represents initialization due to unreliable branch metric calculation. For 16 states trellis diagram, the window size (L) is 32. It is approximately $6K$ where K is the constraint length of a convolutional encoder. The required memory is $192L$ bits when six-bits path metrics are used. It is not greater than the requirement of the conventional Viterbi decoder storage. The output of the LLR calculation can be used for the output of the MAP decoder or produce the extrinsic information.

Now, we estimate the design parameters. If we design the algorithm blocks using a Hardware Description Language and look into a Register-Transfer Level in a specific target hardware environment, we can obtain more accurate estimations of complexity and power consumption. However, we deal with a floating point level design in this chapter. Thus, we estimate them with simple calculation. The power and complexity depend on control signals, memories, and data path calculations (branch metrics, forward path metrics and backward path metrics, and LLRs). When designing a complex system, control signal management is one of tricky parts. Thus, a separate block such as a complex finite state machine, a firmware, or an operating system is required for control signal management. However, the MAP decoder does not require complex data paths or timing. A simple finite state machine using a counter is enough. It controls the data path and memory. In order to estimate the number of memories, we should hold information bits (B_i), parity bits (B_p), LLRs (B_{LLR}), and path metrics (B_{pm}). The simple calculation about memory requirement of the MAP decoder is as follows:

$$\text{Memeory requirement} = 4LM(B_i + B_p + B_{LLR}) + LN_s MB_{pm} \qquad (6.64)$$

where N_s and M are the number of states in the trellis diagram and the number of parallel flows, respectively. The turbo decoder needs two MAP decoders and requires double memories. In addition, a memory for a random interleaver should be considered. The complexity calculation of data paths is approximately estimated when we have the L_{frame} frame length and

$2\,N_s$ branches of the trellis diagram. The complexity calculations are summarized in Table 6.4. The complexity basically increases according to the number of iterations.

The latency of the turbo decoder depends on the delay of the MAP decoder, the number of iteration, and the random interleaver. The delay of the MAP decoder depends on the window size, the number of parallelization, and the frame length. The number of iteration is an important part affecting the turbo decoder latency. Sometimes the number of iteration can be limited if the bit error rate (BER) performance reaches a specific value. We can observe performance saturation after six or seven iterations. The interleaver needs a whole frame to start interleaving. Thus, this is one of key latency parts. Several techniques such as parallel interleaver design or combinational logic design are developed.

Table 6.4   Complexity of the Log-MAP algorithm

Sub-blocks of the Log-MAP decoder	Complexity
Branch metric	$2\,N_s\,L_{\text{frame}}$ (addition)
	$2\,N_s\cdot3\cdot L_{\text{frame}}$ (multiplication)
Forward path metric	$2\,N_s\,L_{\text{frame}}$ (addition)
	$N_s\,L_{\text{frame}}$ (Mux)
	$N_s\,L_{\text{frame}}$ (subtraction)
Backward path metric	$2\,N_s\,L_{\text{frame}}$ (addition)
	$N_s\,L_{\text{frame}}$ (Mux)
	$N_s\,L_{\text{frame}}$ (subtraction)
LLR	$2\cdot2\,N_s\,L_{\text{frame}}$ (addition)
	$2\cdot L_{\text{frame}}$ (Mux)
	L_{frame} (subtraction)

Summary 6.4   Hardware design issues of MAP decoder

1. The optimal quantization width (T) of the soft decision input depends on a noise level. It can be calculated as follows:

$$T=\sqrt{\frac{N_0}{2^b}}.$$

   The four bits or three bits soft decision input is widely used as good tradeoff between performance and complexity.
2. The butterfly structure that two states are paired is useful for simplifying the forward and backward path metric calculation.
3. The ACS is a basic sub-block to implement the forward and backward path metric.
4. The normalization block should be employed to prevent an overflow.
5. The sliding window technique should be used to minimize the memory size. The sliding widow technique defines a window size (L) which is independent of the initial condition.

6.2 Turbo Product Codes

6.2.1 Turbo Product Encoding and Decoding Algorithm

In 1954, P. Elias [7] suggested the construction of simple and powerful linear block codes from simpler linear block codes. We call them product codes. The product code is constructed using a block of information bits (I). Each row of information bits is encoded by a row component code and row parity bits (P_r) are appended to the same row. After finishing row encoding, each column of information bits is encoded by a column component code and column parity bits (P_c and P_{rc}) are appended to the same column. The encoding process is illustrated in Figure 6.21.

Let us consider two systematic linear block codes: C_1 (n_1, k_1, d_1) and C_2 (n_2, k_2, d_2), where n, k, and d are the code length, the information (or message) length, and the minimum distance, respectively. A product code C_p ($n_1 \times n_2$, $k_1 \times k_2$, $d_1 \times d_2$) is obtained from C_1 and C_2. As we can observe from Figure 6.21, one information bit is encoded by both a row code and a column code. When one bit is corrupted in a receiver and cannot be corrected by a row code, a column code may fix it and vice versa. We can expand the two-dimensional product codes to m-dimensional product codes where m is larger than 2. An m-dimensional product code C_p^m ($n_1 \times n_2 \times \cdots \times n_m$, $k_1 \times k_2 \times \cdots \times k_m$, $d_1 \times d_2 \times \cdots \times d_m$) is constructed by the m component codes (n_1, k_1, d_1), (n_2, k_2, d_2), ..., (n_m, k_m, d_m). The component codes can be single parity check codes, Hamming codes, Reed-Muller codes, BCH codes, Reed-Solomon codes, and so on. It is also possible to use a nonlinear block code such as Nordstrom-Robinson code, Preparata codes, and Kerdock codes. R. Pyndiah in 1994 [8] introduced the turbo product codes (or block turbo codes) and presented a new iterative decoding algorithm for decoding product codes based on the Chase algorithm [9]. The turbo product encoding process is same as the product encoding process. The turbo product decoding process is an iterative decoding process based on SISO decoding algorithms such as the Chase algorithm, the symbol by symbol MAP algorithm [10], the SOVA [11], the sliding-window MAP algorithm [12], or the forward recursion only MAP algorithm [13]. The row blocks of the turbo product codes are decoded first and then all of column blocks are decoded. Iteration can be done several times to reduce the error probability. The complexity of SISO algorithms is a very important implementation issue because there is a tradeoff between complexity and performance as we discussed in Section 6.1. When comparing turbo product codes with turbo codes, the turbo product codes have two advantages. The first advantage is that the turbo product codes provide us with a better performance at a high code rate (more than 0.75). The second advantage is that the turbo product codes require smaller frame size. The turbo product decoding process is similar to the turbo code decoding process. The component decoder of the turbo product decoder is illustrated in Figure 6.22. Firstly, each SISO

I_{11}	I_{12}	I_{13}	P_{r1}
I_{21}	I_{22}	I_{23}	P_{r2}
I_{31}	I_{32}	I_{33}	P_{r3}
P_{c1}	P_{c2}	P_{c3}	P_{rc}

Figure 6.21 Product code

decoder calculates the soft output LLR_i from the channel received information R_i. The extrinsic information E_i is obtained from R_i and LLR_i by $w_i^a \mathrm{LLR}_i - R_i$ where w_i^a is a weighting factor.

The SISO decoder is used as the row and column component decoder of the turbo product decoder. The optimal MAP or Log-MAP algorithm has high complexity. Its suboptimal variants such as the Log-MAP algorithm and the Max-log-MAP algorithm are used in practice. The SOVA is another desirable candidate because of its low complexity. However, the SOVA has poorer performance than the Max-log-MAP algorithm.

6.2.2   Example of Turbo Product Encoding and Decoding

Consider a turbo product encoder using a (4, 3, 2) Single Parity Check (SPC) code as a row and column encoder. It is defined as follows:

$$I_1 \oplus I_2 \oplus I_3 = P \tag{6.65}$$

where $\oplus$, I_i, and P denote a modulo 2 adder, an information bit, and a parity bit, respectively. Thus, we obtain a (16, 9, 4) turbo product code. As shown in Figure 6.21, the encoded bits of the turbo product encoder are generated. We transmit the sequence $[I_{11}, I_{12}, I_{13}, P_{r1}, I_{21}, I_{22}, I_{23}, P_{r2}, I_{31}, I_{32}, I_{33}, P_{r3}, P_{c1}, P_{c2}, P_{c3}, P_{rc}]$. Figure 6.23 illustrates the turbo product encoding process.

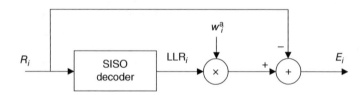

Figure 6.22   Component decoder of the turbo product decoder

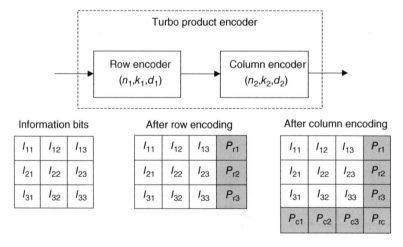

Figure 6.23   Turbo product encoder

It is possible to represent the trellis of block codes. The important advantage is that the trellis representation of block codes enables us to perform MAP or ML decoding. The disadvantage is that the complexity increases when the codeword length increases. Thus, the trellis representation of block codes is suitable for short length codes such as row or column codes of the product codes. The trellis diagram of the (4, 3, 2) SPC code is shown in Figure 6.24.

In Figure 6.24, the information bits are represented from T_0 to T_3 and the parity bit is decided at T_4. Based on this trellis diagram, the turbo product decoder can receive soft decision inputs and MAP decoding is possible. Figure 6.25 illustrates the transmitted symbols, Gaussian noise, and the received symbols.

In the receiver, the turbo product decoder uses the soft decision received symbol sequence as shown in Figure 6.25. Figure 6.26 illustrates the turbo product decoder architecture.

We perform MAP decoding for the row codes and then for the column codes. The branch metrics are calculated as follows:

For the first row decoder,

$$\gamma_1^*(s_0 = 0, s_1 = 0) = (-1) \cdot (-2.12) = -2.12$$

$$\gamma_1^*(s_0 = 0, s_1 = 1) = (+1) \cdot (+2.12) = +2.12$$

$$\gamma_2^*(s_1 = 0, s_2 = 0) = (-1) \cdot (-1.85) = +1.85$$

$$\gamma_2^*(s_1 = 0, s_2 = 1) = (+1) \cdot (-1.85) = -1.85$$

$$\gamma_2^*(s_1 = 1, s_2 = 0) = (+1) \cdot (-1.85) = -1.85$$

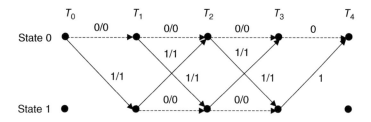

Figure 6.24   Trellis diagram of the (4, 3, 2) SPC code

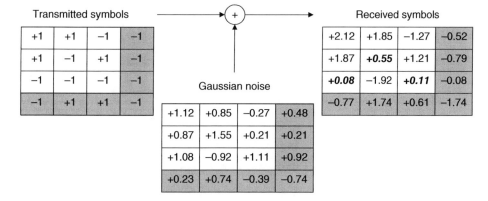

Figure 6.25   Channel noise example

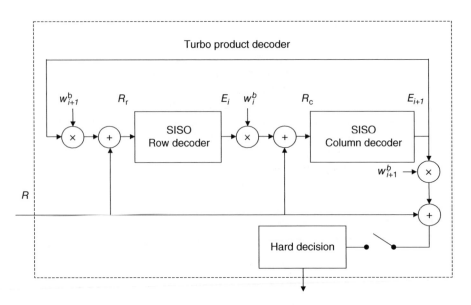

Figure 6.26   Turbo product decoder

$$\gamma_2^*(s_1 = 1, s_2 = 1) = (-1) \cdot (-1.85) = +1.85$$

$$\gamma_3^*(s_2 = 0, s_3 = 0) = (-1) \cdot (+1.27) = -1.27$$

$$\gamma_3^*(s_2 = 0, s_3 = 1) = (+1) \cdot (+1.27) = +1.27$$

$$\gamma_3^*(s_2 = 1, s_3 = 0) = (+1) \cdot (+1.27) = +1.27$$

$$\gamma_3^*(s_2 = 1, s_3 = 1) = (-1) \cdot (+1.27) = -1.27$$

$$\gamma_4^*(s_3 = 0, s_4 = 0) = (-1) \cdot (+0.52) = -0.52$$

$$\gamma_4^*(s_3 = 1, s_4 = 0) = (+1) \cdot (+0.52) = +0.52.$$

For the second row decoder,

$$\gamma_1^*(s_0 = 0, s_1 = 0) = (-1) \cdot (+1.87) = -1.87$$

$$\gamma_1^*(s_0 = 0, s_1 = 1) = (+1) \cdot (+1.82) = +1.87$$

$$\gamma_2^*(s_1 = 0, s_2 = 0) = (-1) \cdot (+0.55) = -0.55$$

$$\gamma_2^*(s_1 = 0, s_2 = 1) = (+1) \cdot (+0.55) = +0.55$$

$$\gamma_2^*(s_1 = 1, s_2 = 0) = (+1) \cdot (+0.55) = +0.55$$

$$\gamma_2^*(s_1 = 1, s_2 = 1) = (-1) \cdot (+0.55) = -0.55$$

$$\gamma_3^*(s_2 = 0, s_3 = 0) = (-1) \cdot (+1.21) = -1.21$$

$$\gamma_3^*(s_2 = 0, s_3 = 1) = (+1) \cdot (+1.21) = +1.21$$

$$\gamma_3^*(s_2 = 1, s_3 = 0) = (+1) \cdot (+1.21) = +1.21$$

$$\gamma_3^*(s_2 = 1, s_3 = 1) = (-1) \cdot (+1.21) = -1.21$$

$$\gamma_4^*(s_3 = 0, s_4 = 0) = (-1) \cdot (-0.79) = +0.79$$
$$\gamma_4^*(s_3 = 1, s_4 = 0) = (+1) \cdot (-0.79) = -0.79.$$

For the third row decoder,

$$\gamma_1^*(s_0 = 0, s_1 = 0) = (-1) \cdot (+0.08) = -0.08$$
$$\gamma_1^*(s_0 = 0, s_1 = 1) = (+1) \cdot (+0.08) = +0.08$$
$$\gamma_2^*(s_1 = 0, s_2 = 0) = (-1) \cdot (-1.92) = +1.92$$
$$\gamma_2^*(s_1 = 0, s_2 = 1) = (+1) \cdot (-1.92) = -1.92$$
$$\gamma_2^*(s_1 = 1, s_2 = 0) = (+1) \cdot (-1.92) = -1.92$$
$$\gamma_2^*(s_1 = 1, s_2 = 1) = (-1) \cdot (-1.92) = +1.92$$
$$\gamma_3^*(s_2 = 0, s_3 = 0) = (-1) \cdot (+0.11) = -0.11$$
$$\gamma_3^*(s_2 = 0, s_3 = 1) = (+1) \cdot (+0.11) = +0.11$$
$$\gamma_3^*(s_2 = 1, s_3 = 0) = (+1) \cdot (+0.11) = +0.11$$
$$\gamma_3^*(s_2 = 1, s_3 = 1) = (-1) \cdot (+0.11) = -0.11$$
$$\gamma_4^*(s_3 = 0, s_4 = 0) = (-1) \cdot (-0.08) = +0.08$$
$$\gamma_4^*(s_3 = 1, s_4 = 0) = (+1) \cdot (-0.08) = -0.08.$$

Figure 6.27 illustrates the branch metric calculations of the row MAP decoders. According to (6.49), the forward path metrics are calculated as follows:
For the first row decoder,

$$\alpha_0^*(s_0 = 0) = 0$$
$$\alpha_1^*(s_1 = 0) = \max_{s_0}(0 + (-2.12)) = -2.12$$
$$\alpha_1^*(s_1 = 1) = \max_{s_0}(0 + (+2.12)) = +2.12$$
$$\alpha_2^*(s_2 = 0) = \max_{s_1}(((-2.12) + (-1.85)), ((+2.12) + (+1.85))) = +3.97$$
$$\alpha_2^*(s_2 = 1) = \max_{s_1}(((-2.12) + (+1.85)), ((+2.12) + (-1.85))) = +0.27$$
$$\alpha_3^*(s_3 = 0) = \max_{s_2}(((+3.97) + (+1.27)), ((+0.27) + (-1.27))) = +5.24$$
$$\alpha_3^*(s_3 = 1) = \max_{s_2}(((+3.97) + (-1.27)), ((+0.27) + (+1.27))) = +2.7$$
$$\alpha_4^*(s_4 = 0) = \max_{s_3}(((+5.24) + (+0.52)), ((+2.7) + (-0.52))) = +5.76$$

For the second row decoder,

$$\alpha_0^*(s_0 = 0) = 0$$
$$\alpha_1^*(s_1 = 0) = \max_{s_0}(0 + (-1.87)) = -1.87$$

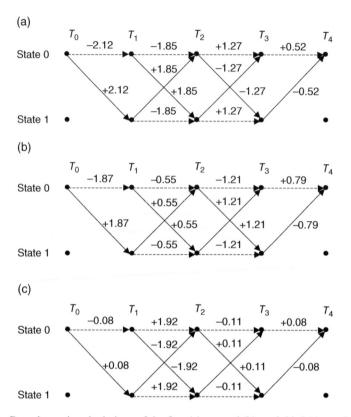

Figure 6.27 Branch metric calculations of the first (a), second (b), and third (c) row MAP decoder

$$\alpha_1^*(s_1 = 1) = \max_{s_0}(0 + (+1.87)) = +1.87$$

$$\alpha_2^*(s_2 = 0) = \max_{s_1}(((-1.87) + (-0.55)),((+1.87) + (+0.55))) = +2.42$$

$$\alpha_2^*(s_2 = 1) = \max_{s_1}(((-1.87) + (+0.55)),((+1.87) + (-0.55))) = +1.32$$

$$\alpha_3^*(s_3 = 0) = \max_{s_2}(((+2.42) + (-1.21)),((+1.32) + (+1.21))) = +2.53$$

$$\alpha_3^*(s_3 = 1) = \max_{s_2}(((+2.42) + (+1.21)),((+1.32) + (-1.21))) = +3.63$$

$$\alpha_4^*(s_4 = 0) = \max_{s_3}(((+2.53) + (+0.79)),((+3.63) + (-0.79))) = +3.32$$

For the third row decoder,

$$\alpha_0^*(s_0 = 0) = 0$$

$$\alpha_1^*(s_1 = 0) = \max_{s_0}(0 + (-0.08)) = -0.08$$

$$\alpha_1^*(s_1 = 1) = \max_{s_0}(0 + (+0.08)) = +0.08$$

$$\alpha_2^*(s_2 = 0) = \max_{s_1}(((-0.08) + (+1.92)),((+0.08) + (-1.92))) = +1.84$$

$$\alpha_2^*(s_2 = 1) = \max_{s_1}(((-0.08) + (-1.92)), ((+0.08) + (+1.92))) = +2$$

$$\alpha_3^*(s_3 = 0) = \max_{s_2}(((+1.84) + (-0.11)), ((+2) + (+0.11))) = +2.11$$

$$\alpha_3^*(s_3 = 1) = \max_{s_2}(((+1.84) + (+0.11)), ((+2) + (-0.11))) = +1.95$$

$$\alpha_4^*(s_4 = 0) = \max_{s_3}(((+2.11) + (+0.08)), ((+1.95) + (-0.08))) = +2.19$$

Figure 6.28 illustrates the forward path metric calculations of the row MAP decoders. According to (6.54), the backward path metrics can be calculated as follows:
For the first row decoder,

$$\beta_4^*(s_4 = 0) = 0$$

$$\beta_3^*(s_3 = 0) = \max_{s_4}(0 + (+0.52)) = +0.52$$

$$\beta_3^*(s_3 = 1) = \max_{s_4}(0 + (-0.52)) = -0.52$$

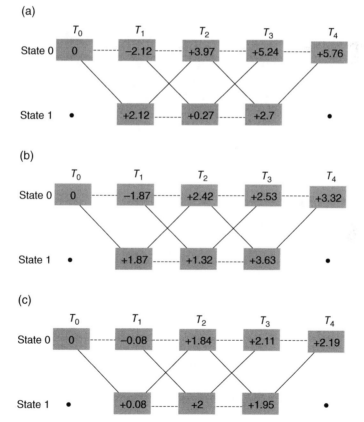

Figure 6.28   Forward path metric calculations of the first (a), second (b), and third (c) row MAP decoder

$$\beta_2^*(s_2 = 0) = \max_{s_3}(((+0.52) + (+1.27)), ((-0.52) + (-1.27))) = +1.79$$

$$\beta_2^*(s_2 = 1) = \max_{s_3}(((+0.52) + (-1.27)), ((-0.52) + (+1.27))) = +0.75$$

$$\beta_1^*(s_1 = 0) = \max_{s_2}(((+1.79) + (-1.85)), ((+0.75) + (+1.85))) = +2.6$$

$$\beta_1^*(s_1 = 1) = \max_{s_2}(((+1.79) + (+1.85)), ((+0.75) + (-1.85))) = +3.64$$

$$\beta_0^*(s_0 = 0) = \max_{s_1}(((+2.6) + (-2.12)), ((+3.64) + (+2.12))) = +5.76$$

For the second row decoder,

$$\beta_4^*(s_4 = 0) = 0$$

$$\beta_3^*(s_3 = 0) = \max_{s_4}(0 + (+0.79)) = +0.79$$

$$\beta_3^*(s_3 = 1) = \max_{s_4}(0 + (-0.79)) = -0.79$$

$$\beta_2^*(s_2 = 0) = \max_{s_3}(((+0.79) + (-1.21)), ((-0.79) + (+1.21))) = +0.42$$

$$\beta_2^*(s_2 = 1) = \max_{s_3}(((+0.79) + (+1.21)), ((-0.79) + (-1.21))) = +2$$

$$\beta_1^*(s_1 = 0) = \max_{s_2}(((+0.42) + (-0.55)), ((+2) + (+0.55))) = +2.55$$

$$\beta_1^*(s_1 = 1) = \max_{s_2}(((+0.42) + (+0.55)), ((+2) + (-0.55))) = +1.45$$

$$\beta_0^*(s_0 = 0) = \max_{s_1}(((+2.55) + (-1.87)), ((+1.45) + (+1.87))) = +3.32$$

For the third row decoder,

$$\beta_4^*(s_4 = 0) = 0$$

$$\beta_3^*(s_3 = 0) = \max_{s_4}(0 + (+0.08)) = +0.08$$

$$\beta_3^*(s_3 = 1) = \max_{s_4}(0 + (-0.08)) = -0.08$$

$$\beta_2^*(s_2 = 0) = \max_{s_3}(((+0.08) + (-0.11)), ((-0.08) + (+0.11))) = +0.03$$

$$\beta_2^*(s_2 = 1) = \max_{s_3}(((+0.08) + (+0.11)), ((-0.08) + (-0.11))) = +0.19$$

$$\beta_1^*(s_1 = 0) = \max_{s_2}(((+0.03) + (+1.92)), ((+0.19) + (-1.92))) = +1.95$$

$$\beta_1^*(s_1 = 1) = \max_{s_2}(((+0.03) + (-1.92)), ((+0.19) + (+1.92))) = +2.11$$

$$\beta_0^*(s_0 = 0) = \max_{s_1}(((+1.95) + (-0.08)), ((+2.11) + (+0.08))) = +2.19$$

Figure 6.29 illustrates the backward path metric calculations of the row MAP decoders. From branch, forward path, and backward path metrics, we calculate the soft decision output of the row MAP decoder as follows:
For the first row decoder,

$$L_r^*(I_{11}) = \max_{(s_0,s_1)\in S^p}((0) + (+2.12) + (+3.64)) - \max_{(s_0,s_1)\in S^n}((0) + (-2.12) + (+2.6)) = +5.28$$

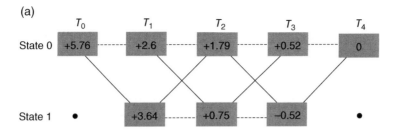

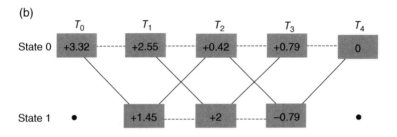

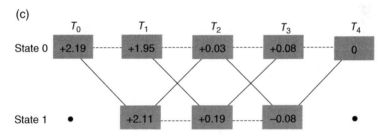

Figure 6.29 Backward path metric calculations of the first (a), second (b), and third (c) row MAP decoder

$$L_r^*(I_{12}) = \max_{(s_1,s_2)\in S^P} (((-2.12)+(+1.85)+(+0.75)),((+2.12)+(+1.85)+(+1.79)))$$

$$- \max_{(s_1,s_2)\in S^n} (((-2.12)+(-1.85)+(+1.79)),((+2.12)+(-1.85)+(+0.75))) = +4.74$$

$$L_r^*(I_{13}) = \max_{(s_2,s_3)\in S^P} (((+3.97)+(-1.27)+(-0.52)),((+0.27)+(-1.27)+(+0.52)))$$

$$- \max_{(s_2,s_3)\in S^n} (((+3.97)+(+1.27)+(+0.52)),((+0.27)+(+1.27)+(-0.52))) = -3.58$$

$$L_r^*(P_{r1}) = \max_{(s_3,s_4)\in S^P} ((+2.7)+(-0.52)+(0)) - \max_{(s_3,s_4)\in S^n} ((+5.24)+(+0.52)+(0)) = -3.58$$

For the second row decoder,

$$L_r^*(I_{21}) = \max_{(s_0,s_1)\in S^P} ((0)+(+1.87)+(+1.45)) - \max_{(s_0,s_1)\in S^n} ((0)+(-1.87)+(+2.55)) = +2.64$$

$$L_r^*(I_{22}) = \max_{(s_1,s_2)\in S^p} (((-1.87)+(+0.55)+(+2)),((+1.87)+(+0.55)+(+0.42)))$$

$$- \max_{(s_1,s_2)\in S^n} (((-1.87)+(-0.55)+(+0.42)),((+1.87)+(-0.55)+(+2))) = -0.48$$

$$L_r^*(I_{23}) = \max_{(s_2,s_3)\in S^p} (((+2.42)+(+1.21)+(-0.79)),((+1.32)+(+1.21)+(+0.79)))$$

$$- \max_{(s_2,s_3)\in S^n} (((+2.42)+(-1.21)+(+0.79)),((+1.32)+(-1.21)+(-0.79))) = +1.32$$

$$L_r^*(P_{r2}) = \max_{(s_3,s_4)\in S^p} ((+3.63)+(-0.79)+(0)) - \max_{(s_3,s_4)\in S^n} ((+2.53)+(+0.79)+(0)) = -0.48$$

For the third row decoder,

$$L_r^*(I_{31}) = \max_{(s_0,s_1)\in S^p} ((0)+(+0.08)+(+2.11)) - \max_{(s_0,s_1)\in S^n} ((0)+(-0.08)+(+1.95)) = +0.32$$

$$L_r^*(I_{32}) = \max_{(s_1,s_2)\in S^p} (((-0.08)+(-1.92)+(+0.19)),((+0.08)+(-1.92)+(+0.03)))$$

$$- \max_{(s_1,s_2)\in S^n} (((-0.08)+(+1.92)+(+0.03)),((+0.08)+(+1.92)+(+0.19))) = -4$$

$$L_r^*(I_{33}) = \max_{(s_2,s_3)\in S^p} (((+1.84)+(+0.11)+(-0.08)),((+2)+(+0.11)+(+0.08)))$$

$$- \max_{(s_2,s_3)\in S^n} (((+1.84)+(-0.11)+(+0.08)),((+2)+(-0.11)+(-0.08))) = +0.38$$

$$L_r^*(P_{r3}) = \max_{(s_3,s_4)\in S^p} ((+1.95)+(-0.08)+(0)) - \max_{(s_3,s_4)\in S^n} ((+2.11)+(+0.08)+(0)) = -0.32$$

Finally, we obtain the row decoding results in the first iteration as shown in Figure 6.30. We had three corrupted bits (I_{22}, I_{31}, and I_{33}) in the received symbol sequence. I_{22} is fixed after the row decoding but I_{31} and I_{33} are not fixed. The output of the row decoder is multiplied by w_i^a which is the weighting factor to adjust the effect of the LLR value. The extrinsic information E_i can be obtained by subtracting the soft input of the MAP decoder from the soft output of the MAP decoder as follows:

$$E_i = w_i^a L_r(R_i) - R_i. \qquad (6.66)$$

The input of the column decoder is obtained by adding the received sequence R with the extrinsic information E_i as follows:

$$R_{i+1} = w_i^b E_i + R \qquad (6.67)$$

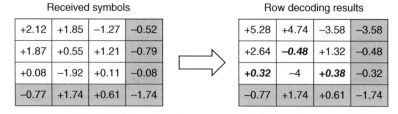

Received symbols

+2.12	+1.85	−1.27	−0.52
+1.87	+0.55	+1.21	−0.79
+0.08	−1.92	+0.11	−0.08
−0.77	+1.74	+0.61	−1.74

Row decoding results

+5.28	+4.74	−3.58	−3.58
+2.64	**−0.48**	+1.32	−0.48
+0.32	−4	**+0.38**	−0.32
−0.77	+1.74	+0.61	−1.74

Figure 6.30   Row decoding results in the first iteration

Extrinsic information E_i

+2.632	+2.416	−1.952	−2.702
+0.506	−0.982	−0.022	+0.358
+0.208	−1.68	+0.232	−0.208
−0.77	+1.74	+0.61	−1.74

Column decoder inputs

+4.964	+4.451	−3.349	−3.274
+2.563	−0.377	+1.309	−0.511
+0.296	−3.792	+0.353	−0.296
−0.77	+1.74	+0.61	−1.74

Figure 6.31   Column decoder inputs in the first iteration

where w_i^b is the weighting factor to adjust the extrinsic information. The experimental results indicate that the performance of the turbo product decoder is very sensitive to the weighting factors [14]. We define $w_i^a = 0.9$ and $w_i^b = 0.9$ in this example. Thus, we obtain the column decoder inputs in the first iteration as shown in Figure 6.31.

The column decoder can be implemented in the same manner as the row MAP decoder. The branch metrics for the column decoder are calculated as follows:

For the first column decoder,

$$\gamma_1^*(s_0 = 0, s_1 = 0) = (-1) \cdot (+4.964) = -2.12$$
$$\gamma_1^*(s_0 = 0, s_1 = 1) = (+1) \cdot (+4.964) = +2.12$$
$$\gamma_2^*(s_1 = 0, s_2 = 0) = (-1) \cdot (+2.563) = -2.563$$
$$\gamma_2^*(s_1 = 0, s_2 = 1) = (+1) \cdot (+2.563) = +2.563$$
$$\gamma_2^*(s_1 = 1, s_2 = 0) = (+1) \cdot (+2.563) = +2.563$$
$$\gamma_2^*(s_1 = 1, s_2 = 1) = (-1) \cdot (+2.563) = -2.563$$
$$\gamma_3^*(s_2 = 0, s_3 = 0) = (-1) \cdot (+0.296) = -0.296$$
$$\gamma_3^*(s_2 = 0, s_3 = 1) = (+1) \cdot (+0.296) = +0.296$$
$$\gamma_3^*(s_2 = 1, s_3 = 0) = (+1) \cdot (+0.296) = +0.296$$
$$\gamma_3^*(s_2 = 1, s_3 = 1) = (-1) \cdot (+0.296) = -0.296$$
$$\gamma_4^*(s_3 = 0, s_4 = 0) = (-1) \cdot (-0.77) = +0.77$$
$$\gamma_4^*(s_3 = 1, s_4 = 0) = (+1) \cdot (-0.77) = -0.77.$$

For the second column decoder,

$$\gamma_1^*(s_0 = 0, s_1 = 0) = (-1) \cdot (+4.451) = -4.451$$
$$\gamma_1^*(s_0 = 0, s_1 = 1) = (+1) \cdot (+4.451) = +4.451$$
$$\gamma_2^*(s_1 = 0, s_2 = 0) = (-1) \cdot (-0.377) = +0.377$$
$$\gamma_2^*(s_1 = 0, s_2 = 1) = (+1) \cdot (-0.377) = -0.377$$
$$\gamma_2^*(s_1 = 1, s_2 = 0) = (+1) \cdot (-0.377) = -0.377$$
$$\gamma_2^*(s_1 = 1, s_2 = 1) = (-1) \cdot (-0.377) = +0.377$$
$$\gamma_3^*(s_2 = 0, s_3 = 0) = (-1) \cdot (-3.792) = +3.792$$
$$\gamma_3^*(s_2 = 0, s_3 = 1) = (+1) \cdot (-3.792) = -3.792$$

$$\gamma_3^*(s_2 = 1, s_3 = 0) = (+1) \cdot (-3.792) = -3.792$$
$$\gamma_3^*(s_2 = 1, s_3 = 1) = (-1) \cdot (-3.792) = +3.792$$
$$\gamma_4^*(s_3 = 0, s_4 = 0) = (-1) \cdot (+1.74) = -1.74$$
$$\gamma_4^*(s_3 = 1, s_4 = 0) = (+1) \cdot (+1.74) = +1.74.$$

For the third column decoder,

$$\gamma_1^*(s_0 = 0, s_1 = 0) = (-1) \cdot (-3.349) = +3.349$$
$$\gamma_1^*(s_0 = 0, s_1 = 1) = (+1) \cdot (-3.349) = -3.349$$
$$\gamma_2^*(s_1 = 0, s_2 = 0) = (-1) \cdot (+1.309) = -1.309$$
$$\gamma_2^*(s_1 = 0, s_2 = 1) = (+1) \cdot (+1.309) = +1.309$$
$$\gamma_2^*(s_1 = 1, s_2 = 0) = (+1) \cdot (+1.309) = +1.309$$
$$\gamma_2^*(s_1 = 1, s_2 = 1) = (-1) \cdot (+1.309) = -1.309.$$
$$\gamma_3^*(s_2 = 0, s_3 = 0) = (-1) \cdot (+0.353) = -0.353$$
$$\gamma_3^*(s_2 = 0, s_3 = 1) = (+1) \cdot (+0.353) = +0.353$$
$$\gamma_3^*(s_2 = 0, s_3 = 1) = (+1) \cdot (+0.353) = +0.353$$
$$\gamma_3^*(s_2 = 1, s_3 = 1) = (-1) \cdot (+0.353) = -0.353$$
$$\gamma_4^*(s_3 = 0, s_4 = 0) = (-1) \cdot (+0.61) = -0.61$$
$$\gamma_4^*(s_3 = 1, s_4 = 0) = (+1) \cdot (+0.61) = +0.61.$$

For the fourth column decoder,

$$\gamma_1^*(s_0 = 0, s_1 = 0) = (-1) \cdot (-3.274) = +3.274$$
$$\gamma_1^*(s_0 = 0, s_1 = 1) = (+1) \cdot (-3.274) = -3.274$$
$$\gamma_2^*(s_1 = 0, s_2 = 0) = (-1) \cdot (-0.511) = +0.511$$
$$\gamma_2^*(s_1 = 0, s_2 = 1) = (+1) \cdot (-0.511) = -0.511$$
$$\gamma_2^*(s_1 = 1, s_2 = 0) = (+1) \cdot (-0.511) = -0.511$$
$$\gamma_2^*(s_1 = 1, s_2 = 1) = (-1) \cdot (-0.511) = +0.511$$
$$\gamma_3^*(s_2 = 0, s_3 = 0) = (-1) \cdot (-0.296) = +0.296$$
$$\gamma_3^*(s_2 = 0, s_3 = 1) = (+1) \cdot (-0.296) = -0.296$$
$$\gamma_3^*(s_2 = 1, s_3 = 0) = (+1) \cdot (-0.296) = -0.296$$
$$\gamma_3^*(s_2 = 1, s_3 = 1) = (-1) \cdot (-0.296) = +0.296$$
$$\gamma_4^*(s_3 = 0, s_4 = 0) = (-1) \cdot (-1.74) = +1.74$$
$$\gamma_4^*(s_3 = 1, s_4 = 0) = (+1) \cdot (-1.74) = -1.74.$$

Figure 6.32 illustrates the branch metric calculations of the column MAP decoders. According to (6.49), the forward path metrics are calculated as follows: For the first column decoder,

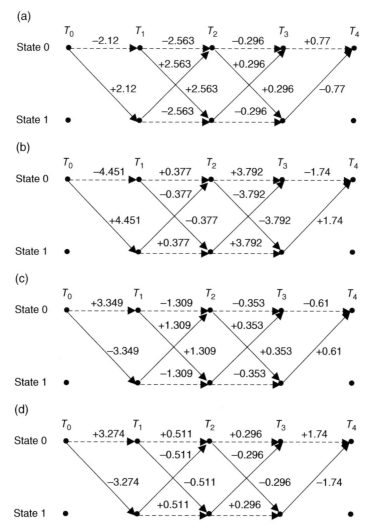

Figure 6.32 Branch metric calculations of the first (a), second (b), third (c), and fourth (d) column MAP decoder

$$\alpha_0^*(s_0 = 0) = 0$$
$$\alpha_1^*(s_1 = 0) = \max_{s_0}(0 + (-2.12)) = -2.12$$
$$\alpha_1^*(s_1 = 1) = \max_{s_0}(0 + (+2.12)) = +2.12$$
$$\alpha_2^*(s_2 = 0) = \max_{s_1}(((-2.12) + (-2.563)), ((+2.12) + (+2.563))) = +4.683$$
$$\alpha_2^*(s_2 = 1) = \max_{s_1}(((-2.12) + (+2.563)), ((+2.12) + (-2.563))) = +0.443$$
$$\alpha_3^*(s_3 = 0) = \max_{s_2}(((+4.683) + (-0.296)), ((+0.443) + (+0.296))) = +4.387$$

$$\alpha_3^*(s_3 = 1) = \max_{s_2}(((+4.683) + (+0.296)), ((+0.443) + (-0.296))) = +4.979$$

$$\alpha_4^*(s_4 = 0) = \max_{s_3}(((+4.387) + (+0.77)), ((+4.979) + (-0.77))) = +5.157$$

For the second column decoder,

$$\alpha_0^*(s_0 = 0) = 0$$

$$\alpha_1^*(s_1 = 0) = \max_{s_0}(0 + (-4.451)) = -4.451$$

$$\alpha_1^*(s_1 = 1) = \max_{s_0}(0 + (+4.451)) = +4.451$$

$$\alpha_2^*(s_2 = 0) = \max_{s_1}(((-4.451) + (+0.377)), ((+4.451) + (-0.377))) = +4.074$$

$$\alpha_2^*(s_2 = 1) = \max_{s_1}(((-4.451) + (-0.377)), ((+4.451) + (+0.377))) = +4.828$$

$$\alpha_3^*(s_3 = 0) = \max_{s_2}(((+4.074) + (+3.792)), ((+4.828) + (-3.792))) = +7.866$$

$$\alpha_3^*(s_3 = 1) = \max_{s_2}(((+4.074) + (-3.792)), ((+4.828) + (+3.792))) = +8.62$$

$$\alpha_4^*(s_4 = 0) = \max_{s_3}(((+7.866) + (-1.74)), ((+8.62) + (+1.74))) = +10.36$$

For the third column decoder,

$$\alpha_0^*(s_0 = 0) = 0$$

$$\alpha_1^*(s_1 = 0) = \max_{s_0}(0 + (+3.349)) = +3.349$$

$$\alpha_1^*(s_1 = 1) = \max_{s_0}(0 + (-3.349)) = -3.349$$

$$\alpha_2^*(s_2 = 0) = \max_{s_1}(((+3.349) + (-1.309)), ((-3.349) + (+1.309))) = +2.04$$

$$\alpha_2^*(s_2 = 1) = \max_{s_1}(((+3.349) + (+1.309)), ((-3.349) + (-1.309))) = +4.658$$

$$\alpha_3^*(s_3 = 0) = \max_{s_2}(((+2.04) + (-0.353)), ((+4.658) + (+0.353))) = +5.011$$

$$\alpha_3^*(s_3 = 1) = \max_{s_2}(((+2.04) + (+0.353)), ((+4.658) + (-0.353))) = +4.305$$

$$\alpha_4^*(s_4 = 0) = \max_{s_3}(((+5.011) + (-0.61)), ((+4.305) + (+0.61))) = +4.915$$

For the fourth column decoder,

$$\alpha_0^*(s_0 = 0) = 0$$

$$\alpha_1^*(s_1 = 0) = \max_{s_0}(0 + (+3.274)) = +3.274$$

$$\alpha_1^*(s_1 = 1) = \max_{s_0}(0 + (-3.274)) = -3.274$$

$$\alpha_2^*(s_2 = 0) = \max_{s_1}(((+3.274) + (+0.511)), ((-3.274) + (-0.511))) = +3.785$$

$$\alpha_2^*(s_2 = 1) = \max_{s_1}(((+3.274) + (-0.511)), ((-3.274) + (+0.511))) = +2.763$$

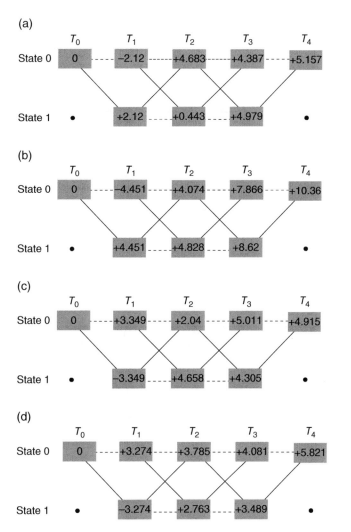

Figure 6.33 Forward path metric calculations of the first (a), second (b), third (c), and fourth (d) column MAP decoder

$$\alpha_3^*(s_3=0) = \max_{s_2}(((+3.785)+(+0.296)),((+2.763)+(-0.296))) = +4.081$$

$$\alpha_3^*(s_3=1) = \max_{s_2}(((+3.785)+(-0.296)),((+2.763)+(+0.296))) = +3.489$$

$$\alpha_4^*(s_4=0) = \max_{s_3}(((+4.081)+(+1.74)),((+3.489)+(-1.74))) = +5.821$$

Figure 6.33 illustrates the forward path metric calculations of the column MAP decoders. According to (6.54), the backward path metrics are calculated as follows:
For the first column decoder,

$$\beta_4^*(s_4 = 0) = 0$$

$$\beta_3^*(s_3 = 0) = \max_{s_4}(0 + (+0.77)) = +0.77$$

$$\beta_3^*(s_3 = 1) = \max_{s_4}(0 + (-0.77)) = -0.77$$

$$\beta_2^*(s_2 = 0) = \max_{s_3}(((+0.77) + (-0.296)), ((-0.77) + (+0.296))) = +0.474$$

$$\beta_2^*(s_2 = 1) = \max_{s_3}(((+0.77) + (+0.296)), ((-0.77) + (-0.296))) = +1.066$$

$$\beta_1^*(s_1 = 0) = \max_{s_2}(((+0.474) + (-2.563)), ((+1.066) + (+2.563))) = +3.629$$

$$\beta_1^*(s_1 = 1) = \max_{s_2}(((+0.474) + (+2.563)), ((+1.066) + (-2.563))) = +3.037$$

$$\beta_0^*(s_0 = 0) = \max_{s_1}(((+3.629) + (-2.12)), ((+3.037) + (+2.12))) = +5.157$$

For the second column decoder,

$$\beta_4^*(s_4 = 0) = 0$$

$$\beta_3^*(s_3 = 0) = \max_{s_4}(0 + (-1.74)) = -1.74$$

$$\beta_3^*(s_3 = 1) = \max_{s_4}(0 + (+1.74)) = +1.74$$

$$\beta_2^*(s_2 = 0) = \max_{s_3}(((-1.74) + (+3.792)), ((+1.74) + (-3.792))) = +2.052$$

$$\beta_2^*(s_2 = 1) = \max_{s_3}(((-1.74) + (-3.792)), ((+1.74) + (+3.792))) = +5.532$$

$$\beta_1^*(s_1 = 0) = \max_{s_2}(((+2.052) + (+0.377)), ((+5.532) + (-0.377))) = +5.155$$

$$\beta_1^*(s_1 = 1) = \max_{s_2}(((+2.052) + (-0.377)), ((+5.532) + (+0.377))) = +5.909$$

$$\beta_0^*(s_0 = 0) = \max_{s_1}(((+5.155) + (-4.451)), ((+5.909) + (+4.451))) = +10.36$$

For the third column decoder,

$$\beta_4^*(s_4 = 0) = 0$$

$$\beta_3^*(s_3 = 0) = \max_{s_4}(0 + (-0.61)) = -0.61$$

$$\beta_3^*(s_3 = 1) = \max_{s_4}(0 + (+0.61)) = +0.61$$

$$\beta_2^*(s_2 = 0) = \max_{s_3}(((-0.61) + (-0.353)), ((+0.61) + (+0.353))) = +0.963$$

$$\beta_2^*(s_2 = 1) = \max_{s_3}(((-0.61) + (+0.353)), ((+0.61) + (-0.353))) = +0.257$$

$$\beta_1^*(s_1 = 0) = \max_{s_2}(((+0.963) + (-1.309)), ((+0.257) + (+1.309))) = +1.566$$

$$\beta_1^*(s_1 = 1) = \max_{s_2}(((+0.963) + (+1.309)), ((+0.257) + (-1.309))) = +2.272$$

$$\beta_0^*(s_0 = 0) = \max_{s_1}(((+1.566) + (+3.349)), ((+2.272) + (-3.349))) = +4.915$$

For the fourth column decoder,

$$\beta_4^*(s_4 = 0) = 0$$

$$\beta_3^*(s_3 = 0) = \max_{s_4}(0 + (+1.74)) = +1.74$$

$$\beta_3^*(s_3 = 1) = \max_{s_4}(0 + (-1.74)) = -1.74$$

$$\beta_2^*(s_2 = 0) = \max_{s_3}(((+1.74)+(+0.296)),((-1.74)+(-0.296))) = +2.036$$

$$\beta_2^*(s_2 = 1) = \max_{s_3}(((+1.74)+(-0.296)),((-1.74)+(+0.296))) = +1.444$$

$$\beta_1^*(s_1 = 0) = \max_{s_2}(((+2.036)+(+0.511)),((+1.444)+(-0.511))) = +2.547$$

$$\beta_1^*(s_1 = 1) = \max_{s_2}(((+2.036)+(-0.511)),((+1.444)+(+0.511))) = +1.955$$

$$\beta_0^*(s_0 = 0) = \max_{s_1}(((+2.547)+(+3.274)),((+1.955)+(-3.274))) = +5.821$$

Figure 6.34 illustrates the backward path metric calculations of the column MAP decoders. From branch, forward path, and backward path metrics, we calculate the soft decision outputs of the column MAP decoder as follows:

For the first column decoder,

$$L_c^*(I_{11}) = \max_{(s_0,s_1)\in S^p}((0)+(+2.12)+(+3.037)) - \max_{(s_0,s_1)\in S^n}((0)+(-2.12)+(+3.629)) = +3.648$$

$$L_c^*(I_{21}) = \max_{(s_1,s_2)\in S^p}(((-2.12)+(+2.563)+(+1.066)),((+2.12)+(+2.563)+(+0.474)))$$

$$- \max_{(s_1,s_2)\in S^n}(((-2.12)+(-2.563)+(+0.474)),((+2.12)+(-2.563)+(+1.066))) = +4.534$$

$$L_c^*(I_{31}) = \max_{(s_2,s_3)\in S^p}(((+4.683)+(+0.296)+(-0.77)),((+0.443)+(+0.296)+(+0.77)))$$

$$- \max_{(s_2,s_3)\in S^n}(((+4.683)+(-0.296)+(+0.77)),((+0.443)+(-0.296)+(-0.77))) = -0.948$$

$$L_c^*(P_{c1}) = \max_{(s_3,s_4)\in S^p}((+4.979)+(-0.77)+(0)) - \max_{(s_3,s_4)\in S^n}((+4.387)+(+0.77)+(0)) = -0.948$$

For the second column decoder,

$$L_c^*(I_{12}) = \max_{(s_0,s_1)\in S^p}((0)+(+4.451)+(+5.909)) - \max_{(s_0,s_1)\in S^n}((0)+(-4.451)+(+5.155)) = +9.656.$$

$$L_c^*(I_{22}) = \max_{(s_1,s_2)\in S^p}(((-4.451)+(-0.377)+(+5.532)),((+4.451)+(-0.377)+(+2.052)))$$

$$- \max_{(s_1,s_2)\in S^n}(((-4.451)+(+0.377)+(+2.052)),((+4.451)+(+0.377)+(+5.532))) = -4.234$$

$$L_c^*(I_{32}) = \max_{(s_2,s_3)\in S^p}(((+4.074)+(-3.792)+(+1.74)),((+4.828)+(-3.792)+(-1.74)))$$

$$- \max_{(s_2,s_3)\in S^n}(((+4.074)+(+3.792)+(-1.74)),((+4.828)+(+3.792)+(+1.74))) = -8.338$$

$$L_c^*(P_{c2}) = \max_{(s_3,s_4)\in S^p}((+8.62)+(+1.74)+(0)) - \max_{(s_3,s_4)\in S^n}((+7.866)+(-1.74)+(0)) = +4.234$$

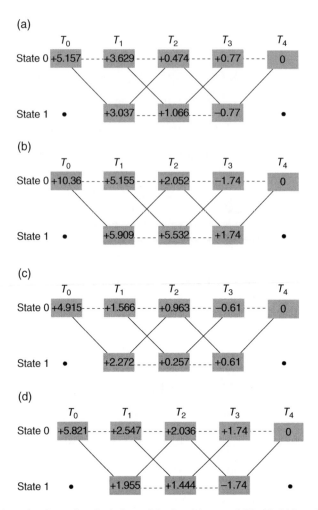

Figure 6.34 Backward path metric calculations of the first (a), second (b), third (c), and fourth (d) column MAP decoder

For the third column decoder,

$$L_c^*(I_{13}) = \max_{(s_0,s_1)\in S^p} ((0)+(-3.349)+(+2.272)) - \max_{(s_0,s_1)\in S^n} ((0)+(+3.349)+(+1.566)) = -5.992$$

$$L_c^*(I_{23}) = \max_{(s_1,s_2)\in S^p} (((+3.349)+(+1.309)+(+0.257)),((-3.349)+(+1.309)+(+0.963)))$$
$$- \max_{(s_1,s_2)\in S^n} (((+3.349)+(-1.309)+(+0.963)),((-3.349)+(-1.309)+(+0.257))) = +1.912$$

$$L_c^*(I_{33}) = \max_{(s_2,s_3)\in S^p} (((+2.04)+(+0.353)+(+0.61)),((+4.658)+(+0.353)+(-0.61)))$$
$$- \max_{(s_2,s_3)\in S^n} (((+2.04)+(-0.353)+(-0.61)),((+4.658)+(-0.353)+(+0.61))) = -0.514$$

$$L_c^*(P_{c3}) = \max_{(s_3,s_4)\in S^p} ((+4.305)+(+0.61)+(0)) - \max_{(s_3,s_4)\in S^n} ((+5.011)+(-0.61)+(0)) = +0.514$$

For the fourth column decoder,

$$L_c^*(P_{r1}) = \max_{(s_0,s_1)\in S^P} ((0)+(-3.274)+(+1.955)) - \max_{(s_0,s_1)\in S^n} ((0)+(+3.274)+(+2.547)) = -7.14$$

$$L_c^*(P_{r2}) = \max_{(s_1,s_2)\in S^P} (((+3.274)+(-0.511)+(+1.444)),((-3.274)+(-0.511)+(+2.036)))$$
$$- \max_{(s_1,s_2)\in S^n} (((+3.274)+(+0.511)+(+2.036)),((-3.274)+(+0.511)+(+1.444))) = -1.614$$

$$L_c^*(P_{r3}) = - \max_{(s_1,s_2)\in S^n} (((+3.274)+(+0.511)+(+2.036)),((-3.274)+(+0.511)+(+1.444))) = -1.614$$
$$- \max_{(s_2,s_3)\in S^n} (((+3.785)+(+0.296)+(+1.74)),((+2.763)+(+0.296)+(-1.74))) = -1.614$$

$$L_c^*(P_{rc}) = \max_{(s_3,s_4)\in S^P} ((+3.489)+(-1.74)+(0)) - \max_{(s_3,s_4)\in S^n} ((+4.081)+(+1.74)+(0)) = -4.072$$

Finally, we obtain the column decoding results in the first iteration as shown in Figure 6.35. As we can observe in Figure 6.35, I_{31} and I_{33} were not fixed after the row decoding. However, they are fixed in the column decoder. In the same way, the second iteration is performed or the output of the turbo product decoder is determined by hard decision.

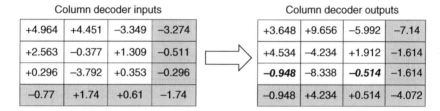

Figure 6.35 Column decoding results in the first iteration

Summary 6.5 Turbo product codes

1. The turbo product encoding process is same as the product encoding process. The turbo product decoding process is iterative using the Soft Input Soft Output (SISO) decoding algorithm.
2. The product code C_p $(n_1 \times n_2, k_1 \times k_2, d_1 \times d_2)$ is obtained from systematic linear block codes C_1 (n_1, k_1, d_1) and C_2 (n_2, k_2, d_2) where n, k, and d are the code length, the information length, and minimum distance, respectively.
3. One information bit is encoded by both a row code and a column code. When the received bit is corrupted and cannot be corrected by a row code, a column code may fix it and vice versa.
4. The row blocks of the turbo product code are decoded first and then all of column blocks are decoded. Iteration can be done several times to reduce the error probability.
5. The trellis representation of block codes enables us to perform MAP or ML decoding. It is suitable for short length codes such as row or column codes of the product codes.
6. When comparing with turbo codes, the first advantage is that turbo product codes provide a better performance at a high code rate (more than 0.75). The second advantage is that the turbo product codes require smaller frame size.

6.2.3   Hardware Implementation of Turbo Product Encoding and Decoding

As we investigated in Section 6.2.1, turbo product codes are constructed by systematic block codes. The encoding process of the turbo product codes can be easily implemented by shift registers and modulo 2 adders. The encoding complexity is proportional to the codeword length. Figure 6.36 illustrates an example of a $(4, 3, 2)$ single parity check encoder and a $(7, 4)$ Hamming encoder.

The turbo product decoder is composed of row and column component decoders. Unlike turbo codes, the turbo product codes do not need an interleaver and each component code is independent. Thus, it is easy to build a parallel architecture. If we design a $(n, k, d)^2$ turbo product code with same component codes (N row codes and N column codes), the turbo product decoder can have simple parallel decoding architecture as shown in Figure 6.37.

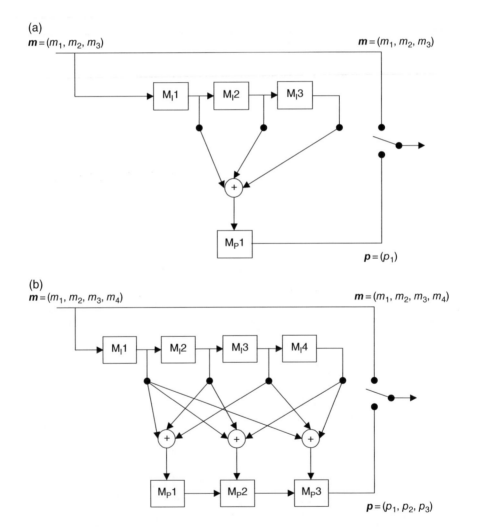

Figure 6.36   Example of a $(4, 3, 2)$ SPC encoder (a) and a $(7, 4, 3)$ Hamming encoder (b)

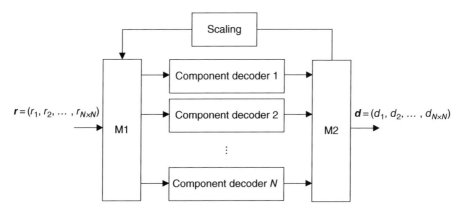

Figure 6.37   Parallel turbo product decoder

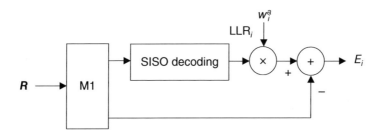

Figure 6.38   Component decoder of the turbo product decoder

In the first iteration, the first memory (M1) holds the received codeword and sends each row codeword to component decoders. The second memory (M2) holds the results of each row decoder and sends extrinsic information to the first memory after scaling it. In the second iteration, the first memory sends the received codeword and extrinsic information as a priori information. Each component decoder performs column decoding. Figure 6.38 illustrates the component decoder.

Since we used the Max-log-MAP algorithm as component decoders of the turbo product decoder, the turbo product decoding process is same as the turbo decoding process. The parallel turbo product decoder enables us to implement a high speed decoding. However, the memory size and the complexity increase. In order to reduce them, we can design a serial turbo product decoder which uses a single component decoder. Since each row and column code can be decoded by a single component, we can approximately reduce the memory size and complexity by a factor of N even if the latency increases.

6.3   Low-Density Parity Check Codes

6.3.1   LDPC Encoding and Decoding Algorithms

In 1962, R. Gallager [Ref. 1 in Chapter 5] originally invented LDPC codes during his PhD. However, the LDPC codes were not widely recognized until D.J.C Mackay and R.M. Neal [Ref. 2 in Chapter 5] rediscovered them as the era of transistors has just started and the hardware technology did not cover the complexity of LDPC encoding and decoding at that time. After the turbo codes emerged in 1993, many researchers have made an effort to prove theoretically how the turbo codes achieve near Shannon limit and have tried to find another new error correction code. In 1996, Mackay and Neal designed a new linear block code including many similar features of turbo codes such as randomness, large block length, and iterative decoding. They soon realized that the new codes are almost same as LDPC codes by Gallager. After that, irregular LDPC codes as generalization of Gallager's LDPC codes are developed by Luby et al. [15] in 1998. The irregular LDPC codes became the most powerful error control codes as of now. When comparing with the turbo codes, LDPC codes have several advantages. Firstly a random interleaver is not required. Secondly, it has a better block error rate and a lower error floor. Thirdly, iterative decoding of LDPC codes is simple even if it requires more iterations. The most highlighted advantage is that it is patent-free.

As the name implies, LDPC codes are linear block codes with a sparse parity check matrix H. The sparseness of the parity check matrix H means that H contains relatively few 1s among many 0s. The sparseness enables LDPC codes to increase the minimum distance. Typically, the minimum distance of LDPC codes linearly increases according to the codeword length. Basically, LDPC codes are same as conventional linear block codes except the sparseness. The difference between LDPC codes and conventional linear block codes is decoding method. The decoder of the conventional linear block codes is generally based on ML decoding which receives n bits codeword and decides the most likely k bits message among the 2^k possible messages. Thus, the codeword length is short and the decoding complexity is low. On the other hands, LDPC codes are iteratively decoded using a graphical representation (Tanner graph) of H. The Tanner graph is consisted of bit (or variable, symbol) nodes and check (or parity check) nodes. The bit nodes and the check nodes represent codeword bits and parity equations, respectively. The edge represents a connection between bit nodes and check nodes if and only if the bit is involved in the corresponding parity check equation. Thus, the number of edges in a Tanner graph means the number of 1s in the parity check matrix H. Figure 6.39 illustrates an example of a Tanner graph. The squares represent check nodes (or parity check equations) and the circles represent bit nodes in the figure.

As we observed in Chapter 5, a linear block code has a $(n-k) \times n$ parity check matrix H. The rows and columns of H represent the parity check equations and the coded bits, respectively.

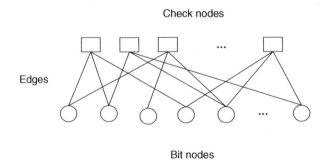

Figure 6.39   Example of Tanner graph

Example 6.1 Parity check matrix and Tanner graph

A (7, 4) linear block code has the following parity check matrix:

$$H = \begin{bmatrix} 1 & 0 & 0 & 1 & 1 & 0 & 1 \\ 0 & 1 & 0 & 1 & 0 & 1 & 1 \\ 0 & 0 & 1 & 0 & 1 & 1 & 1 \end{bmatrix}.$$

When we have the codeword $[c_1, c_2, c_3, c_4, c_5, c_6, c_7]$, find the corresponding parity check equations and draw the corresponding Tanner graph.

Solution

From the equation $c \cdot H^T = 0$,

$$c \cdot H^T = \begin{bmatrix} c_1 & c_2 & c_3 & c_4 & c_5 & c_6 & c_7 \end{bmatrix} \begin{bmatrix} 1 & 0 & 0 \\ 0 & 1 & 0 \\ 0 & 0 & 1 \\ 1 & 1 & 0 \\ 1 & 0 & 1 \\ 0 & 1 & 1 \\ 1 & 1 & 1 \end{bmatrix} = 0.$$

In GF(2), the addition is XOR operation and the multiplication is AND operation. Therefore, we have the following parity check equations:

Parity check equation 1: $c_1 + c_4 + c_5 + c_7 = 0$
Parity check equation 2: $c_2 + c_4 + c_6 + c_7 = 0$
Parity check equation 3: $c_3 + c_5 + c_6 + c_7 = 0$.

The corresponding Tanner graph is illustrated in Figure 6.40. ∎

The regular LDPC code by Gallager is denoted as (n, b_c, b_r) where n is a codeword length, b_c is the number of parity check equations (or 1s per column), and b_r is the number of coded

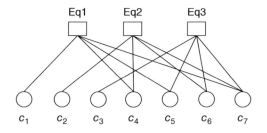

Figure 6.40 Tanner graph of the (7, 4) linear block code

bits (or 1s per row). The regular LDPC code has the following properties: (i) each coded bit is contained in the same number of parity check equations and (ii) each parity check equation contains the same number of coded bits. The regular LDPC code by Gallager is constructed by randomly choosing the locations of 1s with the fixed numbers in each row and column. The rows of the parity check matrix are divided into b_c sets. Each set has n/b_r rows. The first set contains r consecutive 1s descending from left to right. The other sets are obtained by column permutation of the first set.

Example 6.2   Regular LDPC code construction by Gallager
Construct the parity check matrix of a (12, 3, 4) regular Gallager LDPC code.

Solution
Firstly, the rows of the parity check matrix are divided into three sets. The first set contains four consecutive 1s in each row as follows:

$$
\begin{bmatrix} \text{The first set} \\ \text{The second set} \\ \text{The third set} \end{bmatrix} =
\begin{bmatrix}
1 & 1 & 1 & 1 & 0 & 0 & 0 & 0 & 0 & 0 & 0 & 0 \\
0 & 0 & 0 & 0 & 1 & 1 & 1 & 1 & 0 & 0 & 0 & 0 \\
0 & 0 & 0 & 0 & 0 & 0 & 0 & 0 & 1 & 1 & 1 & 1 \\
0 & 0 & 0 & 0 & 0 & 0 & 0 & 0 & 0 & 0 & 0 & 0 \\
0 & 0 & 0 & 0 & 0 & 0 & 0 & 0 & 0 & 0 & 0 & 0 \\
0 & 0 & 0 & 0 & 0 & 0 & 0 & 0 & 0 & 0 & 0 & 0 \\
0 & 0 & 0 & 0 & 0 & 0 & 0 & 0 & 0 & 0 & 0 & 0 \\
0 & 0 & 0 & 0 & 0 & 0 & 0 & 0 & 0 & 0 & 0 & 0 \\
0 & 0 & 0 & 0 & 0 & 0 & 0 & 0 & 0 & 0 & 0 & 0
\end{bmatrix}.
$$

Secondly, the second and third sets are obtained by column permutation of the first set as follows:

$$
H =
\begin{bmatrix}
1 & 1 & 1 & 1 & 0 & 0 & 0 & 0 & 0 & 0 & 0 & 0 \\
0 & 0 & 0 & 0 & 1 & 1 & 1 & 1 & 0 & 0 & 0 & 0 \\
0 & 0 & 0 & 0 & 0 & 0 & 0 & 0 & 1 & 1 & 1 & 1 \\
0 & 1 & 0 & 0 & 1 & 0 & 0 & 0 & 1 & 0 & 0 & 1 \\
1 & 0 & 0 & 1 & 0 & 1 & 0 & 0 & 0 & 1 & 0 & 0 \\
0 & 0 & 1 & 0 & 0 & 0 & 1 & 1 & 0 & 0 & 1 & 0 \\
0 & 1 & 0 & 0 & 0 & 1 & 0 & 0 & 0 & 0 & 1 & 1 \\
0 & 0 & 1 & 0 & 0 & 0 & 1 & 0 & 1 & 1 & 0 & 0 \\
1 & 0 & 0 & 1 & 1 & 0 & 0 & 1 & 0 & 0 & 0 & 0
\end{bmatrix}.
$$

As we can observe from the above parity check matrix, each coded bit is contained in three parity check equations (three bits per column) and each parity check equation contains four coded bits (four bits per row). ∎

Another LDPC code construction by Mackay and Neal fills 1s in the parity check matrix H from left column to right column. The locations of 1s are randomly chosen in each column until each row with the fixed numbers is assigned. If some row is already filled, the row is not assigned and the remaining rows are filled. One important constraint is that the 1s in each column and row should not be square shape. It is important to avoid a cycle of length 4. However, it is not easy to satisfy this constraint. The number of b_r and b_c should be small when comparing with the code length.

Example 6.3 Regular LDPC code construction by Mackay and Neal

Construct the parity check matrix of a (12, 3, 4) regular Mackay and Neal LDPC code.

Solution

Firstly, the location of three bits in each column is randomly chosen from left column to right column with avoiding overlapping until some row with four bits appears as follows:

$$H = \begin{bmatrix}
1 & 0 & 0 & 1 & 0 & 0 & 1 & 0 & 0 & 0 & 0 & 0 \\
0 & 0 & 1 & 1 & 0 & 0 & 0 & 1 & 0 & 0 & 0 & 0 \\
0 & 1 & 0 & 1 & 0 & 0 & 0 & 0 & 1 & 0 & 0 & 0 \\
0 & 0 & 1 & 0 & 1 & 0 & 1 & 0 & 1 & 0 & 0 & 0 \\
0 & 1 & 0 & 0 & 1 & 0 & 0 & 1 & 0 & 0 & 0 & 0 \\
1 & 0 & 0 & 0 & 0 & 1 & 0 & 1 & 1 & 0 & 0 & 0 \\
0 & 0 & 1 & 0 & 0 & 1 & 0 & 0 & 0 & 0 & 0 & 0 \\
0 & 1 & 0 & 0 & 0 & 1 & 1 & 0 & 0 & 0 & 0 & 0 \\
1 & 0 & 0 & 0 & 1 & 0 & 0 & 0 & 0 & 0 & 0 & 0
\end{bmatrix}.$$

When filling in the ninth column, the fourth row and the sixth row of the above matrix contain four bits. Thus, we should not assign 1s in the fourth row and the sixth row. We keep filling in the columns with avoiding overlapping. The fully filled rows are as follows:

$$H = \begin{bmatrix}
1 & 0 & 0 & 1 & 0 & 0 & 1 & 0 & 0 & 0 & 0 & 0 \\
0 & 0 & 1 & 1 & 0 & 0 & 0 & 1 & 0 & 0 & 0 & 0 \\
0 & 1 & 0 & 1 & 0 & 0 & 0 & 0 & 1 & 1 & 0 & 0 \\
0 & 0 & 1 & 0 & 1 & 0 & 1 & 0 & 1 & 0 & 0 & 0 \\
0 & 1 & 0 & 0 & 1 & 0 & 0 & 1 & 0 & 0 & 0 & 0 \\
1 & 0 & 0 & 0 & 0 & 1 & 0 & 1 & 1 & 0 & 0 & 0 \\
0 & 0 & 1 & 0 & 0 & 1 & 0 & 0 & 0 & 1 & 0 & 0 \\
0 & 1 & 0 & 0 & 0 & 1 & 1 & 0 & 0 & 0 & 0 & 0 \\
1 & 0 & 0 & 0 & 1 & 0 & 0 & 0 & 0 & 1 & 0 & 0
\end{bmatrix}.$$

When filling in the 10th column, the third row of the above matrix contains four bits. In the same way, we construct the parity check matrix of a (12, 3, 4) regular LDPC code as follows:

$$H = \begin{bmatrix} 1 & 0 & 0 & 1 & 0 & 0 & 1 & 0 & 0 & 0 & 1 & 0 \\ 0 & 0 & 1 & 1 & 0 & 0 & 0 & 1 & 0 & 0 & 0 & 1 \\ 0 & 1 & 0 & 1 & 0 & 0 & 0 & 0 & 1 & 1 & 0 & 0 \\ 0 & 0 & 1 & 0 & 1 & 0 & 1 & 0 & 1 & 0 & 0 & 0 \\ 0 & 1 & 0 & 0 & 1 & 0 & 0 & 1 & 0 & 0 & 1 & 0 \\ 1 & 0 & 0 & 0 & 0 & 1 & 0 & 1 & 1 & 0 & 0 & 0 \\ 0 & 0 & 1 & 0 & 0 & 1 & 0 & 0 & 0 & 1 & 1 & 0 \\ 0 & 1 & 0 & 0 & 0 & 1 & 1 & 0 & 0 & 0 & 0 & 1 \\ 1 & 0 & 0 & 0 & 1 & 0 & 0 & 0 & 0 & 1 & 0 & 1 \end{bmatrix}.$$

Similar to Gallager's LDPC code construction, the above parity check matrix contains three bits per a column and four bits per a row. ∎

In a Tanner graph, a *cycle* is defined as a node connection starting and ending at same node. The length of the cycle is defined as the number of edges of the cycle. The cycle prevents performance improvement of iterative decoding because it affects the independence of extrinsic information during the iterative process. Thus, it is important to remove cycles in the parity check matrix. Sometimes a cycle breaking technique can be used to remove cycles as shown in Figure 6.41.

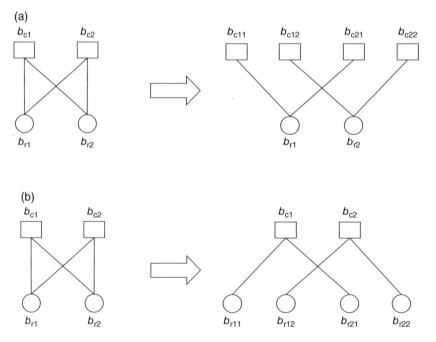

Figure 6.41 Cycle breaking (column splitting (a) and row splitting (b))

This technique changes the Tanner graph and can reduce the number of cycles. However, it increases the number of nodes so that the complexity of iterative decoding increases.

The *girth* of a Tanner graph is defined as the length of the shortest cycle in the Tanner graph. The shortest cycle in a Tanner graph is 4. In the Example 6.3, the girth is 6. The short cycles disturb performance improvement during iterative process of decoding algorithms. Thus, it is one important parameter to maximize the girth when designing a LDPC encoder.

Example 6.4   Cycles of a parity check matrix

Find the cycles of the parity check matrix in Example 6.1.

Solution

In the Tanner graph of Example 6.1, the bit nodes c_4, c_5, c_6, and c_7 are connected with more than two edges and the check nodes Eq1, Eq2, and Eq3 are connected with more than two edges. Thus, there is some possibility to produce cycles. In the parity check matrix H, two 1s in the first row belong to the fourth and seventh columns. Two 1s in the second row belong to the fourth and seventh columns as well. These four 1s produce a cycle of length 4. Figure 6.42 illustrates the cycle and the related edges for them. We can find more cycles of length 4 as shown in Figures 6.43 and 6.44. In Figures 6.42, 6.43 and 6.44, the bold line and the bold 1s of H represent cycles of length 4 and the related edges, respectively.

In the same way, we can find a cycle of length 6 as shown in Figure 6.45. In Figure 6.45, the bold line and the bold 1s of H represent a cycle of length 6 and the related edges, respectively.   ∎

Irregular LDPC codes are generalization of regular LDPC codes. In the parity check matrix of irregular LDPC codes, the degrees of bit nodes and check nodes are not constant. Thus, an irregular LDPC code is represented as degree distribution of bit nodes and check nodes. The bit node degree distribution is defined as follows:

$$\Lambda(x) = \sum_{i=1}^{d_b} \Lambda_i x^i \tag{6.68}$$

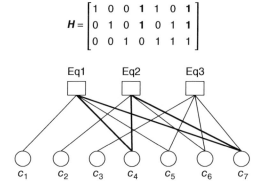

Figure 6.42   Cycle of length 4 and the related edges between the bit nodes (c_4 and c_7) and the check nodes (Eq1 and Eq2)

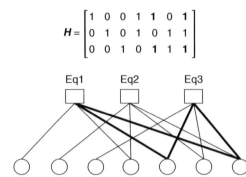

$$H = \begin{bmatrix} 1 & 0 & 0 & 1 & 1 & 0 & 1 \\ 0 & 1 & 0 & 1 & 0 & 1 & 1 \\ 0 & 0 & 1 & 0 & 1 & 1 & 1 \end{bmatrix}$$

Figure 6.43   Cycle of length 4 and the related edges between the bit nodes (c_5 and c_7) and the check nodes (Eq1 and Eq3)

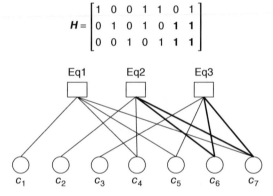

$$H = \begin{bmatrix} 1 & 0 & 0 & 1 & 1 & 0 & 1 \\ 0 & 1 & 0 & 1 & 0 & 1 & 1 \\ 0 & 0 & 1 & 0 & 1 & 1 & 1 \end{bmatrix}$$

Figure 6.44   Cycle of length 4 and the related edges between the bit nodes (c_6 and c_7) and the check nodes (Eq2 and Eq3)

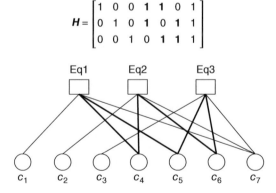

$$H = \begin{bmatrix} 1 & 0 & 0 & 1 & 1 & 0 & 1 \\ 0 & 1 & 0 & 1 & 0 & 1 & 1 \\ 0 & 0 & 1 & 0 & 1 & 1 & 1 \end{bmatrix}$$

Figure 6.45   Cycle of length 6 and the related edges between the bit nodes (c_4, c_5 and c_6) and the check nodes (Eq1, Eq2 and Eq3)

where Λ_i and d_b are the number of bit nodes of degree i and the maximum bit nodes, respectively. The check node degree distribution is defined as follows:

$$P(x) = \sum_{i=1}^{d_c} P_i x^i \tag{6.69}$$

where P_i and d_c are the number of check nodes of degree i and the maximum check nodes, respectively. It is possible to represent in the edge point of view. The bit node degree distribution in the edge point of view is defined as follows:

$$\lambda(x) = \sum_{i=1}^{d_b} \lambda_i x^{i-1} \tag{6.70}$$

where λ_i is the fraction of edges connected to bit nodes of degree i. The check node degree distribution in the edge point of view is defined as follows:

$$\rho(x) = \sum_{i=1}^{d_c} \rho_i x^{i-1} \tag{6.71}$$

where ρ_i is the fraction of edges connected to check nodes of degree i. The design rate of an LDPC code is bounded by

$$R(\lambda,\rho) \geq 1 - \frac{\int_0^1 \rho(x)dx}{\int_0^1 \lambda(x)dx} = 1 - \frac{b_c}{b_r}. \tag{6.72}$$

Example 6.5   Bit and check node degree distribution
Find the bit and check node degree distribution in the Tanner graph of Example 6.1.

Solution
In the Tanner graph of Example 6.1, we can observe that the bit nodes c_1, c_2, and c_3 have one edge, the bit nodes c_4, c_5, and c_6 have two edges, and the bit node c_7 has three edges. Thus, we represent the bit node degree distribution as follows:

$$\Lambda(x) = 3x + 3x^2 + x^3.$$

The check nodes Eq1, Eq2, and Eq3 have four edges. Thus, we represent the check node degree distribution as follows:

$$P(x) = 3x^4.$$

In addition, 12 edges are connected to bit nodes. We can observe that the edges with bit nodes of degree 1 are 3, the edges with bit nodes of degree 2 are 6, and the edges with bit nodes of degree 3 are 3. In the edge point of view, the bit node degree distribution is

$$\lambda(x) = \frac{3}{12} + \frac{6}{12}x + \frac{3}{12}x^2 = \frac{1}{4} + \frac{1}{2}x + \frac{1}{4}x^2.$$

The edges with check nodes of degree 4 are 12. Thus, the check node degree distribution in the edge point of view is

$$\rho(x) = \frac{12}{12}x^3 = x^3.$$ ∎

It is not easy to optimize an irregular LDPC code design. Many approaches to construct an irregular LDPC code are based on computer search. As we discussed in Chapter 5, the generator matrix of a linear block code can be obtained from the parity check matrix. Using Gauss–Jordan elimination, we change the form of the party check matrix H as follows:

$$H = \left[P^T \; I_{n-k} \right] \text{ or } H = \left[I_{n-k} \; P^T \right] \tag{6.73}$$

where P and I_{n-k} are a parity matrix and an identity matrix, respectively. Thus, the generator matrix is

$$G = \left[I_k \; P \right] \text{ or } G = \left[P \, I_k \right]. \tag{6.74}$$

The sparseness of P determines the complexity of LDPC encoders. Unfortunately, P is most likely not spare even if the constructed H is sparse. Basically, a LDPC code requires a long frame size (a large n). The encoding complexity is $O(n^2)$ so that the encoding complexity is one important design issue. There are many approaches to reduce encoding complexity [16].

Summary 6.6   LDPC codes

1. LDPC codes are linear block codes with a sparse parity check matrix H. The sparseness of the parity check matrix H means that H contains relatively few 1s among many 0s.
2. The sparseness enables LDPC codes to increase the minimum distance. Typically, the minimum distance of LDPC codes linearly increases according to the codeword length.
3. LDPC codes are same as conventional linear block codes except the sparseness. The difference between LDPC codes and conventional linear block codes is the decoding method.
4. The regular LDPC code means that each coded bit is contained in the same number of parity check equations and each parity check equation contains the same number of coded bits.
5. The irregular LDPC codes are generalization of regular LDPC codes. They are represented as degree distribution of bit nodes and check nodes.
6. In a Tanner graph, a cycle is defined as a node connection starting and ending at same node. The cycle prevents performance improvement of iterative decoding because it affects the independence of the extrinsic information in iterative process.

The decoding process of LDPC codes is based on the iteration scheme between bit nodes and check nodes in a Tanner graph. A decoding scheme of LDPC codes is known as a message passing algorithm which passes messages forward and backward between the bit nodes and check nodes. There are two types of message passing algorithms: the *bit flipping algorithm* based on a hard decision decoding algorithm and the *belief propagation algorithm* based on a soft decision decoding algorithm. The belief propagation algorithm calculates the maximum a posteriori probability (APP). That is to say, it calculates the probability $P(c_i \mid E)$, which means we find a codeword c_i on the event E (all parity check equations are satisfied). In the belief propagation algorithm, the message represents the belief level (probability) of the received codewords. Each bit node passes a message to each check node connected to the bit node. Each check node passes a message to each bit node connected to the check node. In the final stage, APP of each codeword bit is calculated. We should calculate many multiplication and division operations for the belief propagation algorithm. Thus, the implementation complexity is high. In order to reduce the complexity, it is possible to implement it using log likelihood ratios. Multiplication and division are replaced by addition and subtractions, respectively. We call this a sum-product decoding algorithm. In order to explain the belief propagation algorithm, the Tanner graph is modified. Figure 6.46 illustrates an example of the Tanner graph for LDPC decoding.

In Figure 6.46, v_j, x_i, and y_i represent check nodes, bit nodes, and received codeword bits, respectively. We express the received codeword as follows:

$$y_i = x_i + n_i \tag{6.75}$$

where n_i is Gaussian noise with zero mean and standard deviation σ. We define two messages (estimations): $q_{ij}(x)$ and $r_{ji}(x)$. The message $q_{ij}(x)$ and the message $r_{ji}(x)$ represent the message from the bit node x_i to the check node v_j and the message from the check node v_j to the bit node x_i, respectively. They can be expressed in the Tanner graph as shown in Figure 6.47.

The message $q_{ij}(x)$ means the probability $P(x_i = x \mid y_i)$ or the probability $x_i = x$ satisfying all check node equations except v_j. The message $r_{ji}(x)$ means the probability the parity check node (parity check equation) v_j is satisfied when all bit nodes have x except x_i. Thus, message passing can be described as shown in Figure 6.48.

In the AWGN channel, we have the following initial value of $q_{ij}^{initial}(x)$:

$$q_{ij}^{initial}(x) = P(x_i = x \mid y_i) = \frac{1}{1 + e^{-(2xy_i/\sigma^2)}} \tag{6.76}$$

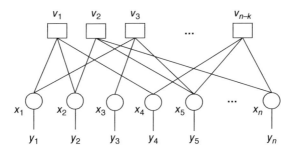

Figure 6.46   Tanner graph example for LDPC decoding

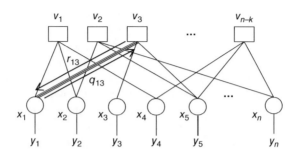

Figure 6.47   Two messages in the Tanner graph

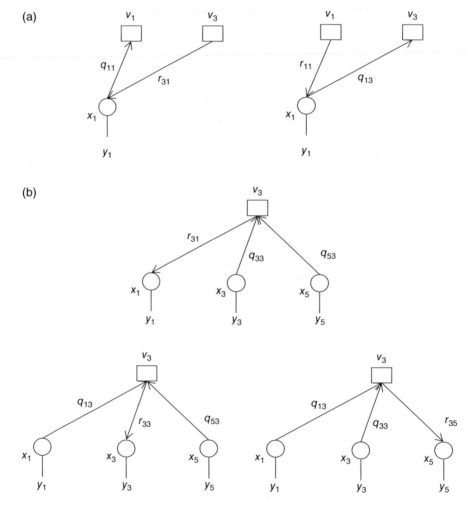

Figure 6.48   Example of bit node (a) and check node (b) message passing in the Tanner graph

where $x = +1$ or -1 (assume BPSK modulation). When the received vector y with the length L is given, Gallager described the probability that the parity check nodes contain an even or odd number of 1s is expressed as follows:

$$P^c = \frac{1}{2} + a\frac{1}{2}\prod_{i=1}^{L}(1-2p_i) \tag{6.77}$$

where

$$a = \begin{cases} +1 & \text{for even parity} \\ -1 & \text{for odd parity} \end{cases} \tag{6.78}$$

and p_i is $P(x_i = -1 | y_i)$ which is the probability that the codeword bit at i is equal to -1. Thus, the message $r_{ji}(+1)$ is expressed as follows:

$$r_{ji}(+1) = \frac{1}{2} + \frac{1}{2}\prod_{i \in V_{ji}}\left(1 - 2q_{ij}(-1)\right) \tag{6.79}$$

where V_{ji} denotes a bit node set connected to the check node v_j except x_i. We express $r_{ji}(-1)$ as follows:

$$r_{ji}(-1) = 1 - r_{ji}(+1). \tag{6.80}$$

The message $q_{ij}(+1)$ is expressed as follows:

$$q_{ij}(+1) = \alpha_{ij}(1-p_i)\prod_{j \in C_{ij}} r_{ji}(+1) \tag{6.81}$$

and the message $q_{ij}(-1)$ is expressed as follows:

$$q_{ij}(-1) = \alpha_{ij} p_i \prod_{j \in C_{ij}} r_{ji}(-1) \tag{6.82}$$

where C_{ij} denotes a check node set connected to the bit node x_i except v_j and the constant α_{ij} are selected to ensure that $q_{ij}(+1) + q_{ij}(-1) = 1$. In this way, we start calculating $r_{ji}(x)$ using the initial values of $q_{ij}(x)$. The message $q_{ij}(x)$ is calculated by $r_{ji}(x)$. Then, we calculate the APP ratio for each codeword bit as follows:

$$Q_i(+1) = \alpha_i(1-p_i)\prod_{j \in C_i} r_{ji}(+1) \tag{6.83}$$

$$Q_i(-1) = \alpha_i p_i \prod_{j \in C_i} r_{ji}(-1) \tag{6.84}$$

where the constant α_{ij} are selected to ensure that $Q_i(+1) + Q_i(-1) = 1$. Lastly, the hard decision outputs are calculated as follows:

$$\hat{x}_i = \begin{cases} +1, & \text{if } Q_i(+1) \geq 0.5 \\ -1, & \text{if } Q_i(+1) < 0.5 \end{cases} \tag{6.85}$$

The iteration process is continued until the estimated codeword bits satisfy the syndrome condition or the maximum number of iteration is finished.

Summary 6.7   Belief propagation algorithm

1. Initialization of $q_{ij}^{\text{initial}}(x)$ is

$$q_{ij}^{\text{initial}}(x) = P(x_i = x \mid y_i) = \frac{1}{1 + e^{-(2xy_i/\sigma^2)}}$$

2. The message $r_{ji}(x)$ from check nodes to bit nodes is

$$r_{ji}(+1) = \frac{1}{2} + \frac{1}{2} \prod_{i \in V_{jii}} (1 - 2q_{ij}(-1))$$

$$r_{ji}(-1) = 1 - r_{ji}(+1)$$

where V_{jii} denotes a bit node set connected to the check node v_j except x_i.

3. The message $q_{ij}(x)$ from bit nodes to check nodes is

$$q_{ij}(+1) = \alpha_{ij}(1 - p_i) \prod_{j \in C_{iij}} r_{ji}(+1)$$

$$q_{ij}(-1) = \alpha_{ij} p_i \prod_{j \in C_{iij}} r_{ji}(-1)$$

where C_{iij} denotes a check node set connected to the bit node x_i except v_j.

4. The APP ratio for each codeword bit is

$$Q_i(+1) = \alpha_i(1 - p_i) \prod_{j \in C_i} r_{ji}(+1)$$

$$Q_i(-1) = \alpha_i p_i \prod_{j \in C_i} r_{ji}(-1).$$

5. The hard decision output is

$$\hat{x}_i = \begin{cases} +1, & \text{if } Q_i(+1) \geq 0.5 \\ -1, & \text{if } Q_i(+1) < 0.5 \end{cases}.$$

Due to the high complexity of the belief propagation algorithm, a sum-product algorithm in the log domain is more suitable for hardware implementation. The initial value of $q_{ij}(x)$ as log likelihood ratio is expressed as follows:

$$L(q_{ij}) = L(p_i) = \log \frac{q_{ij}(+1)}{q_{ij}(-1)} = \log \frac{P(x_i = +1 \mid y_i)}{P(x_i = -1 \mid y_i)} \tag{6.86}$$

$$= \log \frac{1/(1 + e^{-(2y_i/\sigma^2)})}{1/(1 + e^{2y_i/\sigma^2})} = \log \frac{1 + e^{2y_i/\sigma^2}}{1 + e^{-(2y_i/\sigma^2)}} = \frac{2y_i}{\sigma^2}. \tag{6.87}$$

The message $r_{ji}(x)$ as log likelihood ratio is expressed as follows:

$$L(r_{ji}) = \log \frac{r_{ji}(+1)}{r_{ji}(-1)} = \log \frac{1 - r_{ji}(-1)}{r_{ji}(-1)}. \tag{6.88}$$

Using the relationship

$$\tanh\left(\frac{1}{2}\log\left(\frac{1-r_{ji}(-1)}{r_{ji}(-1)}\right)\right) = 1-2r_{ji}(-1),\tag{6.89}$$

we reformulate (6.88) as follows:

$$L(r_{ji}) = 2\tanh^{-1}\left(\prod_{i'\in V_{j/i}}\tanh\left(\frac{1}{2}L(q_{i'j})\right)\right)\tag{6.90}$$

because

$$\tanh\left(\frac{1}{2}L(r_{ji})\right) = 1-2r_{ji}(-1)\tag{6.91}$$

$$= \prod_{i'\in V_{j/i}}\left(1-2q_{i'j}(-1)\right) = \prod_{i'\in V_{j/i}}\tanh\left(\frac{1}{2}L(q_{i'j})\right).\tag{6.92}$$

However, (6.90) still has multiplication. We need to further reformulate as follows:

$$L(r_{ji}) = 2\tanh^{-1}\left(\prod_{i'\in V_{j/i}}\mathrm{sgn}\left(L(q_{i'j})\right)\prod_{i'\in V_{j/i}}\tanh\left(\frac{1}{2}\left|L(q_{i'j})\right|\right)\right)\tag{6.93}$$

$$= \left(\prod_{i'\in V_{j/i}}\mathrm{sgn}\left(L(q_{i'j})\right)\right)\cdot 2\tanh^{-1}\left(\prod_{i'\in V_{j/i}}\tanh\left(\frac{1}{2}\left|L(q_{i'j})\right|\right)\right)\tag{6.94}$$

$$= \left(\prod_{i'\in V_{j/i}}\mathrm{sgn}\left(L(q_{i'j})\right)\right)\cdot f\left(\sum_{i'\in V_{j/i}}f\left(\left|L(q_{i'j})\right|\right)\right)\tag{6.95}$$

where

$$\mathrm{sgn}(x) = \begin{cases} -1 & \text{if } x < 0 \\ 0 & \text{if } x = 0 \\ +1 & \text{if } x > 0 \end{cases}\tag{6.96}$$

and

$$f(x) = -\log\left(\tanh\left(\frac{x}{2}\right)\right) = \ln\left(\frac{e^x+1}{e^x-1}\right).\tag{6.97}$$

Now, $L(r_{ji})$ does not include multiplication but the complexity is still high. The function $f(x)$ is not simple function as shown in Figure 6.49. It can be implemented by a look-up table.

Thus, further simplification using a min-sum algorithm is possible as follows:

$$L(r_{ji}) \approx \left(\prod_{i'\in V_{j/i}}\mathrm{sgn}\left(L(q_{i'j})\right)\right)\cdot\min_{i'\in V_{j/i}}\left|L(q_{i'j})\right|\tag{6.98}$$

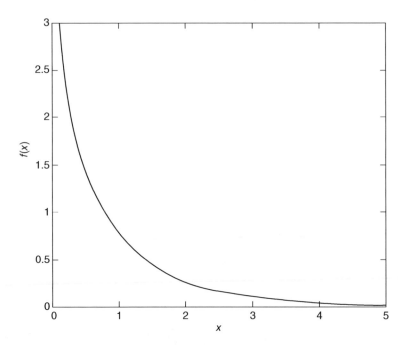

Figure 6.49 Function $f(x)$

because the minimum term of $L\left(q_{i_{j}}\right)$ dominates in the equation. The performance degradation caused by further approximation is about 0.5 dB. However, if a scaling factor (α) [17] is added as follows:

$$L\left(r_{ji}\right) \approx \alpha \left(\prod_{i \in V_{ji}} \mathrm{sgn}\left(L\left(q_{i_{j}}\right)\right)\right) \cdot \min_{i \in V_{ji}} \left|L\left(q_{i_{j}}\right)\right|, \tag{6.99}$$

the performance degradation can be reduced to 0.1 dB. The message $q_{ij}(x)$ as log likelihood ratio is expressed as follows:

$$L(q_{ij}) = \log \frac{q_{ij}(+1)}{q_{ij}(-1)} = \log \frac{\alpha_{ij}(1-p_i) \prod_{j \in C_{i\backslash j}} r_{ji}(+1)}{\alpha_{ij} p_i \prod_{j \in C_{i\backslash j}} r_{ji}(-1)} \tag{6.100}$$

$$= \log \frac{(1-p_i)}{p_i} + \log \frac{\prod_{j \in C_{i\backslash j}} r_{ji}(+1)}{\prod_{j \in C_{i\backslash j}} r_{ji}(-1)} = L(p_i) + \sum_{j \in C_{i\backslash j}} L\left(r_{ji}\right). \tag{6.101}$$

Then, we calculate the log APP ratio for each codeword bit as follows:

$$L(Q_i) = L(p_i) + \sum_{j \in C_i} L(r_{ji}). \tag{6.102}$$

Lastly, the hard decision outputs are calculated as follows:

$$\hat{x}_l = \begin{cases} +1, & \text{if } L(Q_i) > 0 \\ -1, & \text{if } L(Q_i) \le 0 \end{cases}. \tag{6.103}$$

Similar to the belief propagation algorithm, the iteration continues until the syndrome condition is satisfied or the maximum number of iterations is reached.

Summary 6.8   Log domain sum product algorithm

1. Initialization of $L(q_{ij})$ is

$$L(q_{ij}) = L(p_i) = \frac{2y_i}{\sigma^2}.$$

2. The message $L(r_{ji})$ from check nodes to bit nodes is

$$L(r_{ji}) \approx \left(\prod_{i' \in V_{j\setminus i}} \text{sgn}\left(L\left(q_{i'j} \right) \right) \right) \cdot \min_{i' \in V_{j\setminus i}} \left| L\left(q_{i'j} \right) \right|$$

3. The message $L(q_{ij})$ from bit nodes to check nodes is

$$L(q_{ij}) = L(p_i) + \sum_{j' \in C_{i\setminus j}} L\left(r_{j'i} \right).$$

4. The log APP ratio for each codeword bit is

$$L(Q_i) = L(p_i) + \sum_{j \in C_i} L(r_{ji}).$$

5. The hard decision output is

$$\hat{x}_l = \begin{cases} +1, & \text{if } L(Q_i) > 0 \\ -1, & \text{if } L(Q_i) \le 0 \end{cases}.$$

6.3.2   Example of LDPC Encoding and Decoding

We construct an irregular LDPC code having the following parity check matrix H:

$$H = \begin{bmatrix} 1 & 0 & 0 & 1 & 0 & 0 & 0 & 1 & 0 & 0 & 1 & 0 \\ 0 & 0 & 1 & 1 & 1 & 0 & 1 & 0 & 0 & 0 & 0 & 1 \\ 0 & 1 & 0 & 0 & 1 & 1 & 0 & 0 & 0 & 1 & 0 & 1 \\ 0 & 0 & 1 & 0 & 0 & 0 & 1 & 1 & 0 & 1 & 0 & 0 \\ 0 & 1 & 1 & 0 & 0 & 0 & 0 & 1 & 0 & 0 & 1 \\ 1 & 0 & 0 & 0 & 0 & 1 & 0 & 1 & 0 & 0 & 1 & 0 \\ 0 & 1 & 0 & 1 & 0 & 0 & 1 & 0 & 1 & 1 & 0 & 0 \\ 1 & 0 & 0 & 0 & 1 & 1 & 0 & 0 & 1 & 0 & 1 & 0 \end{bmatrix}. \tag{6.104}$$

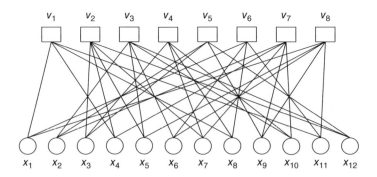

Figure 6.50 Tanner graph of (6.104)

The Tanner graph of the parity check matrix **H** is drawn as shown in Figure 6.50. Using Gauss–Jordan elimination, we change the form of (6.104) as follows:

$$H = \left[I_{n-k} \; P^\mathrm{T} \right] = \begin{bmatrix} 1 & 0 & 0 & 0 & 0 & 0 & 0 & 0 & 0 & 0 & 1 & 1 \\ 0 & 1 & 0 & 0 & 0 & 0 & 0 & 0 & 1 & 1 & 0 & 0 \\ 0 & 0 & 1 & 0 & 0 & 0 & 0 & 0 & 0 & 1 & 0 & 1 \\ 0 & 0 & 0 & 1 & 0 & 0 & 0 & 0 & 1 & 1 & 0 & 1 \\ 0 & 0 & 0 & 0 & 1 & 0 & 0 & 0 & 0 & 1 & 0 & 0 \\ 0 & 0 & 0 & 0 & 0 & 1 & 0 & 0 & 1 & 1 & 0 & 1 \\ 0 & 0 & 0 & 0 & 0 & 0 & 1 & 0 & 1 & 1 & 0 & 1 \\ 0 & 0 & 0 & 0 & 0 & 0 & 0 & 1 & 1 & 1 & 0 & 0 \end{bmatrix}. \tag{6.105}$$

From (6.105), we find the generator matrix of the irregular LDPC code as follows:

$$G = \left[P \; I_k \right] = \begin{bmatrix} 0 & 1 & 0 & 1 & 0 & 1 & 1 & 1 & 1 & 0 & 0 & 0 \\ 0 & 1 & 1 & 1 & 1 & 1 & 1 & 1 & 0 & 1 & 0 & 0 \\ 1 & 0 & 0 & 0 & 0 & 0 & 0 & 0 & 0 & 0 & 1 & 0 \\ 1 & 0 & 1 & 1 & 0 & 1 & 1 & 0 & 0 & 0 & 0 & 1 \end{bmatrix}. \tag{6.106}$$

The message $m = [1\ 0\ 1\ 0]$ generates the codeword $[1\ 1\ 0\ 1\ 0\ 1\ 1\ 1\ 1\ 0\ 1\ 0]$ by mG. The codeword is modulated by BPSK modulation. The modulated symbol sequence $[+1\ +1\ -1\ +1\ -1\ +1\ +1\ +1\ +1\ -1\ +1\ -1]$ is transmitted. After passing through AWGN channel with the standard deviation $\sigma = 0.7$ as shown in Figure 6.51, the LDPC decoder receives the symbol $[+1.11\ +1.20\ -1.44\ +0.84\ -0.20\ -0.65\ +1.64\ +1.15\ +0.95\ -1.22\ +1.13\ -0.98]$. We can observe that an error occurs at the sixth codeword symbol.

We use a sum-product algorithm in the log domain. The initialization of $L(q_{ij})$ is carried out as the first step of the LDPC decoding as follows:

$$L(q_{11}) = L(q_{16}) = L(q_{18}) = L(p_1) = \frac{2 \cdot (+1.11)}{(0.7)^2} = +4.53$$

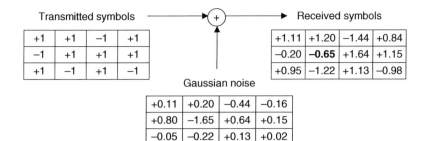

Figure 6.51   Channel noise example

$$L(q_{23}) = L(q_{25}) = L(q_{27}) = L(p_2) = \frac{2\cdot(+1.20)}{(0.7)^2} = +4.90$$

$$L(q_{32}) = L(q_{34}) = L(q_{35}) = L(p_3) = \frac{2\cdot(-1.44)}{(0.7)^2} = -5.88$$

$$L(q_{41}) = L(q_{42}) = L(q_{47}) = L(p_4) = \frac{2\cdot(+0.84)}{(0.7)^2} = +3.43$$

$$L(q_{52}) = L(q_{53}) = L(q_{58}) = L(p_5) = \frac{2\cdot(-0.20)}{(0.7)^2} = -0.82$$

$$L(q_{63}) = L(q_{66}) = L(q_{68}) = L(p_6) = \frac{2\cdot(-0.65)}{(0.7)^2} = -2.65$$

$$L(q_{72}) = L(q_{74}) = L(q_{77}) = L(p_7) = \frac{2\cdot(+1.64)}{(0.7)^2} = +6.69$$

$$L(q_{81}) = L(q_{84}) = L(q_{86}) = L(p_8) = \frac{2\cdot(+1.15)}{(0.7)^2} = +4.69$$

$$L(q_{95}) = L(q_{97}) = L(q_{98}) = L(p_9) = \frac{2\cdot(+0.95)}{(0.7)^2} = +3.88$$

$$L(q_{103}) = L(q_{104}) = L(q_{107}) = L(p_{10}) = \frac{2\cdot(-1.22)}{(0.7)^2} = -4.98$$

$$L\big(q_{111}\big) = L\big(q_{116}\big) = L\big(q_{118}\big) = L\big(p_{11}\big) = \frac{2\cdot(+1.13)}{(0.7)^2} = +4.61$$

$$L(q_{122}) = L(q_{123}) = L(q_{125}) = L(p_{12}) = \frac{2\cdot(-0.98)}{(0.7)^2} = -4.$$

The second step is to calculate $L(r_{ji})$ from check nodes to bit nodes. $L(r_{11})$ is calculated by the initial values of the $L(q_{41})$, $L(q_{81})$, and $L(q_{111})$ as follows:

$$L(r_{11}) \approx \mathrm{sgn}(L(q_{41}))\cdot \mathrm{sgn}(L(q_{81}))\cdot \mathrm{sgn}(L(q_{111}))\cdot \min\big(|L(q_{41})|,|L(q_{81})|,|L(q_{111})|\big)$$
$$= \mathrm{sgn}(+3.43)\cdot \mathrm{sgn}(+4.69)\cdot \mathrm{sgn}(+4.61)\cdot \min\big(|+3.43|,|+4.69|,|+4.61|\big) = +3.43.$$

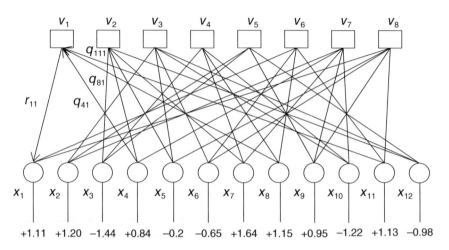

Figure 6.52   Calculations of $L(r_{11})$

Figure 6.52 illustrates the calculation process of $L(r_{11})$.
In the same way, we calculate the remaining $L(r_{ji})$ as follows:

$$L(r_{14}) \approx \text{sgn}(L(q_{11})) \cdot \text{sgn}(L(q_{81})) \cdot \text{sgn}(L(q_{111})) \cdot \min\left(\left|L(q_{11})\right|,\left|L(q_{81})\right|,\left|L(q_{111})\right|\right)$$
$$= \text{sgn}(+4.53) \cdot \text{sgn}(+4.69) \cdot \text{sgn}(+4.61) \cdot \min\left(\left|+4.53\right|,\left|+4.69\right|,\left|+4.61\right|\right) = +4.53$$

$$L(r_{18}) \approx \text{sgn}(L(q_{11})) \cdot \text{sgn}(L(q_{41})) \cdot \text{sgn}(L(q_{111})) \cdot \min\left(\left|L(q_{11})\right|,\left|L(q_{41})\right|,\left|L(q_{111})\right|\right)$$
$$= \text{sgn}(+4.53) \cdot \text{sgn}(+3.43) \cdot \text{sgn}(+4.61) \cdot \min\left(\left|+4.53\right|,\left|+3.43\right|,\left|+4.61\right|\right) = +3.43$$

$$L(r_{111}) \approx \text{sgn}(L(q_{11})) \cdot \text{sgn}(L(q_{41})) \cdot \text{sgn}(L(q_{81})) \cdot \min\left(\left|L(q_{11})\right|,\left|L(q_{41})\right|,\left|L(q_{81})\right|\right)$$
$$= \text{sgn}(+4.53) \cdot \text{sgn}(+3.43) \cdot \text{sgn}(+4.69) \cdot \min\left(\left|+4.53\right|,\left|+3.43\right|,\left|+4.69\right|\right) = +3.43$$

$$L(r_{23}) \approx \text{sgn}(L(q_{42})) \cdot \text{sgn}(L(q_{52})) \cdot \text{sgn}(L(q_{72})) \cdot \text{sgn}(L(q_{122})) \cdot \min\left(\left|L(q_{42})\right|,\left|L(q_{52})\right|,\left|L(q_{72})\right|,\left|L(q_{122})\right|\right)$$
$$= \text{sgn}(+3.43) \cdot \text{sgn}(-0.82) \cdot \text{sgn}(+6.69) \cdot \text{sgn}(-4) \cdot \min\left(\left|+3.43\right|,\left|-0.82\right|,\left|+6.69\right|,\left|-4\right|\right) = +0.82$$

$$L(r_{24}) \approx \text{sgn}(L(q_{32})) \cdot \text{sgn}(L(q_{52})) \cdot \text{sgn}(L(q_{72})) \cdot \text{sgn}(L(q_{122})) \cdot \min\left(\left|L(q_{32})\right|,\left|L(q_{52})\right|,\left|L(q_{72})\right|,\left|L(q_{122})\right|\right)$$
$$= \text{sgn}(-5.88) \cdot \text{sgn}(-0.82) \cdot \text{sgn}(+6.69) \cdot \text{sgn}(-4) \cdot \min\left(\left|-5.88\right|,\left|-0.82\right|,\left|+6.69\right|,\left|-4\right|\right) = -0.82$$

$$L(r_{25}) \approx \text{sgn}(L(q_{32})) \cdot \text{sgn}(L(q_{42})) \cdot \text{sgn}(L(q_{72})) \cdot \text{sgn}(L(q_{122})) \cdot \min\left(\left|L(q_{32})\right|,\left|L(q_{42})\right|,\left|L(q_{72})\right|,\left|L(q_{122})\right|\right)$$
$$= \text{sgn}(-5.88) \cdot \text{sgn}(+3.43) \cdot \text{sgn}(+6.69) \cdot \text{sgn}(-4) \cdot \min\left(\left|-5.88\right|,\left|+3.43\right|,\left|+6.69\right|,\left|-4\right|\right) = +3.43$$

$$L(r_{27}) \approx \text{sgn}(L(q_{32})) \cdot \text{sgn}(L(q_{42})) \cdot \text{sgn}(L(q_{52})) \cdot \text{sgn}(L(q_{122})) \cdot \min\left(\left|L(q_{32})\right|,\left|L(q_{42})\right|,\left|L(q_{52})\right|,\left|L(q_{122})\right|\right)$$
$$= \text{sgn}(-5.88) \cdot \text{sgn}(+3.43) \cdot \text{sgn}(-0.82) \cdot \text{sgn}(-4) \cdot \min\left(\left|-5.88\right|,\left|+3.43\right|,\left|-0.82\right|,\left|-4\right|\right) = -0.82$$

$$L(r_{212}) \approx \text{sgn}(L(q_{32})) \cdot \text{sgn}(L(q_{42})) \cdot \text{sgn}(L(q_{52})) \cdot \text{sgn}(L(q_{72})) \cdot \min\left(\left|L(q_{32})\right|,\left|L(q_{42})\right|,\left|L(q_{52})\right|,\left|L(q_{72})\right|\right)$$
$$= \text{sgn}(-5.88) \cdot \text{sgn}(+3.43) \cdot \text{sgn}(-0.82) \cdot \text{sgn}(+6.69) \cdot \min\left(\left|-5.88\right|,\left|+3.43\right|,\left|-0.82\right|,\left|+6.69\right|\right) = +0.82$$

$$L(r_{32}) \approx \mathrm{sgn}(L(q_{53})) \cdot \mathrm{sgn}(L(q_{63})) \cdot \mathrm{sgn}(L(q_{103})) \cdot \mathrm{sgn}(L(q_{123})) \cdot \min\left(\left|L(q_{53})\right|,\left|L(q_{63})\right|,\left|L(q_{103})\right|,\left|L(q_{123})\right|\right)$$
$$= \mathrm{sgn}(-0.82) \cdot \mathrm{sgn}(-2.65) \cdot \mathrm{sgn}(-4.98) \cdot \mathrm{sgn}(-4) \cdot \min\left(\left|-0.82\right|,\left|-2.65\right|,\left|-4.98\right|,\left|-4\right|\right) = +0.82$$

$$L(r_{35}) \approx \mathrm{sgn}(L(q_{23})) \cdot \mathrm{sgn}(L(q_{63})) \cdot \mathrm{sgn}(L(q_{103})) \cdot \mathrm{sgn}(L(q_{123})) \cdot \min\left(\left|L(q_{23})\right|,\left|L(q_{63})\right|,\left|L(q_{103})\right|,\left|L(q_{123})\right|\right)$$
$$= \mathrm{sgn}(+4.90) \cdot \mathrm{sgn}(-2.65) \cdot \mathrm{sgn}(-4.98) \cdot \mathrm{sgn}(-4) \cdot \min\left(\left|+4.90\right|,\left|-2.65\right|,\left|-4.98\right|,\left|-4\right|\right) = -2.65$$

$$L(r_{36}) \approx \mathrm{sgn}(L(q_{23})) \cdot \mathrm{sgn}(L(q_{53})) \cdot \mathrm{sgn}(L(q_{103})) \cdot \mathrm{sgn}(L(q_{123})) \cdot \min\left(\left|L(q_{23})\right|,\left|L(q_{53})\right|,\left|L(q_{103})\right|,\left|L(q_{123})\right|\right)$$
$$= \mathrm{sgn}(+4.90) \cdot \mathrm{sgn}(-0.82) \cdot \mathrm{sgn}(-4.98) \cdot \mathrm{sgn}(-4) \cdot \min\left(\left|+4.90\right|,\left|-0.82\right|,\left|-4.98\right|,\left|-4\right|\right) = -0.82$$

$$L(r_{310}) \approx \mathrm{sgn}(L(q_{23})) \cdot \mathrm{sgn}(L(q_{53})) \cdot \mathrm{sgn}(L(q_{63})) \cdot \mathrm{sgn}(L(q_{123})) \cdot \min\left(\left|L(q_{23})\right|,\left|L(q_{53})\right|,\left|L(q_{63})\right|,\left|L(q_{123})\right|\right)$$
$$= \mathrm{sgn}(+4.90) \cdot \mathrm{sgn}(-0.82) \cdot \mathrm{sgn}(-2.65) \cdot \mathrm{sgn}(-4) \cdot \min\left(\left|+4.90\right|,\left|-0.82\right|,\left|-2.65\right|,\left|-4\right|\right) = -0.82$$

$$L(r_{312}) \approx \mathrm{sgn}(L(q_{23})) \cdot \mathrm{sgn}(L(q_{53})) \cdot \mathrm{sgn}(L(q_{63})) \cdot \mathrm{sgn}(L(q_{103})) \cdot \min\left(\left|L(q_{23})\right|,\left|L(q_{53})\right|,\left|L(q_{63})\right|,\left|L(q_{103})\right|\right)$$
$$= \mathrm{sgn}(+4.90) \cdot \mathrm{sgn}(-0.82) \cdot \mathrm{sgn}(-2.65) \cdot \mathrm{sgn}(-4) \cdot \min\left(\left|+4.90\right|,\left|-0.82\right|,\left|-2.65\right|,\left|-4\right|\right) = -0.82$$

$$L(r_{43}) \approx \mathrm{sgn}(L(q_{74})) \cdot \mathrm{sgn}(L(q_{84})) \cdot \mathrm{sgn}(L(q_{104})) \cdot \min\left(\left|L(q_{74})\right|,\left|L(q_{84})\right|,\left|L(q_{104})\right|\right)$$
$$= \mathrm{sgn}(+6.69) \cdot \mathrm{sgn}(+4.69) \cdot \mathrm{sgn}(-4.98) \cdot \min\left(\left|+6.69\right|,\left|+4.69\right|,\left|-4.98\right|\right) = -4.69$$

$$L(r_{47}) \approx \mathrm{sgn}(L(q_{34})) \cdot \mathrm{sgn}(L(q_{84})) \cdot \mathrm{sgn}(L(q_{104})) \cdot \min\left(\left|L(q_{34})\right|,\left|L(q_{84})\right|,\left|L(q_{104})\right|\right)$$
$$= \mathrm{sgn}(-5.88) \cdot \mathrm{sgn}(+4.69) \cdot \mathrm{sgn}(-4.98) \cdot \min\left(\left|-5.88\right|,\left|+4.69\right|,\left|-4.98\right|\right) = +4.69$$

$$L(r_{48}) \approx \mathrm{sgn}(L(q_{34})) \cdot \mathrm{sgn}(L(q_{74})) \cdot \mathrm{sgn}(L(q_{104})) \cdot \min\left(\left|L(q_{34})\right|,\left|L(q_{74})\right|,\left|L(q_{104})\right|\right)$$
$$= \mathrm{sgn}(-5.88) \cdot \mathrm{sgn}(+6.69) \cdot \mathrm{sgn}(-4.98) \cdot \min\left(\left|-5.88\right|,\left|+6.69\right|,\left|-4.98\right|\right) = +4.98$$

$$L(r_{410}) \approx \mathrm{sgn}(L(q_{34})) \cdot \mathrm{sgn}(L(q_{74})) \cdot \mathrm{sgn}(L(q_{84})) \cdot \min\left(\left|L(q_{34})\right|,\left|L(q_{74})\right|,\left|L(q_{84})\right|\right)$$
$$= \mathrm{sgn}(-5.88) \cdot \mathrm{sgn}(+6.69) \cdot \mathrm{sgn}(+4.69) \cdot \min\left(\left|-5.88\right|,\left|+6.69\right|,\left|+4.69\right|\right) = -4.69$$

$$L(r_{52}) \approx \mathrm{sgn}\left(L(q_{35})\right) \cdot \mathrm{sgn}(L(q_{95})) \cdot \mathrm{sgn}(L(q_{125})) \cdot \min\left(\left|L(q_{35})\right|,\left|L(q_{95})\right|,\left|L(q_{125})\right|\right)$$
$$= \mathrm{sgn}(-5.88) \cdot \mathrm{sgn}(+3.88) \cdot \mathrm{sgn}(-4) \cdot \min\left(\left|-5.88\right|,\left|+3.88\right|,\left|-4\right|\right) = +3.88$$

$$L(r_{53}) \approx \mathrm{sgn}(L(q_{25})) \cdot \mathrm{sgn}(L(q_{95})) \cdot \mathrm{sgn}(L(q_{125})) \cdot \min\left(\left|L(q_{25})\right|,\left|L(q_{95})\right|,\left|L(q_{125})\right|\right)$$
$$= \mathrm{sgn}(+4.90) \cdot \mathrm{sgn}(+3.88) \cdot \mathrm{sgn}(-4) \cdot \min\left(\left|+4.90\right|,\left|+3.88\right|,\left|-4\right|\right) = -3.88$$

$$L(r_{59}) \approx \mathrm{sgn}(L(q_{25})) \cdot \mathrm{sgn}(L(q_{35})) \cdot \mathrm{sgn}(L(q_{125})) \cdot \min\left(\left|L(q_{25})\right|,\left|L(q_{35})\right|,\left|L(q_{125})\right|\right)$$
$$= \mathrm{sgn}(+4.90) \cdot \mathrm{sgn}(-5.88) \cdot \mathrm{sgn}(-4) \cdot \min\left(\left|+4.90\right|,\left|-5.88\right|,\left|-4\right|\right) = +4$$

$$L(r_{512}) \approx \mathrm{sgn}(L(q_{25})) \cdot \mathrm{sgn}(L(q_{35})) \cdot \mathrm{sgn}(L(q_{95})) \cdot \min\left(\left|L(q_{25})\right|,\left|L(q_{35})\right|,\left|L(q_{95})\right|\right)$$
$$= \mathrm{sgn}(+4.90) \cdot \mathrm{sgn}(-5.88) \cdot \mathrm{sgn}(+3.88) \cdot \min\left(\left|+4.90\right|,\left|-5.88\right|,\left|+3.88\right|\right) = -3.88$$

$$L(r_{61}) \approx \mathrm{sgn}(L(q_{66})) \cdot \mathrm{sgn}(L(q_{86})) \cdot \mathrm{sgn}(L(q_{116})) \cdot \min\left(\left|L(q_{66})\right|,\left|L(q_{86})\right|,\left|L(q_{116})\right|\right)$$
$$= \mathrm{sgn}(-2.65) \cdot \mathrm{sgn}(+4.69) \cdot \mathrm{sgn}(+4.61) \cdot \min\left(\left|-2.65\right|,\left|+4.69\right|,\left|+4.61\right|\right) = -2.65$$

$$L(r_{66}) \approx \text{sgn}(L(q_{16})) \cdot \text{sgn}(L(q_{86})) \cdot \text{sgn}(L(q_{116})) \cdot \min\left(\left|L(q_{16})\right|, \left|L(q_{86})\right|, \left|L(q_{116})\right|\right)$$
$$= \text{sgn}(+4.53) \cdot \text{sgn}(+4.69) \cdot \text{sgn}(+4.61) \cdot \min\left(\left|+4.53\right|, \left|+4.69\right|, \left|+4.61\right|\right) = +4.53$$

$$L(r_{68}) \approx \text{sgn}(L(q_{16})) \cdot \text{sgn}(L(q_{66})) \cdot \text{sgn}(L(q_{116})) \cdot \min\left(\left|L(q_{16})\right|, \left|L(q_{66})\right|, \left|L(q_{116})\right|\right)$$
$$= \text{sgn}(+4.53) \cdot \text{sgn}(-2.65) \cdot \text{sgn}(+4.61) \cdot \min\left(\left|+4.53\right|, \left|-2.65\right|, \left|+4.61\right|\right) = -2.65$$

$$L(r_{611}) \approx \text{sgn}(L(q_{16})) \cdot \text{sgn}(L(q_{66})) \cdot \text{sgn}(L(q_{86})) \cdot \min\left(\left|L(q_{16})\right|, \left|L(q_{66})\right|, \left|L(q_{86})\right|\right)$$
$$= \text{sgn}(+4.53) \cdot \text{sgn}(-2.65) \cdot \text{sgn}(+4.69) \cdot \min\left(\left|+4.53\right|, \left|-2.65\right|, \left|+4.69\right|\right) = -2.65$$

$$L(r_{72}) \approx \text{sgn}(L(q_{47})) \cdot \text{sgn}(L(q_{77})) \cdot \text{sgn}(L(q_{97})) \cdot \text{sgn}(L(q_{107})) \cdot \min\left(\left|L(q_{47})\right|, \left|L(q_{77})\right|, \left|L(q_{97})\right|, \left|L(q_{107})\right|\right)$$
$$= \text{sgn}(+3.43) \cdot \text{sgn}(+6.69) \cdot \text{sgn}(+3.88) \cdot \text{sgn}(-4.98) \cdot \min\left(\left|+3.43\right|, \left|+6.69\right|, \left|+3.88\right|, \left|-4.98\right|\right) = -3.43$$

$$L(r_{74}) \approx \text{sgn}(L(q_{27})) \cdot \text{sgn}(L(q_{77})) \cdot \text{sgn}(L(q_{97})) \cdot \text{sgn}(L(q_{107})) \cdot \min\left(\left|L(q_{27})\right|, \left|L(q_{77})\right|, \left|L(q_{97})\right|, \left|L(q_{107})\right|\right)$$
$$= \text{sgn}(+4.90) \cdot \text{sgn}(+6.69) \cdot \text{sgn}(+3.88) \cdot \text{sgn}(-4.98) \cdot \min\left(\left|+4.90\right|, \left|+6.69\right|, \left|+3.88\right|, \left|-4.98\right|\right) = -3.88$$

$$L(r_{77}) \approx \text{sgn}(L(q_{27})) \cdot \text{sgn}(L(q_{47})) \cdot \text{sgn}(L(q_{97})) \cdot \text{sgn}(L(q_{107})) \cdot \min\left(\left|L(q_{27})\right|, \left|L(q_{47})\right|, \left|L(q_{97})\right|, \left|L(q_{107})\right|\right)$$
$$= \text{sgn}(+4.90) \cdot \text{sgn}(+3.43) \cdot \text{sgn}(+3.88) \cdot \text{sgn}(-4.98) \cdot \min\left(\left|+4.90\right|, \left|+3.43\right|, \left|+3.88\right|, \left|-4.98\right|\right) = -3.43$$

$$L(r_{79}) \approx \text{sgn}(L(q_{27})) \cdot \text{sgn}(L(q_{47})) \cdot \text{sgn}(L(q_{77})) \cdot \text{sgn}(L(q_{107})) \cdot \min\left(\left|L(q_{27})\right|, \left|L(q_{47})\right|, \left|L(q_{77})\right|, \left|L(q_{107})\right|\right)$$
$$= \text{sgn}(+4.90) \cdot \text{sgn}(+3.43) \cdot \text{sgn}(+6.69) \cdot \text{sgn}(-4.98) \cdot \min\left(\left|+4.90\right|, \left|+3.43\right|, \left|+6.69\right|, \left|-4.98\right|\right) = -3.43$$

$$L(r_{710}) \approx \text{sgn}(L(q_{27})) \cdot \text{sgn}(L(q_{47})) \cdot \text{sgn}(L(q_{77})) \cdot \text{sgn}(L(q_{97})) \cdot \min\left(\left|L(q_{27})\right|, \left|L(q_{47})\right|, \left|L(q_{77})\right|, \left|L(q_{97})\right|\right)$$
$$= \text{sgn}(+4.90) \cdot \text{sgn}(+3.43) \cdot \text{sgn}(+6.69) \cdot \text{sgn}(+3.88) \cdot \min\left(\left|+4.90\right|, \left|+3.43\right|, \left|+6.69\right|, \left|+3.88\right|\right) = +3.43$$

$$L(r_{81}) \approx \text{sgn}(L(q_{58})) \cdot \text{sgn}(L(q_{68})) \cdot \text{sgn}(L(q_{98})) \cdot \text{sgn}(L(q_{118})) \cdot \min\left(\left|L(q_{58})\right|, \left|L(q_{68})\right|, \left|L(q_{98})\right|, \left|L(q_{118})\right|\right)$$
$$= \text{sgn}(-0.82) \cdot \text{sgn}(-2.65) \cdot \text{sgn}(+3.88) \cdot \text{sgn}(+4.61) \cdot \min\left(\left|-0.82\right|, \left|-2.65\right|, \left|+3.88\right|, \left|+4.61\right|\right) = +0.82$$

$$L(r_{85}) \approx \text{sgn}(L(q_{18})) \cdot \text{sgn}(L(q_{68})) \cdot \text{sgn}(L(q_{98})) \cdot \text{sgn}(L(q_{118})) \cdot \min\left(\left|L(q_{18})\right|, \left|L(q_{68})\right|, \left|L(q_{98})\right|, \left|L(q_{118})\right|\right)$$
$$= \text{sgn}\left(+4.53\right) \cdot \text{sgn}\left(-2.65\right) \cdot \text{sgn}\left(+3.88\right) \cdot \text{sgn}\left(+4.61\right) \cdot \min\left(\left|+4.53\right|, \left|-2.65\right|, \left|+3.88\right|, \left|+4.61\right|\right) = -2.65$$

$$L(r_{86}) \approx \text{sgn}(L(q_{18})) \cdot \text{sgn}(L(q_{58})) \cdot \text{sgn}(L(q_{98})) \cdot \text{sgn}(L(q_{118})) \cdot \min\left(\left|L(q_{18})\right|, \left|L(q_{58})\right|, \left|L(q_{98})\right|, \left|L(q_{118})\right|\right)$$
$$= \text{sgn}(+4.53) \cdot \text{sgn}(-0.82) \cdot \text{sgn}(+3.88) \cdot \text{sgn}(+4.61) \cdot \min\left(\left|+4.53\right|, \left|-0.82\right|, \left|+3.88\right|, \left|+4.61\right|\right) = -0.82$$

$$L(r_{89}) \approx \text{sgn}(L(q_{18})) \cdot \text{sgn}(L(q_{58})) \cdot \text{sgn}(L(q_{68})) \cdot \text{sgn}(L(q_{118})) \cdot \min\left(\left|L(q_{18})\right|, \left|L(q_{58})\right|, \left|L(q_{68})\right|, \left|L(q_{118})\right|\right)$$
$$= \text{sgn}(+4.53) \cdot \text{sgn}(-0.82) \cdot \text{sgn}(-2.65) \cdot \text{sgn}(+4.61) \cdot \min\left(\left|+4.53\right|, \left|-0.82\right|, \left|-2.65\right|, \left|+4.61\right|\right) = +0.82$$

$$L(r_{811}) \approx \text{sgn}(L(q_{18})) \cdot \text{sgn}(L(q_{58})) \cdot \text{sgn}(L(q_{68})) \cdot \text{sgn}(L(q_{98})) \cdot \min\left(\left|L(q_{18})\right|, \left|L(q_{58})\right|, \left|L(q_{68})\right|, \left|L(q_{98})\right|\right)$$
$$= \text{sgn}(+4.53) \cdot \text{sgn}(-0.82) \cdot \text{sgn}(-2.65) \cdot \text{sgn}(+3.88) \cdot \min\left(\left|+4.53\right|, \left|-0.82\right|, \left|-2.65\right|, \left|+3.88\right|\right) = +0.82.$$

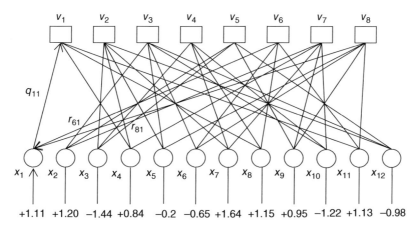

Figure 6.53 Calculations of $L(q_{11})$

The third step is to calculate $L(q_{ij})$ from bit nodes to check nodes. $L(q_{11})$ is calculated by $L(p_1)$, $L(r_{61})$, and $L(r_{81})$ as follows:

$$L(q_{11}) = L(p_1) + L(r_{61}) + L(r_{81}) = +4.53 - 2.65 + 0.82 = +2.7.$$

Figure 6.53 illustrates the calculation process of $L(q_{11})$.
In the same way, we calculate the remaining $L(q_{ij})$ as follows:

$$L(q_{16}) = L(p_1) + L(r_{11}) + L(r_{81}) = +4.53 + 3.43 + 0.82 = +8.78$$
$$L(q_{18}) = L(p_1) + L(r_{11}) + L(r_{61}) = +4.53 + 3.43 - 2.65 = +5.31$$

$$L(q_{23}) = L(p_2) + L(r_{52}) + L(r_{72}) = +4.90 + 3.88 - 3.43 = +5.35$$
$$L(q_{25}) = L(p_2) + L(r_{32}) + L(r_{72}) = +4.90 + 0.82 - 3.43 = +2.29$$
$$L(q_{27}) = L(p_2) + L(r_{32}) + L(r_{52}) = +4.90 + 0.82 + 3.88 = +9.6$$

$$L(q_{32}) = L(p_3) + L(r_{43}) + L(r_{53}) = -5.88 - 4.69 - 3.88 = -14.45$$
$$L(q_{34}) = L(p_3) + L(r_{23}) + L(r_{53}) = -5.88 + 0.82 - 3.88 = -8.94$$
$$L(q_{35}) = L(p_3) + L(r_{23}) + L(r_{43}) = -5.88 + 0.82 - 4.69 = -9.75$$

$$L(q_{41}) = L(p_4) + L(r_{24}) + L(r_{74}) = +3.43 - 0.82 - 3.88 = -1.27$$
$$L(q_{42}) = L(p_4) + L(r_{14}) + L(r_{74}) = +3.43 + 4.53 - 3.88 = +4.08$$
$$L(q_{47}) = L(p_4) + L(r_{14}) + L(r_{24}) = +3.43 + 4.53 - 0.82 = +7.14$$

$$L(q_{52}) = L(p_5) + L(r_{35}) + L(r_{85}) = -0.82 - 2.65 - 2.65 = -6.12$$
$$L(q_{53}) = L(p_5) + L(r_{25}) + L(r_{85}) = -0.82 + 3.43 - 2.65 = -0.04$$
$$L(q_{58}) = L(p_5) + L(r_{25}) + L(r_{35}) = -0.82 + 3.43 - 2.65 = -0.04$$

$$L(q_{63}) = L(p_6) + L(r_{66}) + L(r_{86}) = -2.65 + 4.53 - 0.82 = +1.06$$
$$L(q_{66}) = L(p_6) + L(r_{36}) + L(r_{86}) = -2.65 - 0.82 - 0.82 = -4.29$$
$$L(q_{68}) = L(p_6) + L(r_{36}) + L(r_{66}) = -2.65 - 0.82 + 4.53 = +1.06$$

$$L(q_{72}) = L(p_7) + L(r_{47}) + L(r_{77}) = +6.69 + 4.69 - 3.43 = +7.95$$
$$L(q_{74}) = L(p_7) + L(r_{27}) + L(r_{77}) = +6.69 - 0.82 - 3.43 = +2.44$$
$$L(q_{77}) = L(p_7) + L(r_{27}) + L(r_{47}) = +6.69 - 0.82 + 4.69 = +10.56$$

$$L(q_{81}) = L(p_8) + L(r_{48}) + L(r_{68}) = +4.69 + 4.98 - 2.65 = +7.02$$
$$L(q_{84}) = L(p_8) + L(r_{18}) + L(r_{68}) = +4.69 + 3.43 - 2.65 = +5.47$$
$$L(q_{86}) = L(p_8) + L(r_{18}) + L(r_{48}) = +4.69 + 3.43 + 4.98 = +13.1$$

$$L(q_{95}) = L(p_9) + L(r_{79}) + L(r_{89}) = +3.88 - 3.43 + 0.82 = +1.27$$
$$L(q_{97}) = L(p_9) + L(r_{59}) + L(r_{89}) = +3.88 + 4 + 0.82 = +8.7$$
$$L(q_{98}) = L(p_9) + L(r_{59}) + L(r_{79}) = +3.88 + 4 - 3.43 = +4.45$$

$$L(q_{103}) = L(p_{10}) + L(r_{410}) + L(r_{710}) = -4.98 - 4.69 + 3.43 = -6.24$$
$$L(q_{104}) = L(p_{10}) + L(r_{310}) + L(r_{710}) = -4.98 - 0.82 + 3.43 = -2.37$$
$$L(q_{107}) = L(p_{10}) + L(r_{310}) + L(r_{410}) = -4.98 - 0.82 - 4.69 = -10.49$$

$$L(q_{111}) = L(p_{11}) + L(r_{611}) + L(r_{811}) = +4.61 - 2.65 + 0.82 = +2.78$$
$$L(q_{116}) = L(p_{11}) + L(r_{111}) + L(r_{811}) = +4.61 + 3.43 + 0.82 = +8.86$$
$$L(q_{118}) = L(p_{11}) + L(r_{111}) + L(r_{611}) = +4.61 + 3.43 - 2.65 = +5.39$$

$$L(q_{122}) = L(p_{12}) + L(r_{312}) + L(r_{512}) = -4 - 0.82 - 3.88 = -8.7$$
$$L(q_{123}) = L(p_{12}) + L(r_{212}) + L(r_{512}) = -4 + 0.82 - 3.88 = -7.06$$
$$L(q_{125}) = L(p_{12}) + L(r_{212}) + L(r_{312}) = -4 + 0.82 - 0.82 = -4$$

The fourth step is to calculate $L(Q_i)$ as follows:

$$L(Q_1) = L(p_1) + L(r_{11}) + L(r_{61}) + L(r_{81}) = +4.53 + 3.43 - 2.65 + 0.82 = +6.13$$
$$L(Q_2) = L(p_2) + L(r_{32}) + L(r_{52}) + L(r_{72}) = +4.90 + 0.82 + 3.88 - 3.43 = +6.17$$
$$L(Q_3) = L(p_3) + L(r_{23}) + L(r_{43}) + L(r_{53}) = -5.88 + 0.82 - 4.69 - 3.88 = -13.63$$
$$L(Q_4) = L(p_4) + L(r_{14}) + L(r_{24}) + L(r_{74}) = +3.43 + 4.53 - 0.82 - 3.88 = +3.26$$
$$L(Q_5) = L(p_5) + L(r_{25}) + L(r_{35}) + L(r_{85}) = -0.82 + 3.43 - 2.65 - 2.65 = -2.69$$
$$L(Q_6) = L(p_6) + L(r_{36}) + L(r_{66}) + L(r_{86}) = -2.65 - 0.82 + 4.53 - 0.82 = +0.24$$
$$L(Q_7) = L(p_7) + L(r_{27}) + L(r_{47}) + L(r_{77}) = +6.69 - 0.82 + 4.69 - 3.43 = +7.13$$
$$L(Q_8) = L(p_8) + L(r_{18}) + L(r_{48}) + L(r_{68}) = +4.69 + 3.43 + 4.98 - 2.65 = +10.45$$
$$L(Q_9) = L(p_9) + L(r_{59}) + L(r_{79}) + L(r_{89}) = +3.88 + 4 - 3.43 + 0.82 = +5.27$$
$$L(Q_{10}) = L(p_{10}) + L(r_{310}) + L(r_{410}) + L(r_{710}) = -4.98 - 0.82 - 4.69 + 3.43 = -7.06$$
$$L(Q_{11}) = L(p_{11}) + L(r_{111}) + L(r_{611}) + L(r_{811}) = +4.61 + 3.43 - 2.65 + 0.82 = +6.21$$
$$L(Q_{12}) = L(p_{12}) + L(r_{212}) + L(r_{312}) + L(r_{512}) = -4 + 0.82 - 0.82 - 3.88 = -7.88.$$

After the hard decision of the APPs, we obtain [+1 +1 −1 +1 −1 +1 +1 +1 +1 −1 +1 −1]. Thus, the estimated codeword bits $\hat{x}$ are [1 1 0 1 0 1 1 1 1 0 1 0]. The syndrome test is carried out as follows:

$$\hat{x}\,H^{\mathrm{T}} = [110101111010] \begin{bmatrix} 1\,0\,0\,0\,0\,1\,0\,1 \\ 0\,0\,1\,0\,1\,0\,1\,0 \\ 0\,1\,0\,1\,1\,0\,0\,0 \\ 1\,1\,0\,0\,0\,0\,1\,0 \\ 0\,1\,1\,0\,0\,0\,0\,1 \\ 0\,0\,1\,0\,0\,1\,0\,1 \\ 0\,1\,0\,1\,0\,0\,1\,0 \\ 1\,0\,0\,1\,0\,1\,0\,0 \\ 0\,0\,0\,0\,1\,0\,1\,1 \\ 0\,0\,1\,1\,0\,0\,1\,0 \\ 1\,0\,0\,0\,0\,1\,0\,1 \\ 0\,1\,1\,0\,1\,0\,0\,0 \end{bmatrix} = [00000000].$$

Since the syndrome result is all zero, the LDPC decoder stops iteration and estimates [1 1 0 1 0 1 1 1 0 1 0] is the transmitted codeword. Thus, the decoded message $\hat{m}$ is [1 0 1 0]. If the syndrome result is not all zero, the iteration continues until the syndrome result is all zero or the maximum number of iteration is reached.

6.3.3   Hardware Implementation of LDPC Encoding and Decoding

As we discussed in the previous section and Chapter 5, a codeword of a LPDC code can be constructed by performing $c = mG$. The generator matrix G ($= [I_k\ P]$ or $[P\ I_k]$) is created by a parity check matrix H. We should make the parity check matrix in a systematic form H ($= [P^{\mathrm{T}}\ I_{n-k}]$ or $[I_{n-k}\ P^{\mathrm{T}}]$) using Gauss–Jordan elimination. However, it is very difficult to have a sparse P matrix. A codeword construction by $c = mG$ is not simple. Especially, the complexity of a LDPC code construction is one critical problem in a large codeword length. Thus, many researchers developed a low complexity LDPC encoding. In [18], two steps encoding method with the low complexity ($0.017^2 n^2 + O(n)$) is developed. As the first step is called preprocessing, the parity check matrix is rearranged into several submatrices as follows:

$$H = \begin{bmatrix} A & B & T \\ C & D & E \end{bmatrix} \tag{6.107}$$

where A, B, T, C, D, and E are of size $(m-g) \times (n-m)$, $(m-g) \times g$, $(m-g) \times (m-g)$, $g \times (n-m)$, $g \times g$, and $g \times (m-g)$, respectively. T is a lower triangular sub-matrix with 1s along the diagonal. All sub-matrices are spares because the parity check matrix is rearranged by only

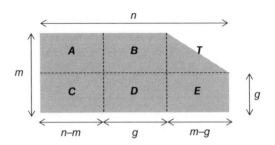

Figure 6.54   Rearranged parity check matrix

row and column permutations. Figure 6.54 illustrates the rearranged parity check matrix in approximate lower triangular form.

In the lower triangular form, a gap g should be as small as possible. The sub-matrix E is cleared by performing the pre-multiplication as follows:

$$\begin{bmatrix} I & 0 \\ ET^{-1} & I \end{bmatrix}\begin{bmatrix} A & B & T \\ C & D & E \end{bmatrix} = \begin{bmatrix} A & B & T \\ -ET^{-1}A+C & -ET^{-1}B+D & 0 \end{bmatrix}. \qquad (6.108)$$

We check whether or not $-ET^{-1}B+D$ is singular. If it is singular, the singularity should be removed. The second step is to construct codeword $c=(u, p_1, p_2)$ where u is an information part and p_1 (length g) and p_2 (length $m-g$) are parity parts. Since $Hc^T=0$, we have the following equations:

$$\begin{bmatrix} A & B & T \\ -ET^{-1}A+C & -ET^{-1}B+D & 0 \end{bmatrix}\begin{bmatrix} u^T \\ p_1^T \\ p_2^T \end{bmatrix} = \begin{bmatrix} 0 \\ 0 \end{bmatrix} \qquad (6.109)$$

$$Au^T + Bp_1^T + Tp_2^T = 0 \qquad (6.110)$$

$$(-ET^{-1}A+C)u^T + (-ET^{-1}B+D)p_1^T = 0. \qquad (6.111)$$

From (6.111), we obtain the first parity part p_1 as follows:

$$p_1^T = -\Phi^{-1}(-ET^{-1}A+C)u^T \qquad (6.112)$$

where $\Phi = -ET^{-1}B+D$. In the same way, we obtain the second parity part p_2 from (6.110) as follows:

$$p_2^T = -T^{-1}\left(Au^T + Bp_1^T\right). \qquad (6.113)$$

Example 6.6   LDPC encoding
Consider the following parity check matrix:

$$H = \begin{bmatrix} 1 & 0 & 1 & 0 & 0 & 0 & 0 & 0 & 1 & 1 & 1 & 1 \\ 1 & 1 & 0 & 0 & 0 & 0 & 1 & 1 & 1 & 1 & 0 & 0 \\ 0 & 0 & 1 & 1 & 0 & 1 & 0 & 1 & 0 & 0 & 1 & 1 \\ 0 & 1 & 0 & 1 & 1 & 1 & 0 & 1 & 1 & 0 & 0 & 0 \\ 0 & 1 & 1 & 1 & 1 & 0 & 1 & 0 & 0 & 1 & 0 & 0 \\ 1 & 0 & 0 & 0 & 1 & 1 & 1 & 0 & 0 & 0 & 1 & 1 \end{bmatrix}$$

and find the codeword for message $u = [1\ 0\ 1\ 0\ 1\ 0]$.

Solution
As the first step, we permute the columns in the order of [1 2 3 4 7 9 10 11 12 8 6 5]. Thus, the parity check matrix is rearranged into several sub-matrices as follows:

$$H_r = \begin{bmatrix} A & B & T \\ C & D & E \end{bmatrix} = \begin{bmatrix} 1 & 0 & 1 & 0 & 0 & 1 & 1 & 1 & 1 & 0 & 0 & 0 \\ 1 & 1 & 0 & 0 & 1 & 1 & 1 & 0 & 0 & 1 & 0 & 0 \\ 0 & 0 & 1 & 1 & 0 & 0 & 0 & 1 & 1 & 1 & 1 & 0 \\ 0 & 1 & 0 & 1 & 0 & 1 & 0 & 0 & 0 & 1 & 1 & 1 \\ 0 & 1 & 1 & 1 & 1 & 0 & 1 & 0 & 0 & 0 & 0 & 1 \\ 1 & 0 & 0 & 0 & 1 & 0 & 0 & 1 & 1 & 0 & 1 & 1 \end{bmatrix}.$$

We calculate the following matrix for pre-multiplication:

$$\begin{bmatrix} I & 0 \\ ET^{-1} & I \end{bmatrix} = \begin{bmatrix} 1 & 0 & 0 & 0 & 0 & 0 \\ 0 & 1 & 0 & 0 & 0 & 0 \\ 0 & 0 & 1 & 0 & 0 & 0 \\ 0 & 0 & 0 & 1 & 0 & 0 \\ 1 & 0 & 1 & 1 & 1 & 0 \\ 1 & 1 & 0 & 1 & 0 & 1 \end{bmatrix}.$$

Therefore, we clear sub-matrix E in H_r as follows:

$$\begin{bmatrix} A & B & T \\ -ET^{-1}A+C & -ET^{-1}B+D & 0 \end{bmatrix} = \begin{bmatrix} 1 & 0 & 1 & 0 & 0 & 1 & 1 & 1 & 1 & 0 & 0 & 0 \\ 1 & 1 & 0 & 0 & 1 & 1 & 1 & 0 & 0 & 1 & 0 & 0 \\ 0 & 0 & 1 & 1 & 0 & 0 & 0 & 1 & 1 & 1 & 1 & 0 \\ 0 & 1 & 0 & 1 & 0 & 1 & 0 & 0 & 0 & 1 & 1 & 1 \\ \hline 1 & 0 & 1 & 1 & 1 & 0 & 0 & 0 & 0 & 0 & 0 & 0 \\ 1 & 0 & 1 & 1 & 0 & 1 & 0 & 0 & 0 & 0 & 0 & 0 \end{bmatrix}.$$

In the above matrix, $\Phi = -ET^{-1}B + D$ is singular. Thus, we exchange Column 4 and 5 with Column 7 and 8 as follows:

$$
\begin{bmatrix}
1 & 0 & 1 & 1 & 1 & 1 & 0 & 0 & 1 & 0 & 0 & 0 \\
1 & 1 & 0 & 1 & 0 & 1 & 0 & 1 & 0 & 1 & 0 & 0 \\
0 & 0 & 1 & 0 & 1 & 0 & 1 & 0 & 1 & 1 & 1 & 0 \\
0 & 1 & 0 & 0 & 0 & 1 & 1 & 0 & 0 & 1 & 1 & 1 \\
1 & 0 & 1 & 0 & 0 & 0 & 1 & 1 & 0 & 0 & 0 & 0 \\
1 & 0 & 1 & 0 & 0 & 1 & 1 & 0 & 0 & 0 & 0 & 0
\end{bmatrix}.
$$

We have the following equivalent parity check matrix through reordering the columns:

$$
H_e = \begin{bmatrix} A & B & T \\ C & D & E \end{bmatrix} =
\begin{bmatrix}
1 & 0 & 1 & 1 & 1 & 1 & 0 & 0 & 1 & 0 & 0 & 0 \\
1 & 1 & 0 & 1 & 0 & 1 & 0 & 1 & 0 & 1 & 0 & 0 \\
0 & 0 & 1 & 0 & 1 & 0 & 1 & 0 & 1 & 1 & 1 & 0 \\
0 & 1 & 0 & 0 & 0 & 1 & 1 & 0 & 0 & 1 & 1 & 1 \\
0 & 1 & 1 & 1 & 0 & 0 & 1 & 1 & 0 & 0 & 0 & 1 \\
1 & 0 & 0 & 0 & 1 & 0 & 0 & 1 & 1 & 0 & 1 & 1
\end{bmatrix}.
$$

From (6.112), we find the first parity part as follows:

$$
p_1^{\mathrm{T}} = -\Phi^{-1}\left(-ET^{-1}A + C\right)u^{\mathrm{T}} = \begin{bmatrix} 0 & 1 \\ 1 & 1 \end{bmatrix}\begin{bmatrix} 1 & 0 & 1 & 0 & 0 & 0 \\ 1 & 0 & 1 & 0 & 0 & 1 \end{bmatrix}u^{\mathrm{T}}
$$

$$
= \begin{bmatrix} 1 & 0 & 1 & 0 & 0 & 1 \\ 0 & 0 & 0 & 0 & 0 & 1 \end{bmatrix}u^{\mathrm{T}} = \begin{bmatrix} 1 & 0 & 1 & 0 & 0 & 1 \\ 0 & 0 & 0 & 0 & 0 & 1 \end{bmatrix}\begin{bmatrix} 1 \\ 0 \\ 1 \\ 0 \\ 1 \\ 0 \end{bmatrix} = \begin{bmatrix} 0 \\ 0 \end{bmatrix}.
$$

From (6.113), we find the second parity part as follows:

$$
p_2^{\mathrm{T}} = -T^{-1}\left(Au^{\mathrm{T}} + Bp_1^{\mathrm{T}}\right) = \begin{bmatrix} 1 & 0 & 0 & 0 \\ 0 & 1 & 0 & 0 \\ 1 & 1 & 1 & 0 \\ 1 & 0 & 1 & 1 \end{bmatrix}\left(\begin{bmatrix} 1 \\ 1 \\ 0 \\ 0 \end{bmatrix} + \begin{bmatrix} 0 \\ 0 \\ 0 \\ 0 \end{bmatrix}\right) = \begin{bmatrix} 1 \\ 1 \\ 0 \\ 0 \\ 1 \end{bmatrix}.
$$

Finally, we find the codeword $c = [u \; p_1 \; p_2] = [1\ 0\ 1\ 0\ 1\ 0\ 0\ 0\ 1\ 1\ 0\ 1]$ and then confirm that $H_e c^{\mathrm{T}} = 0$. ∎

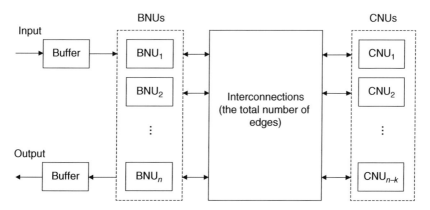

Figure 6.55   Fully parallel LDPC decoder architecture

Another easy encoding of LDPC codes is to use cyclic or quasi cyclic code characteristics. A Quasi Cyclic (QC) LDPC code [19] has sub-matrices with cyclic or quasi cyclic characteristics. It can be constructed by shifting sub-matrices.

In the LDPC decoder, there are many design issues such as complexity, power consumption, latency, interconnection between nodes, scheduling, error floor reduction, reconfigurable design for multiple code rates and codeword lengths. According to interconnection between nodes, LDPC decoder architectures can be classified into fully parallel LDPC decoders, partially parallel LDPC decoders, and serial LDPC decoders. In the fully parallel LDPC decoder, each node of the Tanner graph is directly mapped into a processing unit as shown in Figure 6.55. The number of Check processing Node Units (CNUs) and Bit processing Node Units (BNUs) is same as the number of rows and columns, respectively. The number of interconnections is same as the number of edges. Each CNU and BNU calculates the probability (or logarithms of probabilities) and passes a message to opposite nodes through interconnections. The fully parallel LDPC decoder architecture provides us with the highest throughput. We don't need scheduling and a large size memory. However, it requires the maximum number of calculations and interconnections. For a large codeword, the interconnection is very complex and difficult to implement. Another disadvantage is that it is not flexible. CNU, BNU, and interconnection are fixed. It could not support multiple code rates and codeword lengths. The complexity of the fully parallel LDPC decoder is proportional to the codeword length.

The partially parallel LDPC decoder architecture as shown in Figure 6.56 provides us with a lower complexity and a better scalability than the fully parallel LDPC decoder. The number of CNUs, BNUs, and interconnections in the partially parallel LDPC decoder is much smaller than them in the fully parallel LDPC decoder because they are shared. The grouping of CNUs and BNUs can be carried out according to cycles. The interconnection and scheduling of messages depend on the grouping. However, the partially parallel LDPC decoder has a lower throughput and needs scheduling. The throughput is proportional to the number of CNUs and BNUs. Thus, the number of CNUs and BNUs should be selected in terms of the required throughput.

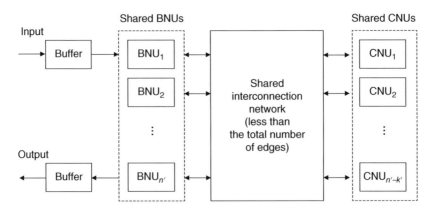

Figure 6.56   Partially parallel LDPC decoder architecture

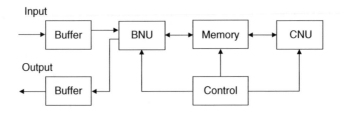

Figure 6.57   Serial LDPC decoder architecture

The serial LDPC decoder architecture as shown in Figure 6.57 has only one CNU and BNU. Thus, it provides us with the minimum hardware complexity. The CNU and BNU calculate one row and column at a time. In addition, it is highly scalable and can satisfy any code rates or codeword lengths. However, the throughput is extremely low. Thus, it is diffi-cult to apply a practical system because the most of wireless communication system require a high throughput.

The latency of a LDPC decoder is one of the disadvantages. Typically, LPDC codes need about 25 iterations to converge but turbo codes converge the performance in about 6 or 7 iterations. The latency is related to node calculations, interconnections, and scheduling. In order to reduce the node calculation, approximations such as the min-sum algorithm can be used. Most of the low complexity algorithms [20] have a near optimum performance. The quantization level of the input signal in a LDPC decoder is one of the important design parameters. This is highly related to decoding performance, memory size, and complexity. They are a trade-off between decoding performance and complexity. About four to six bits level brings a good balance between them [21, 22]. Another disadvantage of LDPC decoder is that it is not easy to design reconfigurable decoders because it naturally has a random structure. Thus, a new class of LDPC codes such as QC-LDPC and structured LDPC code [23] are widely used in many wireless communication systems. The QC-LDPC code struc-ture composed of square sub-matrices allows us to efficiently design the partially parallel LDPC decoder.

6.4 Problems

6.1. Consider the $(2, 1, 1)$ RSC code with generator matrix $G(D)=[1 \; 1/(1+D)]$. Draw the trellis diagram.

6.2. Consider the turbo encoder composed of two RSC codes of Problem 6.1 and a random interleaver. Describe the turbo encoding and decoding process when the code rate is 1/2 (Puncturing) and 1/3 (no puncturing).

6.3. Find the minimum free distance of the turbo code in Problem 6.2.

6.4. In one of 3GPPP standards, the turbo code is defined as follows:
   - Code rate $= 1/2, 1/3, 1/4$, and $1/5$
   - The transfer function of the component code is

   $$G(D)=[1 \; n_0(D)/d(D) \; n_1(D)/d(D)]$$

   where $d(D)=1+D2+D3$, $n_0(D)=1+D+D3$ and $n_1(D)=1+D+D2+D3$.
   - Puncturing patterns

Output	Code rate			
	1/2	1/3	1/4	1/5
X	11	11	11	11
Y_0	10	11	11	11
Y_1	00	00	10	11
X'	00	00	00	00
Y_0'	01	11	01	11
Y_1'	00	00	11	11

   - Random interleaver
   Design the turbo encoder and decoder.

6.5. Consider the turbo product encoder composed of two $(7, 4, 3)$ Hamming codes as row and column encoder. Design the turbo product decoder using Chase algorithm.

6.6. Explain why turbo product codes have a lower error floor than turbo codes.

6.7. Consider the following parity check matrix

$$H = \begin{bmatrix} 110100 \\ 101010 \\ 011001 \end{bmatrix}.$$

6.8. Find the parity check equations and draw the corresponding Tanner graph.

6.9. Construct the parity check matrix of a (20, 3, 4) regular LDPC code by (i) Gallager and (ii) Mackay and Neal.

6.10. Find the cyclic of the parity check matrix in Problem 6.8 and perform cycle breaking.

6.11. Extrinsic Information Transfer (EXIT) charts provide us with a graphical depiction of density evolution. Draw the EXIT chart for (20, 3, 4) regular LDPC code in Problem 6.8.

6.12. Describe the encoding and decoding process of a (20, 3, 4) regular LDPC code.

References

[1] D. Divslar and R. J. McEliece, "Effective Free Distance of Turbo Codes," *Electronic Letters*, vol. **32**, no. 5, pp. 445–446, 1996.

[2] D. Divsalar and F. Pollara, "One the Design of Turbo Codes," TDA Progress Report 42-123, Jet Propulsion Lab, Pasadena, CA, pp. 99–121, 1995.

[3] P. Robertson, P. Hoeher, and E. Villebrun, "Optimal and Sub-optimal Maximum a Posteriori Algorithms Suitable for Turbo Decoding," *European Transactions on Telecommunications*, vol. **8**, pp. 119–125, 1997.

[4] 3rd Generation Partnership Project (3GPP), Technical Specification Group Radio Access Network, "Multiplexing and Channel Coding (FDD)," W-CDMA 3GPP TS 25.212 v6.3, December 2004.

[5] CDMA2000 3GPP2 C.S0002-C Version 2.0, "Physical Layer Standard for cdma200 Spread Spectrum Systems" July 23, 2004.

[6] A. J. Viterbi, "An Intuitive Justification and a Simplified Implementation of the MAP Decoder for Convolutional Codes," *IEEE Journal on Selected Areas in Communications*, vol. **16**, no. 2, pp. 260–264, 1998.

[7] P. Elias, "Error-Free Coding," *IRE Transactions on Information Theory*, vol. **PGIT-4**, pp. 29–37, 1954.

[8] R. Pyndiah, A. Glavieux, A. Picart, and S. Jacq, "Near Optimum Decoding of Product Codes," *IEEE Global Telecommunications Conference (GLOBECOM'94)*, San Francisco, CA, vol. **1**, pp. 339–343, November 28, 1994.

[9] D. Chase, "A Class of Algorithms for Decoding Block Codes with Channel Measurement Information," *IEEE Transactions on Information Theory*, vol. **IT-18**, pp. 170–182, 1972.

[10] L. R. Bahl, J. Cocke, F. Jelinek, and J. Raviv, "Optimal Decoding of Linear Codes for Minimizing Symbol Error Rate," *IEEE Transactions on Information Theory*, vol. IT-**20**, pp. 284–297, 1974.

[11] J. Hagenauer and P. Hoeher, "A Viterbi Algorithm with Soft-Decision Output and its Applications," *IEEE Global Telecommunications Conference (GLOBECOM'89)*, Dallas, TX, vol. **3**, pp. 1680–1686, November 27–30, 1989.

[12] S. S. Pietrobon and S. A. Barbulescu, "A Simplification of the Modified Bahl Algorithm for Systematic Convolutional Codes," *Proceedings of IEEE International Symposium on Information Theory and Its Applications (ISITA'94)*, pp. 1071–1077, Sydney, Australia, November 20–25, 1994.

[13] Y. Li, B. Vucetic, and Y. Sato, "Optimum Soft-Output Detection for Channels with Intersymbol Interference," *IEEE Transactions on Information Theory*, vol. **41**, no. 3, pp. 704–713, 1995.

[14] H. Kim, G. Markarian, and V. Rocha, "Low Complexity Iterative Decoding of Product Codes Using a Generalized Array Code Form of the Nordstrom-Robinson Code," *Proceedings of the 6th International Symposium on Turbo Codes and Iterative Information Processing*, pp. 88–92, Brest, France, September 6–10, 2010.

[15] M. G. Luby, M. Mitzenmacher, M. A. Shokrollahi, and D. A. Speilman, "Improved Low-Density Parity Check Codes Using Irregular Graphs and Belief Propagation," *Proceedings of IEEE International Symposium on Information Theory (ISIT'98)*, pp. 16–21, Cambridge, MA, USA, August 16–21, 1998.

[16] T. Richardson and R. Urbanke, "Efficient Encoding of Low-Density Parity Check Codes," *IEEE Transactions on Information Theory*, vol. **47**, pp. 638–656, 2011.

[17] J. Heo, "Analysis of Scaling Soft Information on Low Density Parity Check Code," *IEE Electronics Letters*, vol. **39**, no. 2, pp. 219–220, 2003.

[18] T. J. Richardson and R. L. Urbanke, "Efficient Encoding of Low-Density Parity-Check Codes," *IEEE Transactions on Information Theory*, vol. **47**, no. 2, pp. 638–656, 2001.

[19] M. Fossorier, "Quasi-Cyclic Low-Density Parity-Check Codes from Circulant Permutation Matrices," *IEEE Transactions on Information Theory*, vol. **50**, no. 8, pp. 1788–1793, 2004.

[20] E. Elefthriou, T. Mittelholzer, and A. Dholakia, "Reduced Complexity Decoding Algorithm for Low-Density Parity Check Codes," *IEE Electronics Letter*, vol.**37**, no. 2, pp. 102–103, 2011.

[21] L. Ping and W. K. Leung, "Decoding Low Density Parity Check Codes with Finite Quantization Bits," *IEEE Communications Letters*, vol. **4**, no. 2, pp. 62–64, 2000.

[22] Z. Zhang, L. Dolecek, M. Wainwright, V. Anantharam, and B. Nikolic, "Quantization Effects in Low-Density Parity-Check Decoder," *Proceedings of IEEE International Conference on Communications (ICC'07)*, pp. 6231–6237, Glasgow, UK, June 28, 2007.

[23] R. M. Tanner, D. Sridhara, T. E. Fuja, and D. J. Costello, "LDPC Block and Convolutional Codes Based on Circulant Matrices," *IEEE Transactions on Information Theory*, vol. **50**, no. 12, pp. 2966–2984, 2004.

7

Orthogonal Frequency-Division Multiplexing

The Orthogonal Frequency-Division Multiplexing (OFDM) became the most popular transmission technique for a broadband communication system because it has many advantages such as robustness to multipath fading channels and high spectral efficiency. The Orthogonal Frequency-Division Multiple Access (OFDMA) is used for a multiuser system based on the OFDM technique. It allows multiple users to receive information at the same time on different parts of the channels. The OFDMA typically allocates multiple subcarriers to an individual user. Many recent standards use the OFDM/OFDMA technique. For example, European Telecommunications Standards Institute (ETSI) included the OFDM technique in the Digital Video Broadcasting-Terrestrial (DVB-T) system in 1997. The WiFi included the OFDM technique for its physical layer in 1999. In addition, IEEE802.16e/m and LTE adopted the OFDMA technology. In this chapter, we design the OFDM system and discuss hardware implementation issues.

7.1 OFDM System Design

We should consider many design parameters of the OFDM system. The first design parameters of the OFDM system for wireless communications are *Inter-Carrier Interference (ICI) power*, *the size of the Discrete Fourier Transform (DFT)*, and *the subcarrier spacing (or the OFDM symbol duration)*. It is important to determine the optimal DFT size balancing protection against multipath fading, the Doppler shift, and complexity [1]. The DFT size and subcarrier spacing are related to the performance and complexity of the OFDM system. For example, when 20 MHz bandwidth is given to design an OFDM system, we can consider two designs: one OFDM system can have 64 subcarriers and 312.5 kHz subcarrier spacing and another OFDM system can have 2048 subcarriers and 15 kHz subcarrier spacing. The OFDM symbol duration is the inverse of the subcarrier spacing. Thus, their OFDM symbol durations are

Wireless Communications Systems Design, First Edition. Haesik Kim.
© 2015 John Wiley & Sons, Ltd. Published 2015 by John Wiley & Sons, Ltd.

3.2 μs (=1/312.5 kHz) and 66.7 μs (=1/15 kHz). Generally speaking, the short OFDM symbol duration and wide subcarrier spacing is suitable when the channel is rapidly varying in time domain and slowly varying in frequency domain. Thus, this OFDM system configuration is suitable for short-range communications. On the other hand, the long OFDM symbol duration and narrow subcarrier spacing is suitable when the channel is slowly varying in time domain and rapidly varying in frequency domain. Thus, this OFDM system configuration is suitable for cellular systems. In order to find suitable OFDM parameters, two main elements (the Doppler spread (coherence time) and multipath delay (coherence bandwidth)) should be considered. As we discussed in Chapter 3, $T_s < T_c$ and $B_s < B_c$ should be satisfied so that OFDM symbols experience slow and flat fading. For example, a mobile station speed 100 km/h (=28 m/s) is considered in this analysis for mobility support. This maximum speed basically provides us with good coverage of vehicular speed. The maximum Doppler shift (f_m) corresponding to the operation at 3.5 GHz (wavelength=0.086 m=3×10^8 m/s/3.5×10^9/s) is given as follows:

$$f_m = \frac{v}{\lambda} = \frac{28 \text{ m/s}}{0.086 \text{ m}} = 326 \text{ Hz} \tag{7.1}$$

where v and λ are velocity and wavelength, respectively. The symbol duration affects the ICI by the Doppler spread. The ICI degrades the OFDM system performance. A longer symbol duration basically means a higher ICI in wireless channels. In Ref. [2], a simple upper bound on the ICI power is defined as follows:

$$P_{ici} \le \frac{1}{12} \left(2\pi f_m T_s \right)^2 . \tag{7.2}$$

Figure 7.1 illustrates the relationship between the ICI power and $f_m T_s$.

The symbol duration T_s should be carefully selected to satisfy the condition that $f_m T_s$ is small enough. Thus, the Doppler effect to the symbol must be negligible. When considering 10 kHz subcarrier spacing, the ICI power corresponding to the Doppler shift is calculated as follows:

$$P_{ici} \le \frac{1}{12} \left(2\pi \left(326 \right) \left(\frac{1}{10 \times 10^3} \right) \right)^2 = 0.0035 = -24.5 \text{ dB}. \tag{7.3}$$

This ICI power is less than a noise level. Thus, we can ignore the ICI power. The coherence time T_c of the wireless channel corresponding to the Doppler shift is calculated from (3.24) as follows:

$$T_c = \sqrt{\frac{9}{16\pi f_m^2}} = \frac{0.423}{f_m} = \frac{0.423}{326 \text{ Hz}} = 1.3 \text{ ms}. \tag{7.4}$$

This means the update rate of 770 Hz (=1/1.3 ms) is required for channel estimation and equalization. The International Telecommunications Union—Radio communication Sector

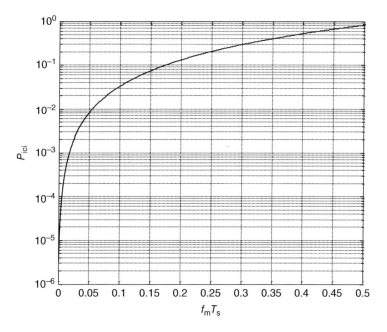

Figure 7.1   Upper bound on the ICI power

Vehicular Channel Model B (ITU-VB channel model) [3, 4] shows us the rms delay spread values of up to 20 μs for mobile environments. The subcarrier spacing design requires that each subcarrier experiences flat fading for the worst-case delay spread values of 20 μs with a guard time overhead of no more than 10% for a target delay spread of 10 μs. The coherence bandwidth (B_c) of the wireless channel for correlation greater than 0.5 corresponding to the 20 μs delay spread (τ_{rms}) is calculated from (3.17) as follows:

$$B_c \approx \frac{1}{5\tau_{rms}} = \frac{1}{5 \cdot 20 \ \mu s} = 10 \ \text{kHz}. \tag{7.5}$$

This means that multipath fading for the delay spread values of up to 20 μs is considered to be flat fading over 10 kHz subcarrier width. The above analysis based on the coherence time, the Doppler shift, and the coherence bandwidth of the channel is the basis for the consideration of the OFDM system design. For example, when a standard defines 10 MHz bandwidth at 3.5 GHz carrier frequency and the OFDM system should support a mobile user with speed 100 km/h, the subcarrier spacing should be chosen as less than 10 kHz.

The second design parameter is *the guard interval and the cyclic prefix* (CP). The guard interval is used to overcome the Inter-Symbol Interference (ISI) among OFDM symbols. It should be selected as a larger interval than the maximum delay. If the maximum delay exceeds the guard interval, the signal constellation is seriously distorted. In Ref. [5], the guard interval is chosen as at least fourfold rms delay spread. The OFDM symbol duration should be much larger than the guard interval in order to prevent the SNR loss by the guard interval.

However, a larger OFDM symbol duration affects the DFT size, phase noise, frequency offset, and Peak-to-Average Ratio (PAPR). Thus, the OFDM symbol duration is practically defined as at least fivefold or sixfold guard interval. For example, the rms delay spread is 100 ns. The guard interval is 400 ns (=100 ns×4) and the OFDM symbol duration is 2.4 μs (=400 ns×6). A cyclic prefix is used for eliminating the ISI. The guard interval consists of null signals but the cyclic prefix is composed of the OFDM symbol extension cyclically. The CP length should be larger than the channel impulse response. Thus, all distributed energy caused by multipath delay is recovered and the orthogonality is maintained. In LTE standard, two CP lengths are defined. The normal CP is used in urban (small) cells and high data rates and the extended CP is used in rural (large) cells and low data rates. The normal CP and the extended CP are defined as about 7.5 and 25% of the OFDM symbol duration, respectively. For example, the OFDM symbol duration is 2 μs. The normal CP is 150 ns and the extended CP is 500 ns. The CP is a copy of the last subcarriers in the OFDM symbol and appended at the guard interval in front of the OFDM symbol. Figure 7.2 illustrates the effect of the guard interval and cyclic prefix in the OFDM symbol. Figure 7.2a and b shows us signal distortion when there is no guard interval. Figure 7.2c and d represents that the guard interval (or Zero padding) prevents signal distortion. Figure 7.2e and f shows us overcoming multipath delay due to the CP.

The third design parameter is *the pilot subcarrier allocation*. The pilot signals are used for channel estimation and frequency offset/phase noise compensation. In an OFDM symbol, some subcarriers are allocated for pilots. For example, when we consider 256 DFT size of the OFDM system, 56 subcarriers are used for guard bands. Among 56 subcarriers, a low frequency guard band, a high frequency guard band, and a Direct Current (DC) signal are 28 subcarriers, 27 subcarriers, and 1 subcarrier, respectively. They are null signals. Among the remaining 200 subcarriers, data signals are 192 subcarriers and pilots are 8 subcarriers. The goal of pilot design is to find an optimum pilot allocation balancing between the spectral efficiency and the channel estimation performance. As we discussed the pilot spacing in frequency and time domain in Chapter 5, pilot spacing in frequency domain is related to the maximum delay spread (τ_{max}). It should be smaller than the variation of the channel in frequency domain $\left(f_p \leq 1/2\tau_{max}\Delta f \ (5.68) \right)$. If pilot spacing is larger than this condition, the channel cannot be accurately estimated. A wireless channel is varying during OFDM transmission. The channel variation in time domain should be estimated. The maximum pilot spacing in time domain is described by the Doppler spread and the OFDM symbol duration. It is expressed as $T_p \leq 1/2f_m T_{OFDM}$ (5.67). In a practical system, double sampling is often used to estimate the channel more accurately [6]. Thus, (5.67) and (5.68) are modified as follows:

$$f_p \leq \frac{1}{4\tau_{max}\Delta f} \tag{7.6}$$

and

$$T_p \leq \frac{1}{4f_m T_{OFDM}}. \tag{7.7}$$

However, there is trade-off relationship between the accuracy of the channel estimation and the spectral efficiency. Thus, pilot spacing should be designed according to system requirements.

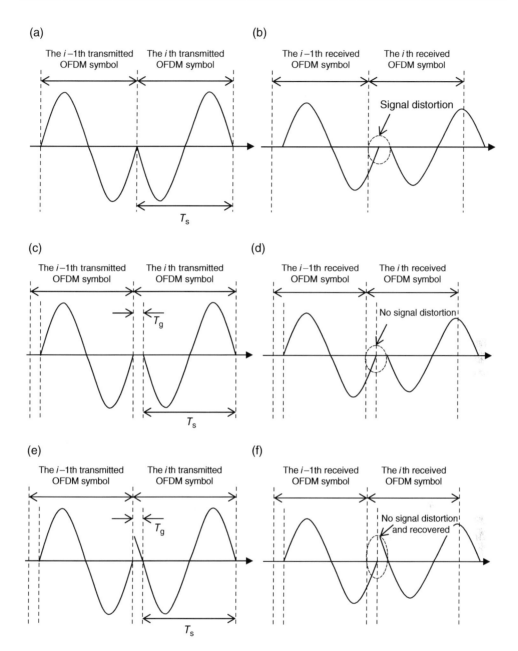

Figure 7.2   Guard interval and cyclic prefix effect: No guard interval (a) and (b), guard interval (c) and (d), and guard interval with cyclic prefix (e) and (f)

For example, we assume the maximum delay spread, the Doppler spread and the subcarrier spacing are 20 μs, 326 Hz, and 10 kHz, respectively. Using (5.67) and (5.68), we calculate the sampling interval of pilot spacing in frequency domain and the maximum pilot spacing in time domain as follows:

$$f_p \le \frac{1}{2 \cdot 20 \ \mu s \cdot 10 \ kHz} = 2.5 \tag{7.8}$$

and

$$T_p \le \frac{1}{2 \cdot 326 \ Hz \cdot 100 \ \mu s} = 15.3. \tag{7.9}$$

The first calculation means that the channel frequency response estimation of an OFDM system is sampled with 2.5 subcarrier spacing. Namely, the pilot symbols are needed for every two data subcarriers. If we assume L tap channel model, the DFT size (N) is equal to $f_p L$. The second calculation means that the pilot symbol should be inserted at least every 15 OFDM symbols in order to track the channel variation. To sum up, a block-type pilot structure is suitable for this system. An optimum pilot allocation for a certain channel may not be optimal allocation for another channel due to channel variation. Thus, we should find pilot allocation for each channel in order to find an optimal pilot allocation. However, it is not efficient way in complexity point of view. Thus, constant pilot allocation in OFDM symbols is used in practical OFDM systems. This is trade-off relationship between the channel estimation performance and the complexity. In Ref. [7], the optimal pilot subcarrier allocation for AWGN channel is described. The pilot subcarrier set is expressed as follows:

$$\left\{ i, i + \frac{N}{N_p}, i + \frac{2N}{N_p}, \dots, i + \frac{(N_p - 1)N}{N_p} \right\} \tag{7.10}$$

where $i = 0, 1, 2, \dots, N/N_p - 1$ and N_p is the number of pilot subcarriers. In addition, we should insert a pilot subcarrier around both ends of the OFDM symbol.

The fourth design parameter is *windowing*. An OFDM symbol is composed of unfiltered subcarriers. Thus, the out-of-band spectrum decreases slowly depending on the number of subcarriers. A larger number of subcarriers decrease more rapidly. A windowing technique is used to make the spectrum go down faster. Several windowing techniques are used such as raised cosine, Hann, Hamming, Blackman, and Kaiser. Among them, the Blackman windowing technique provides us with the best performance. However, the raised cosine function is widely used due to the trade-off between performance and complexity. It is defined as follows:

$$w_T(t) = \begin{cases} \dfrac{1 + \cos\left(\pi + \dfrac{\pi}{\beta T_t} t\right)}{2}, & 0 \le t \le \beta T_t \\ 1 & \beta T_t \le t \le T_t \\ \dfrac{1 + \cos\left(\dfrac{\pi}{\beta T_t}(t - T_t)\right)}{2}, & T_t \le t \le (1 + \beta)T_t \end{cases} \tag{7.11}$$

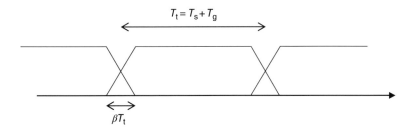

Figure 7.3   OFDM windowing

where β and T_t are the roll-off factor and $T_s + T_g$, respectively. Figure 7.3 illustrates the windowing.

In the figure, βT_t is the transition time between two consecutive OFDM symbols transmission. The roll-off factor means the excess bandwidth of the filter. A large roll-off factor can improve the spectrum. Besides windowing, it is possible to use null subcarriers at the edge of the spectrum.

Example 7.1   OFDM system design
Consider the following design requirements and design the OFDM system:

Requirements	Values
Speed	80 km/h
Carrier frequency	2 GHz
Tolerable delay spread	10 μs
Bandwidth	500 kHz
Target data rate	1 Mbps

Solution
From the required maximum speed (80 km/h (=22 m/s)) and the carrier frequency (2 GHz (wavelength = 0.15 m = 3×10^8 m/s / 2×10^9/s)), we calculate the maximum Doppler shift (f_m) as follows:

$$f_m = \frac{v}{\lambda} = \frac{22 \text{ m/s}}{0.15 \text{ m}} = 146 \text{ Hz}.$$

From this result and the tolerable delay spread (10 μs), the coherence time is

$$T_c = \frac{0.423}{146 \text{ Hz}} = 2.9 \text{ ms}$$

and the coherence bandwidth is

$$B_c \approx \frac{1}{5\tau_{rms}} = \frac{1}{5 \cdot 10 \text{ μs}} = 20 \text{ kHz}.$$

From the tolerable delay spread, the guard interval is $40\,\mu s$ ($=10\,\mu s \times 4$) and the OFDM symbol duration is $0.24\,ms$ ($=40\,\mu s \times 6$). The DFT period is $0.2\,ms$ ($=0.24\,ms$ to $40\,\mu s$). The OFDM symbol duration is the inverse of the subcarrier spacing. Thus, the subcarrier spacing is calculated by the DFT period as follows:

$$\text{Subcarrier spacing} = \frac{1}{0.2\ ms} = 5\ kHz.$$

This value satisfies the condition that subcarrier spacing should be less than $20\,kHz$. The ICI power corresponding to the Doppler shift is calculated as follows:

$$P_{ici} \le \frac{1}{12}\left(2\pi\left(146\right)\left(\frac{1}{5\times 10^3}\right)\right)^2 = 0.0028 = -25.5\ dB.$$

This ICI power is less than noise and cochannel interference. The required total number of data subcarriers is calculated as follows:

$$\frac{500\ kHz}{5\ kHz} = 100.$$

This result means the total number of data subcarriers should be smaller than 100. The sampling rate for pilot spacing is

$$f_p \le \frac{1}{2 \cdot 10\ \mu s \cdot 5\ kHz} = 10$$

and the normal CP is $18\,\mu s$ ($=0.24\,ms \times 0.075$). Thus, we can configure the parameters of the OFDM system as follows: Since the target data rate is $1\,Mbps$, we need to contain 200 bits per one OFDM symbol ($=1\,Mbps \times 0.2\,ms$). We can choose 100 data subcarriers. The modulation scheme is QPSK modulation without coding or 16 QAM with code rate ½. Thus, the DFT size can be chosen as 128. The number of pilot subcarriers can be selected as 12. When we choose 10 guard subcarriers and 6 null subcarriers, subcarriers of the OFDM system can be composed of data, pilots, guard subcarriers, and null subcarrier as follows:

128 DFT size = 100 data subcarriers + 12 pilots subcarriers + 10 guard subcarriers + 6 null carriers.

From (7.10), one possible pilot allocation set is [0 10 21 32 42 53 64 74 85 96 106 117]. The interval of pilot subcarriers is 10 or 11. We should insert pilot subcarriers around both edges of the OFDM symbol and also consider guard carriers and null carriers. Thus, one possible subcarrier allocation of the OFDM symbol is illustrated in Figure 7.4.

The parameters of the OFDM system are summarized in Table 7.1.   ■

Summary 7.1 OFDM design

1. The OFDM symbol duration is the inverse of the subcarrier spacing.
2. The ICI degrades the OFDM system performance. A simple upper bound on the ICI power is defined as follows:

$$P_{\text{ici}} \leq \frac{1}{12}\left(2\pi f_{\text{m}} T_{\text{s}}\right)^2.$$

3. The guard interval is chosen as at least fourfold rms delay spread.
4. The OFDM symbol duration is practically defined as at least fivefold or sixfold guard interval.
5. The pilot spacing in frequency domain should be smaller than the variation of channel in frequency domain $\left(f_{\text{p}} \leq 1/2\tau_{\text{max}}\Delta f\right)$.
6. The maximum pilot spacing in time domain is described as $T_{\text{p}} \leq 1/2 f_{\text{m}} T_{\text{OFDM}}$.
7. The raised cosine function as windowing is defined as follows:

$$w_T(t) = \begin{cases} \dfrac{1+\cos\left(\pi + \dfrac{\pi}{\beta T_{\text{t}}}t\right)}{2}, & 0 \leq t \leq \beta T_{\text{t}} \\ 1 , & \beta T_{\text{t}} \leq t \leq T_{\text{t}} \\ \dfrac{1+\cos\left(\dfrac{\pi}{\beta T_{\text{t}}}(t - T_{\text{t}})\right)}{2}, & T_{\text{t}} \leq t \leq (1+\beta)T_{\text{t}} \end{cases}$$

where β and T_{t} are the roll-off factor and $T_{\text{s}} + T_{\text{g}}$, respectively.

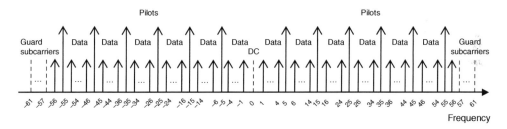

Figure 7.4 Subcarrier allocation in the OFDM symbol

7.2 FFT Design

The OFDM system transmits multiple subcarriers which are orthogonal to each other. The DFT/IDFT is used to avoid many local oscillators. Thus, we can easily transmit and receive the multicarrier signals. In the OFDM transmitter, the modulated signals such as QPSK or 16QAM signals are defined in frequency domain and mapped into each subcarrier as the input

Table 7.1 OFDM system parameters

Parameters	Description	Value
N	DFT size	128
N_d	Number of data subcarriers	100
N_p	Number of pilot subcarriers	12
N_g	Number of guard subcarriers	10
Δf	Subcarrier spacing	5 kHz
T_{OFDM}	OFDM symbol duration	0.24 ms
T_g	Guard interval	40 μs
T_{cp}	Normal CP	18 μs
M-ary/R	Modulation and code rate	QPSK/No coding or 16QAM/1/2

of the IDFT. The signal is transformed into time domain signal by the IDFT. In the OFDM receiver, the inverse process of the OFDM transmitter is carried out. The DFT/IDFT is an essential part of the OFDM system. The Fast Fourier Transform (FFT) and the Inverse Fast Fourier Transform (IFFT) are efficient algorithms to compute the DFT and IDFT. They are widely used when implementing the DFT and IDFT. Consider a sequence of N samples x_n where $n = 0, 1, \ldots, N-1$. The DFT of this sequence is given as follows:

$$X_k = \sum_{n=0}^{N-1} x_n e^{-j\frac{2\pi}{N}kn}, \quad k = 0,1,\ldots,N-1 \tag{7.12}$$

and the IDFT is given as follows:

$$x_n = \frac{1}{N}\sum_{k=0}^{N-1} X_k e^{j\frac{2\pi}{N}kn}, \quad n = 0,1,\ldots,N-1. \tag{7.13}$$

We define twiddle factors as follows:

$$W_N = e^{-j\frac{2\pi}{N}} \tag{7.14}$$

and rewrite the DFT using twiddle factors as follows:

$$X_k = \sum_{n=0}^{N-1} x_n W_N^{kn}, \quad k = 0,1,\ldots,N-1. \tag{7.15}$$

The DFT is rewritten in matrix form as follows:

$$\begin{bmatrix} X_0 \\ X_1 \\ \vdots \\ X_{N-1} \end{bmatrix} = \begin{bmatrix} W_N^0 & W_N^0 & W_N^0 & \cdots & W_N^0 \\ W_N^0 & W_N^1 & W_N^2 & & W_N^{N-1} \\ \vdots & \vdots & & \ddots & \vdots \\ W_N^0 & W_N^{N-1} & W_N^{2(N-1)} & \cdots & W_N^{(N-1)(N-1)} \end{bmatrix} \begin{bmatrix} x_0 \\ x_1 \\ \vdots \\ x_{N-1} \end{bmatrix}. \tag{7.16}$$

When observing (7.16), the number of complex multiplications is proportional to N^2. If the input sequence is composed of 1024 symbols, a million arithmetic operation (1024^2) should be carried out. It is not acceptable. Thus, we need more efficient algorithm to compute the DFT. In order to drive the FFT, we firstly split (7.15) into two parts as follows:

$$X_k = \sum_{m=0}^{N/2-1} x_{2m} W_N^{2mk} + \sum_{m=0}^{N/2-1} x_{2m+1} W_N^{(2m+1)k}, \quad k = 0,1,\ldots,N-1. \tag{7.17}$$

Using $W_N^{2mk} = W_{N/2}^{mk}$ (or $e^{-j\frac{2\pi}{N}2mk} = e^{-j\frac{2\pi}{N/2}mk}$), we consider the first half of the DFT outputs ($k=0, 1, \ldots, N/2-1$) and rewrite (7.17) in terms of $N/2$ length DFT of even and odd samples as follows:

$$X_k = \sum_{m=0}^{N/2-1} x_{2m} W_{N/2}^{mk} + W_N^k \sum_{m=0}^{N/2-1} x_{2m+1} W_{N/2}^{mk}, \quad k = 0,1,\ldots,\frac{N}{2}-1 \tag{7.18}$$

$$= X_k^{\text{even}} + W_N^k X_k^{\text{odd}}, k = 0,1,\ldots,\frac{N}{2}-1 \tag{7.19}$$

where X_k^{even} is the DFT of even samples ($x_{2m} = [x_0, x_2, \ldots, x_{N-4}, x_{N-2}]$) and X_k^{odd} is the DFT of odd samples ($x_{2m+1} = [x_1, x_3, \ldots, x_{N-3}, x_{N-1}]$). The second half of the DFT outputs is represented as follows:

$$X_{k+N/2} = \sum_{m=0}^{N/2-1} x_{2m} W_{N/2}^{m(k+(N/2))} + W_N^{k+(N/2)} \sum_{m=0}^{N/2-1} x_{2m+1} W_{N/2}^{m(k+(N/2))}, \quad k = 0,1,\ldots,\frac{N}{2}-1. \tag{7.20}$$

Further simplification is possible due to periodic and symmetric properties of the DFT as follows:

$$W_{N/2}^{m(k+(N/2))} = W_{N/2}^{mk} W_{N/2}^{mN/2} = W_{N/2}^{mk} \tag{7.21}$$

$$W_N^{(k+(N/2))} = W_N^k W_N^{N/2} = -W_N^k. \tag{7.22}$$

Thus, (7.20) is rewritten as follows:

$$X_{k+N/2} = \sum_{m=0}^{N/2-1} x_{2m} W_{N/2}^{mk} - W_N^k \sum_{m=0}^{N/2-1} x_{2m+1} W_{N/2}^{mk}, \quad k = 0,1,\ldots,\frac{N}{2}-1. \tag{7.23}$$

$$= X_k^{\text{even}} - W_N^k X_k^{\text{odd}}, \quad k = 0,1,\ldots,\frac{N}{2}-1. \tag{7.24}$$

Now, we define the FFT butterfly structure consisting of multiplication, addition, and subtraction as shown in Figure 7.5.

Simpler form is represented as shown in Figure 7.6.

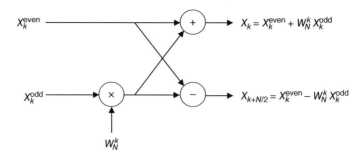

Figure 7.5   FFT butterfly

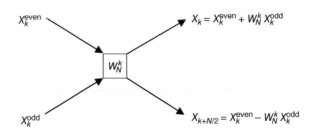

Figure 7.6   Simple form of FFT butterfly

As we can observe from Figures 7.5 and 7.6, an N point DFT is divided into two $N/2$ point DFTs. They are composed of the terms X_k^{even} and X_k^{odd}. Thus, we priorly calculate and store the term W_N^k. Then, we calculate the terms X_k^{even} and X_k^{odd} and use them for X_k and $X_{k+N/2}$ calculation. This butterfly architecture reduces the complexity of the DFT. Two $N/2$ point DFTs require $2(N/2)^2$ complex multiplies and $2(N/2)$ complex additions. Thus, the total number of the complexity is approximately $N^2/2$. This is very meaningful result because the complexity has approximately halved through rearranging the equations. Figure 7.7 illustrates an example of a 8-point FFT structure based on butterfly.

We can further reduce the complexity using periodic and symmetric properties. In Equations 7.19 and 7.24, the $N/2$-point DFT X_k^{even} is represented as follows:

$$X_k^{\text{even}} = X'^{\text{even}}_k + W_{N/2}^k X'^{\text{odd}}_k, \ k = 0,1,\ldots,\frac{N}{4}-1 \tag{7.25}$$

$$X_{k+N/4}^{\text{even}} = X'^{\text{even}}_k - W_{N/2}^k X'^{\text{odd}}_k, \ k = 0,1,\ldots,\frac{N}{4}-1 \tag{7.26}$$

and the $N/2$-point DFT X_k^{odd} is represented as follows:

$$X_k^{\text{odd}} = X''^{\text{even}}_k + W_{N/2}^k X''^{\text{odd}}_k, k = 0,1,\ldots,\frac{N}{4}-1 \tag{7.27}$$

$$X_{k+N/4}^{\text{odd}} = X''^{\text{even}}_k - W_{N/2}^k X''^{\text{odd}}_k, k = 0,1,\ldots,\frac{N}{4}-1 \tag{7.28}$$

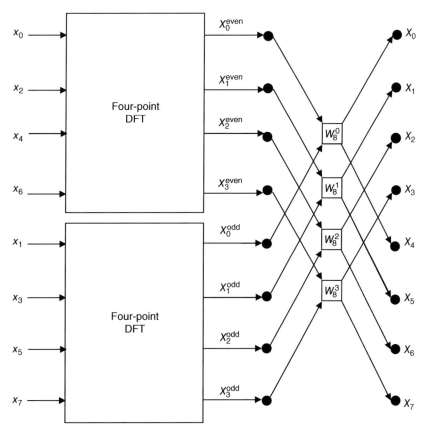

Figure 7.7   8-point FFT structure based on butterfly

where the twiddle factor $W_{N/2}^k$ is written as follows:

$$W_{N/2}^k = W_N^{2k}.$$
(7.29)

Thus, (7.25) to (7.28) are rewritten as follows:

$$X_k^{\text{even}} = X'^{\text{even}}_k + W_N^{2k} X'^{\text{odd}}_k, \quad k = 0,1,\ldots,\frac{N}{4}-1$$
(7.30)

$$X_{k+N/4}^{\text{even}} = X'^{\text{even}}_k - W_N^{2k} X'^{\text{odd}}_k, \quad k = 0,1,\ldots,\frac{N}{4}-1$$
(7.31)

$$X_k^{\text{odd}} = X''^{\text{even}}_k + W_N^{2k} X''^{\text{odd}}_k, \quad k = 0,1,\ldots,\frac{N}{4}-1$$
(7.32)

$$X_{k+N/4}^{\text{odd}} = X''^{\text{even}}_k - W_N^{2k} X''^{\text{odd}}_k, \quad k = 0,1,\ldots,\frac{N}{4}-1.$$
(7.33)

Figure 7.8 illustrates an example of 8-point FFT structure with further reduction.

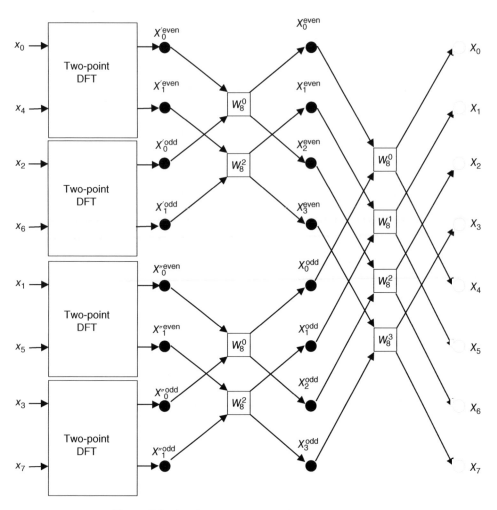

Figure 7.8   8-point FFT structure with further reduction

In the same way, we keep splitting into even and odd terms M times ($N=2^M$). Thus, the N-point DFT is reconstructed by butterflies. Figure 7.9 illustrates an example of 8-point FFT structure consisting of butterflies only.

In the N-point DFT, N is represented as follows:

$$N = r^M \tag{7.34}$$

where both r and M are integers. r is called the radix. When N is a power of 2, the type of the FFT is called a radix-2 FFT. In Figure 7.9, the radix-2 FFT is called decimation-in-time FFT. When observing the input of Figure 7.9, the input sequences are shuffled as $[x_0, x_4, x_2, x_6, x_1, x_5, x_3, x_7]$. We can implement this address order using the bit reversal as shown in Table 7.2.

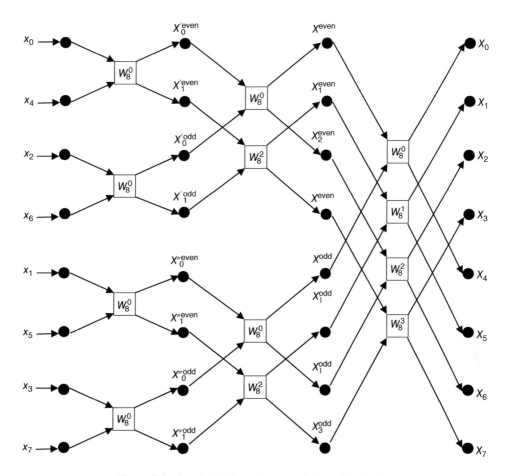

Figure 7.9   8-point FFT structure consisting of butterflies

Table 7.2   Bit reversal

Binary (decimal)	Bit reversal (decimal)
000 (0)	000 (0)
001 (1)	100 (4)
010 (2)	010 (2)
011 (3)	110 (6)
100 (4)	001 (1)
101 (5)	101 (5)
110 (6)	011 (3)
111 (7)	111 (7)

For IFFT derivation, (7.12) is rewritten using twiddle factors as follows:

$$x_n = \frac{1}{N}\sum_{k=0}^{N-1} X_k W_N^{-kn}, \quad n = 0,1,\ldots,N-1. \tag{7.35}$$

The difference is a negative power of the twiddle factor and a multiplication factor ($1/N$). Thus, the FFT can be simply modified to the IFFT and the complexity of the IFFT increases as much as additional multiplication ($1/N$).

Example 7.2   Radix-2 8-point FFT
Consider the following input sequence:

$$x_n = \left[x_0,x_1,x_2,x_3,x_4,x_5,x_6,x_7\right]$$
$$= \left[1+i,-1-i,1-i,-1+i,1-i,-1-i,1+i,-1-i\right].$$

Calculate the output sequence of the radix-2 8-point FFT and the 8-point DFT and then compare them.

Solution
We can construct the decimation-in-time 8-point FFT with following parameters:

Number of stages $= \log_2 N = \log_2 8 = 3$
Number of blocks per stage $= N/2^{stage}$
   The first stage: Number of blocks $= 8/2^1 = 4$
   The second stage: Number of blocks $= 8/2^2 = 2$
   The third stage: Number of blocks $= 8/2^3 = 1$
Number of butterflies per block at each stage $= 2^{stage-1}$
   The first stage: Number of butterflies per block $= 2^{1-1} = 1$
   The second stage: Number of butterflies per block $= 2^{2-1} = 2$
   The third stage: Number of butterflies per block $= 2^{3-1} = 4$.

The radix-2 8-point FFT architecture is same as Figure 7.8. The output sequence of the radix-2 8-point FFT is calculated by following steps: Firstly, the input sequence should be readdressed in order to become the input of FFT. Thus, we use bit reversal in Table 7.2 and have the following input sequence:

$$x_n = \left[x_0,x_4,x_2,x_6,x_1,x_5,x_3,x_7\right].$$
$$= \left[1+i,1-i,1-i,1+i,-1-i,-1-i,-1+i,-1-i\right]$$

Secondly, we need to calculate and store the twiddle factors $\left(W_8^0, W_8^1, W_8^2, \text{and } W_8^3\right)$ as follows:

$$W_8^0 = e^{-j\frac{2\pi}{8}0} = 1$$

$$W_8^1 = e^{-j\frac{2\pi}{8}1} = 0.7 - 0.7i$$

$$W_8^2 = e^{-j\frac{2\pi}{8}2} = -i$$

$$W_8^3 = e^{-j\frac{2\pi}{8}3} = -0.7 - 0.7i.$$

Thirdly, we calculate each stage using butterflies. In the first stage, we have the following values:

$$(1+i)+1\times(1-i)=2$$
$$(1+i)-1\times(1-i)=2i$$
$$(1-i)+1\times(1+i)=2$$
$$(1-i)-1\times(1+i)=-2i$$
$$(-1-i)+1\times(-1-i)=-2-2i$$
$$(-1-i)-1\times(-1-i)=0$$
$$(-1+i)+1\times(-1-i)=-2$$
$$(-1+i)-1\times(-1-i)=2i.$$

In the second stage, we have the following values:

$$(2)+1\times(2)=4$$
$$(2i)+(-i)\times(-2i)=-2+2i$$
$$(2)-1\times(2)=0$$
$$(2i)-(-i)\times(-2i)=2+2i$$

$$(-2-2i)+1\times(-2)=-4-2i$$
$$(0)+(-i)\times(2i)=2$$
$$(-2-2i)-1\times(-2)=-2i$$
$$(0)-(-i)\times(2i)=-2.$$

In the third stage, we have the following values:

$$(4)+1\times(-4-2i)=-2i$$
$$(-2+2i)+(0.7-0.7i)\times(2)=-0.6+0.6i$$
$$(0)+(-i)\times(-2i)=-2$$
$$(2+2i)+(-0.7-0.7i)\times(-2)=3.4+3.4i$$
$$(4)-1\times(-4-2i)=8+2i$$
$$(-2+2i)-(0.7-0.7i)\times(2)=-3.4+3.4i$$
$$(0)-(-i)\times(-2i)=2$$
$$(2+2i)-(-0.7-0.7i)\times(-2)=0.6+0.6i.$$

Thus, the output sequence of radix-2 8-point FFT is

$$X_k=[-2i, -0.6+0.6i, -2, 3.4+3.4i, 8+2i, -3.4+3.4i, 2, 0.6+0.6i].$$

Figure 7.10 illustrates each stage calculation.

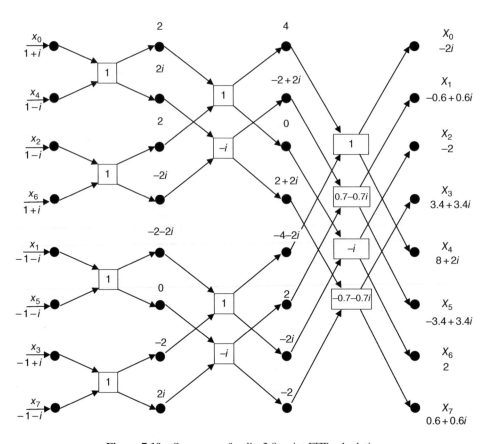

Figure 7.10   Summary of radix-2 8-point FFT calculation

From (7.12), the output sequence of the 8-point DFT is calculated as follows:

$$X_0 = (1+i)e^{-j\frac{2\pi}{8}0\cdot0} + (-1+i)e^{-j\frac{2\pi}{8}0\cdot1} + (1-i)e^{-j\frac{2\pi}{8}0\cdot2} + (-1+i)e^{-j\frac{2\pi}{8}0\cdot3}$$
$$+ (1-i)e^{-j\frac{2\pi}{8}0\cdot4} + (-1-i)e^{-j\frac{2\pi}{8}0\cdot5} + (1+i)e^{-j\frac{2\pi}{8}0\cdot6} + (-1-i)e^{-j\frac{2\pi}{8}0\cdot7} = -2i$$

$$X_1 = (1+i)e^{-j\frac{2\pi}{8}1\cdot0} + (-1-i)e^{-j\frac{2\pi}{8}1\cdot1} + (1-i)e^{-j\frac{2\pi}{8}1\cdot2} + (-1+i)e^{-j\frac{2\pi}{8}1\cdot3}$$
$$+ (1-i)e^{-j\frac{2\pi}{8}1\cdot4} + (-1-i)e^{-j\frac{2\pi}{8}1\cdot5} + (1+i)e^{-j\frac{2\pi}{8}1\cdot6} + (-1-i)e^{-j\frac{2\pi}{8}1\cdot7} = -0.6+0.6i$$

$$X_2 = (1+i)e^{-j\frac{2\pi}{8}2\cdot0} + (-1-i)e^{-j\frac{2\pi}{8}2\cdot1} + (1-i)e^{-j\frac{2\pi}{8}2\cdot2} + (-1+i)e^{-j\frac{2\pi}{8}2\cdot3}$$
$$+ (1-i)e^{-j\frac{2\pi}{8}2\cdot4} + (-1-i)e^{-j\frac{2\pi}{8}2\cdot5} + (1+i)e^{-j\frac{2\pi}{8}2\cdot6} + (-1-i)e^{-j\frac{2\pi}{8}2\cdot7} = -2$$

$$X_3 = (1+i)e^{-j\frac{2\pi}{8}3\cdot0} + (-1-i)e^{-j\frac{2\pi}{8}3\cdot1} + (1-i)e^{-j\frac{2\pi}{8}3\cdot2} + (-1+i)e^{-j\frac{2\pi}{8}3\cdot3}$$
$$+ (1-i)e^{-j\frac{2\pi}{8}3\cdot4} + (-1-i)e^{-j\frac{2\pi}{8}3\cdot5} + (1+i)e^{-j\frac{2\pi}{8}3\cdot6} + (-1-i)e^{-j\frac{2\pi}{8}3\cdot7} = 3.4 + 3.4i$$

$$X_4 = (1+i)e^{-j\frac{2\pi}{8}4\cdot0} + (-1-i)e^{-j\frac{2\pi}{8}4\cdot1} + (1-i)e^{-j\frac{2\pi}{8}4\cdot2} + (-1+i)e^{-j\frac{2\pi}{8}4\cdot3}$$
$$+ (1-i)e^{-j\frac{2\pi}{8}4\cdot4} + (-1-i)e^{-j\frac{2\pi}{8}4\cdot5} + (1+i)e^{-j\frac{2\pi}{8}4\cdot6} + (-1-i)e^{-j\frac{2\pi}{8}4\cdot7} = 8 + 2i$$

$$X_5 = (1+i)e^{-j\frac{2\pi}{8}5\cdot0} + (-1-i)e^{-j\frac{2\pi}{8}5\cdot1} + (1-i)e^{-j\frac{2\pi}{8}5\cdot2} + (-1+i)e^{-j\frac{2\pi}{8}5\cdot3}$$
$$+ (1-i)e^{-j\frac{2\pi}{8}5\cdot4} + (-1-i)e^{-j\frac{2\pi}{8}5\cdot5} + (1+i)e^{-j\frac{2\pi}{8}5\cdot6} + e^{-j\frac{2\pi}{8}5\cdot7} = -3.4 + 3.4i$$

$$X_6 = (1+i)e^{-j\frac{2\pi}{8}6\cdot0} + (-1-i)e^{-j\frac{2\pi}{8}6\cdot1} + (1-i)e^{-j\frac{2\pi}{8}6\cdot2} + (-1+i)e^{-j\frac{2\pi}{8}6\cdot3}$$
$$+ (1-i)e^{-j\frac{2\pi}{8}6\cdot4} + (-1-i)e^{-j\frac{2\pi}{8}6\cdot5} + (1+i)e^{-j\frac{2\pi}{8}6\cdot6} + (-1-i)e^{-j\frac{2\pi}{8}6\cdot7} = 2$$

$$X_7 = (1+i)e^{-j\frac{2\pi}{8}7\cdot0} + (-1-i)e^{-j\frac{2\pi}{8}7\cdot1} + (1-i)e^{-j\frac{2\pi}{8}7\cdot2} + (-1+i)e^{-j\frac{2\pi}{8}7\cdot3}$$
$$+ (1-i)e^{-j\frac{2\pi}{8}7\cdot4} + (-1-i)e^{-j\frac{2\pi}{8}7\cdot5} + (1+i)e^{-j\frac{2\pi}{8}7\cdot6} + (-1-i)e^{-j\frac{2\pi}{8}7\cdot7} = 0.6 + 0.6i.$$

Thus, the output sequence of the 8-point DFT is

$$X_k = [-2i, -0.6 + 0.6i, -2, 3.4 + 3.4i, 8 + 2i, -3.4 + 3.4i, 2, 0.6 + 0.6i].$$

The output sequence of the 8-point DFT is same as the output sequence of the radix-2 8-point FFT. ∎

For decimation-in-frequency FFT derivation, we firstly divide the input sequence into two groups ($[x_0, x_1, \ldots, x_{N/2-1}]$ and $[x_{N/2}, x_{N/2+1}, \ldots, x_{N-1}]$) and split (7.15) into two parts as follows:

$$X_{2m} = \sum_{n=0}^{N/2-1} x_n W_N^{2mn} + \sum_{n=N/2}^{N-1} x_n W_N^{2mn}, \quad m = 0,1,\ldots,\frac{N}{2}-1 \tag{7.36}$$

and

$$X_{2m+1} = \sum_{n=0}^{N/2-1} x_n W_N^{(2m+1)n} + \sum_{n=N/2}^{N-1} x_n W_N^{(2m+1)n}, \quad m = 0,1,\ldots,\frac{N}{2}-1. \tag{7.37}$$

For even index ($k = 2m$), we rewrite (7.34) using index shifting and periodic property as follows:

$$X_{2m} = \sum_{n=0}^{N/2-1} x_n W_N^{2mn} + \sum_{n=0}^{N/2-1} x_{n+N/2} W_N^{2m(n+(N/2))}, \quad m = 0,1,\ldots,\frac{N}{2}-1 \tag{7.38}$$

$$= \sum_{n=0}^{N/2-1} \left(x_n + x_{n+N/2} \right) W_{N/2}^{mn}, \quad m = 0, 1, \ldots, \frac{N}{2} - 1. \tag{7.39}$$

As we can observe from (7.39), the DFT output sequence with even index can be obtained as $N/2$ DFT of $x_n + x_{n+N/2}$. For odd index ($k = 2m + 1$), we rewrite (7.37) using index shifting and periodic and symmetric properties as follows:

$$X_{2m+1} = \sum_{n=0}^{N/2-1} x_n W_N^{(2m+1)n} + \sum_{n=0}^{N/2-1} x_{n+N/2} W_N^{(2m+1)(n+(N/2))}, \quad m = 0, 1, \ldots, \frac{N}{2} - 1 \tag{7.40}$$

$$= \sum_{n=0}^{N/2-1} \left(x_n - x_{n+N/2} \right) W_N^n W_{N/2}^{mn}, \quad m = 0, 1, \ldots, \frac{N}{2} - 1. \tag{7.41}$$

As we can observe from (7.41), the DFT output sequence with odd index can be obtained as $N/2$ DFT of $\left(x_n - x_{n+N/2} \right) W_N^n$. Similar to decimation-in-time butterfly, we can define the decimation-in-frequency FFT butterfly structure consisting of multiplication, addition, and subtraction as shown in Figure 7.11.

Simpler form is represented as shown in Figure 7.12.

Figure 7.13 illustrates an example of the decimation-in-frequency 8-point FFT structure based on butterfly.

In the same way, we keep dividing and splitting so that the N-point DFT is reconstructed by butterflies. Figure 7.14 illustrates an example of the decimation-in-frequency 8-point FFT structure consisting of butterflies only.

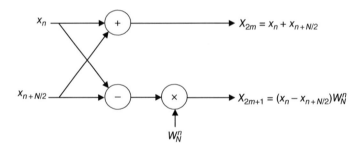

Figure 7.11   Decimation-in-frequency FFT butterfly

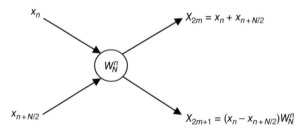

Figure 7.12   Simple form of decimation-in-frequency FFT butterfly

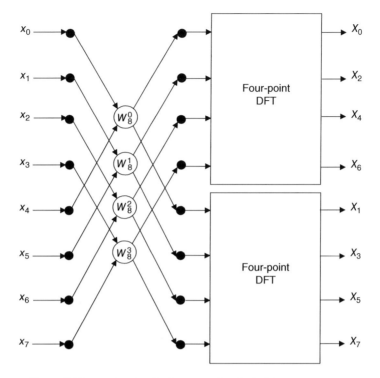

Figure 7.13   Decimation-in-frequency 8-point FFT structure

Example 7.3   Radix-2 decimation-in-frequency 8-point FFT

Consider the following input sequence:

$$x_n = \left[x_0, x_1, x_2, x_3, x_4, x_5, x_6, x_7 \right]$$
$$= \left[1+i, -1-i, 1-i, -1+i, 1-i, -1-i, 1+i, -1-i \right].$$

Calculate the output sequence of the radix-2 decimation-in-frequency 8-point FFT and then compare with the result of Example 7.2.

Solution

We construct the decimation-in-frequency 8-point FFT in a similar way to the decimation-in-time 8-point FFT. The radix-2 decimation-in-frequency 8-point FFT architecture is same as Figure 7.14. The output sequence of the radix-2 decimation-in-frequency 8-point FFT is calculated as following steps: Firstly, we need to calculate and store the twiddle factors $\left(W_8^0, W_8^1, W_8^2, \text{ and } W_8^3 \right)$ as follows:

$$W_8^0 = e^{-j\frac{2\pi}{8}0} = 1$$

$$W_8^1 = e^{-j\frac{2\pi}{8}1} = 0.7 - 0.7i$$

$$W_8^2 = e^{-j\frac{2\pi}{8}2} = -i$$

$$W_8^3 = e^{-j\frac{2\pi}{8}3} = -0.7 - 0.7i.$$

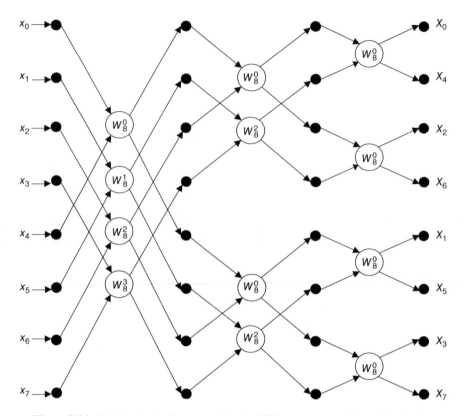

Figure 7.14   Decimation-in-frequency 8-point FFT structure consisting of butterflies

Secondly, we calculate each stage using butterflies. In the first stage, we have the following values:

$$((1+i)+(1-i))\times(1)=2$$
$$((-1-i)+(-1-i))\times(0.7-0.7i)=-2-2i$$
$$((1-i)+(1+i))\times(-i)=2$$
$$((-1+i)+(-1-i))\times(-0.7-0.7i)=-2$$
$$((1+i)-(1-i))\times(1)=2i$$
$$((-1-i)-(-1-i))\times(0.7-0.7i)=0$$
$$((1-i)-(1+i))\times(-i)=-2$$
$$((-1+i)-(-1-i))\times(-0.7-0.7i)=1.4-1.4i$$

In the second stage, we have the following values:

$$((2)+(2))\times(1)=4$$
$$((-2-2i)+(-2))\times(-i)=-4-2i$$
$$((2)-(2))\times(1)=0$$
$$((-2-2i)-(-2))\times(-i)=-2$$
$$((2i)+(-2))\times(1)=-2+2i$$
$$((0)+(1.4-1.4i))\times(-i)=1.4-1.4i$$
$$((2i)-(-2))\times(1)=2+2i$$
$$((0)-(1.4-1.4i))\times(-i)=1.4+1.4i$$

In the third stage, we have the following values:

$$((4)+(-4-2i))\times(1)=-2i$$
$$((4)-(-4-2i))\times(1)=8+2i$$
$$((0)+(-2))\times(1)=-2$$
$$((0)-(-2))\times(1)=2$$
$$((-2+2i)+(1.4-1.4i))\times(1)=-0.6+0.6i$$
$$((-2+2i)-(1.4-1.4i))\times(1)=-3.4+3.4i$$
$$((2+2i)+(1.4+1.4i))\times(1)=3.4+3.4i$$
$$((2+2i)-(1.4+1.4i))\times(1)=0.6+0.6i.$$

Thus, the output sequence of the radix-2 decimation-in-frequency 8-point FFT is

$$X_k=[X_0, X_4, X_2, X_6, X_1, X_5, X_3, X_7]$$
$$= [-2i, 8+2i, -2, 2, -0.6+0.6i, -3.4+3.4i, 3.4+3.4i, 0.6+0.6i].$$

Lastly, the output sequence should be readdressed. We use bit reversal in Table 7.2 and have the following output sequence:
$$X_k=[-2i, -0.6+0.6i, -2, 3.4+3.4i, 8+2i, -3.4+3.4i, 2, 0.6+0.6i].$$
This result is same as the result of Example 7.2. Figure 7.15 illustrates each stage calculation.   ∎

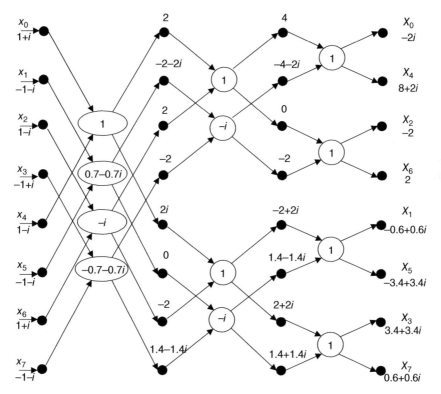

Figure 7.15   Summary of radix-2 decimation-in-frequency 8-point FFT calculation

Summary 7.2   FFT design

1. Radix-2 decimation-in-time FFT
   The first half and the second half of the DFT outputs ($k = 0, 1, \ldots, N/2 - 1$) in terms of $N/2$ length DFT of even and odd samples are represented as follows:

$$X_k = X_k^{\text{even}} + W_N^k X_k^{\text{odd}}, \quad k = 0, 1, \ldots, \frac{N}{2} - 1$$

$$X_{k+N/2} = X_k^{\text{even}} - W_N^k X_k^{\text{odd}}, \quad k = 0, 1, \ldots, \frac{N}{2} - 1$$

   where X_k^{even} and X_k^{odd} are the DFT of even samples (x_{2m}) and odd samples $\left(x_{2m+1}\right)$, respectively.

2. Radix-2 decimation-in-frequency FFT
   The DFT is presented into two parts as follows:

$$X_{2m} = \sum_{n=0}^{N/2-1} \left(x_n + x_{n+N/2}\right) W_{N/2}^{mn}, \quad m = 0, 1, \ldots, \frac{N}{2} - 1.$$

$$X_{2m+1} = \sum_{n=0}^{N/2-1} \left(x_n - x_{n+N/2}\right) W_N^n W_{N/2}^{mn}, \quad m = 0, 1, \ldots, \frac{N}{2} - 1.$$

   The DFT output sequence with even and odd index is obtained as $N/2$ DFT of $x_n + x_{n+N/2}$ and $\left(x_n - x_{n+N/2}\right) W_N^n$, respectively.

7.3   Hardware Implementations of FFT

When implementing the FFT block, many design issues such as complexity, latency, power, and memory should be considered. In 1970s, intensive researches about the FFT architecture were carried out. The Cooley–Tukey's FFT algorithm [8] is widely used for developing FFT architectures. Since then, many different types of FFT architectures have been developed. Pipelined architectures such as the radix-2 single-path delay feedback (R2SDF) [9] and the radix-2 multi-path delay commutator (R2MDC) [10] allow us to develop a high-speed FFT. Thus, it is suitable for a broadband communication system. The pipelined FFT architecture can be classified in parallelism and data flow. According to both parallelism and data flow, the complexity and latency of the FFT are decided. As we discussed in the previous section, there are two types of FFT architectures: the decimation-in-time FFT and the decimation-in-frequency FFT. Their hardware implementation is different each other. We can intuitively build two architectures (the decimation-in-time FFT and the decimation-in-frequency FFT) by signal flow. Figure 7.16 illustrates the radix-2 decimation-in-time 8-point FFT signal flow graph and hardware architecture. In the first stage, signal elements are paired ((x_0 x_4), (x_2 x_6), (x_1 x_5), and (x_3 x_7)). One element (x_4, x_6, x_5, and x_7) of the pair is multiplied by the twiddle factor. Then, they become inputs of the butterfly. In the second stage, the outputs of the butterfly in the first stage are rearranged and paired to become inputs of the second stage butterfly $\left(\left(x_0' \; x_2'\right), \left(x_4' \; x_6'\right), \left(x_1' \; x_3'\right), \text{and} \left(x_5' \; x_7'\right)\right)$. In the third stage, the outputs of the

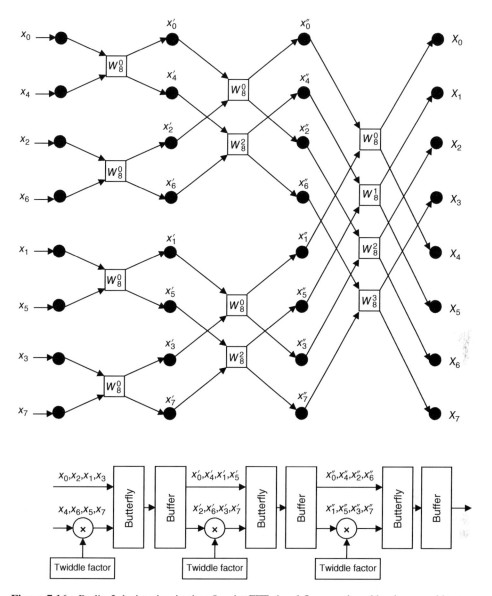

Figure 7.16   Radix-2 decimation-in-time 8-point FFT signal flow graph and hardware architecture

FFT (X_0, X_1, X_2, X_3, X_4, X_5, X_6, and X_7) can be obtained by rearranging the inputs $\left(\left(x_0'' \; x_1''\right),\left(x_4'' \; x_5''\right),\left(x_2'' \; x_3''\right), \text{and} \left(x_6'' \; x_7''\right)\right)$ and calculating the butterfly. In the same way, we calculate the radix-2 decimation-in-frequency 8-point FFT as shown in Figure 7.17. In the first stage, one element (x_4, x_5, x_6, and x_7) of the signal pair is multiplied by twiddle factor. The paired signals ((x_0 x_4), (x_1 x_5), (x_2 x_6), and (x_3 x_7)) become inputs of the butterfly. In the second stage, the rearranged signal pairs $\left(x_0' \; x_2'\right),\left(x_1' \; x_3'\right),\left(x_4' \; x_6'\right), \text{and} \left(x_5' \; x_7'\right)$ are calculated by the butterfly. In the third stage, the output of the FFT (X_0, X_4, X_2, X_6, X_1, X_5, X_3, and X_7) can be

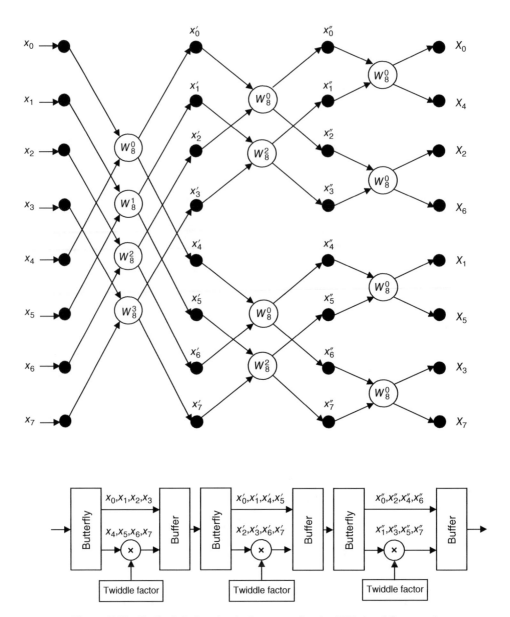

Figure 7.17 Radix-2 decimation-in-frequency 8-point FFT signal flow graph

obtained by rearranging the inputs $\left(\left(x_0'' \ x_1'' \right), \left(x_2'' \ x_3'' \right), \left(x_4'' \ x_5'' \right), \text{and} \left(x_6'' \ x_7'' \right) \right)$ and calculating the butterfly. These two architectures are designed in signal flow point of view. We need to consider efficiently using parallelism and memory usage when implementing FFT architectures. Thus, we look into the R2SDF and R2MDC in this section.

Figure 7.18 illustrates the radix-2 single-path delay feedback.

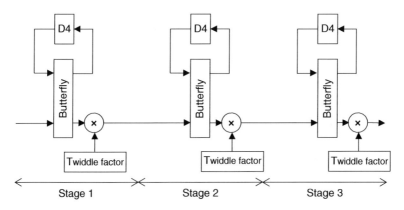

Figure 7.18   Radix-2 single-path delay feedback.

Table 7.3   Signal flow of R2SDF

	Stage 1	
	Delay 4	**Butterfly**
$x_7, x_6, x_5, x_4, x_3, x_2, x_1, x_0$	$x_7, \mathbf{x_6, x_5, x_4}, x_3, x_2, x_1, x_0$ $x_7, x_6, x_5, x_4, \mathbf{x_3, x_2, x_1, x_0}$	$x_7', x_6', x_5', x_4', x_3', x_2', x_1', x_0'$

	Stage 2	
	Delay 2	**Butterfly**
$x_7', x_6', x_5', x_4', x_3', x_2', x_1', x_0'$	$\mathbf{x_7', x_6'}, x_5', x_4', \mathbf{x_3', x_2'}, x_1', x_0'$ $x_7', x_6', \mathbf{x_5', x_4'}, x_3', x_2', \mathbf{x_1', x_0'}$	$x_7'', x_6'', x_5'', x_4'', x_3'', x_2'', x_1'', x_0''$

	Stage 3	
	Delay 1	**Butterfly**
$x_7'', x_6'', x_5'', x_4'', x_3'', x_2'', x_1'', x_0''$	$x_7'', \mathbf{x_6''}, x_5'', \mathbf{x_4''}, x_3'', \mathbf{x_2''}, \mathbf{x_1''}, x_0''$ $x_7'', \mathbf{x_6''}, x_5'', \mathbf{x_4''}, x_3'', \mathbf{x_2''}, x_1'', \mathbf{x_0''}$	$X_0, X_4, X_2, X_6, X_1, X_5, X_3, X_7$

The R2SDF has a single data path going through the butterfly and multiplier and a delay line arranging input pairs of the butterfly. Each stage has different delays (Delay 4, Delay 2, and Delay 1) which arrange the input pairs as follows: In the first stage, the first half of the delayed signal (x_7, x_6, x_5, and x_4) is paired with the second half of the original signal (x_3, x_2, x_1, and x_0). In the second stage, four elements of the delayed signal (x_7', x_6', x_3', and x_2') are paired with the four elements of the original signal (x_5', x_4', x_1', and x_0'). In the third stage, four elements of the delayed signal (x_7'', x_5'', x_3'', and x_1'') are paired with the four elements of the original signal (x_6'', x_4'', x_2'', and x_0''). Table 7.3 summarizes the signal flow of the R2SDF.

The delay line is simply implemented by shift registers. In order to reduce power consumption and area, SRAM can be used. Thus, it can perform the read and write operation in one clock cycle [11]. The R2SDF can efficiently use the registers. However, it is not convenient to implement a large size FFT because it requires a large size shift register and a high power is consumed by shifting a large size data.

The R2MDC uses two parallel data paths which are calculated in one butterfly. Figure 7.19 illustrates the radix-2 multi-path delay commutator. As we can observe from the figure, there is no feedback. Thus, it can be implemented using more pipelining and parallel processing techniques.

Table 7.4 summarizes the signal flow of the R2MDC. In the first stage, the switch sends the first signal (x_3, x_2, x_1, and x_0) to upper signal line and is delayed. The upper signal line is aligned with the second half of the lower signal line (x_7, x_6, x_5, and x_4). In the second stage, the lower signal line is delayed and the second half is switched with the upper signal line ((x_3', x_5') and (x_2', x_4')). Then, the upper signal line is delayed and aligned with the lower signal line. In the third stage, the lower signal line is delayed and two signal elements are switched with the upper signal line ((x_5', x_6') and (x_1', x_2')). Then, the upper signal line is delayed and aligned with the lower signal line. Finally, we obtain the output of the FFT. As we can observe Figure 7.19 and the Table 7.4, the butterfly in the first stage is not working until signals are paired. Thus, the hardware utilization rate of the R2MDC is 50%. For increasing the data throughput, the radix-4 multipath delay commutator can be used. This architecture can achieve

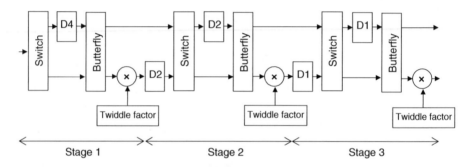

Figure 7.19   Radix-2 multi-path delay commutator

Table 7.4   Signal flow of radix-2 multi-path delay commutator

Stage 1

	Switch	Delay 4	Butterfly
$x_7, x_6, x_5, x_4, x_3, x_2, x_1, x_0$	x_3, x_2, x_1, x_0	x_3, x_2, x_1, x_0	x_3', x_2', x_1', x_0'
	x_7, x_6, x_5, x_4	x_7, x_6, x_5, x_4	x_7', x_6', x_5', x_4'

Stage 2

Delay 2	Switch	Delay 2	Butterfly
x_3', x_2', x_1', x_0'	x_5', x_4', x_1', x_0'	x_5', x_4', x_1', x_0'	$x_5'', x_4'', x_1'', x_0''$
x_7', x_6', x_5', x_4'	x_7', x_6', x_3', x_2'	x_7', x_6', x_3', x_2'	$x_7'', x_6'', x_3'', x_2''$

Stage 3

Delay 1	Switch	Delay 1	Butterfly
$x_5'', x_4'', x_1'', x_0''$	$x_6'', x_4'', x_2'', x_0''$	$x_6'', x_4'', x_2'', x_0''$	$X_0, X_4, X_2, X_6, X_1, X_5, X_3, X_7$
$x_7'', x_6'', x_3'', x_2''$	$x_7'', x_5'', x_3'', x_1''$	$x_7'', x_5'', x_3'', x_1''$	

higher data throughput due to four parallel data path but the hardware utilization rate is only 25%. Thus, the data throughput and the hardware utilization are trade-off relationship. One possible solution to increase both metrics is to use the duplicated input. That is to say the input signal is duplicated. The duplicated signal is aligned with the original signal. They are put into the FFT simultaneously with delay.

7.4 Problems

7.1. Consider an OFDM system with 1024 subcarriers (768 data, 83 pilots, and 173 guard subcarriers), subcarrier spacing = 125 Hz, guard interval = 1 ms, and QPSK modulation. Find the bandwidth and data rate of the OFDM system.

7.2. Consider an OFDM system with 128 subcarriers (96 data, 9 pilots, and 23 guard subcarriers), bandwidth = 1 MHz, bandwidth efficiency = 0.9 and 16QAM modulation. Find the data rate of the OFDM system.

7.3. Compare the PAPRs of an OFDM system with 128 and 1024 DFT size.

7.4. There are many PAPR reduction techniques such as clipping and filtering, block coding, interleaving, tone reservation, tone injection, selective mapping, and partial transmit sequences. Compare their advantages and disadvantages.

7.5. Compare the performance of zero padding and cyclic prefix.

7.6. Mobile WiMAX supports scalable channel bandwidths and adopts scalable OFDMA (SOFDMA). SOFDMA adjusts the DFT/IDFT size while subcarrier spacing is fixed. Describe the advantages and disadvantages of the SOFDMA.

7.7. Consider the following design requirements and design the OFDM system:

Requirements	Values
Speed	120 km/h
Carrier frequency	3.5 GHz
Tolerable delay spread	20 µs
Bandwidth	1 MHz
Target data rate	2 Mbps

7.8. Consider the following input sequence:

$$x_n = \left[x_0, x_1, x_2, x_3, x_4, x_5, x_6, x_7 \right]$$
$$= \left[1-i, 1+i, 1+i, 1-i, -1-i, -1+i, 1-i, 1-i \right].$$

7.9. Calculate the output sequences of the radix-2 8-point FFT, radix-2 decimation-in-frequency 8-point FFT and DFT and then compare them.

References

[1] H. Yaghoobi, "Scalable OFDMA Physical Layer in IEEE802.16 Wireless MAN," *Intel Technology Journal*, vol.**8**, no. 3, pp. 201–212, 2004.

[2] Y. Li and L. J. Cimini, "Bounds on the Interchannel Interference of OFDM in Time Varying Impairments," *IEEE Transactions on Communications*, vol. **49**, no.3, pp. 401–404, 2001.

[3] IEEE802.16 Broadband Wireless Access Working Group, "Channel Models for Fixed Wireless Application." Revision 4.0, IEEE802.16.3c-01/29r4, July 2001.

[4] ITU-R Recommendation M.1225, "Guidelines for Evaluation of Radio Transmission Technologies for IMT-2000, 1997." Feburary 1997. http://www.itu.int/rec/R-REC-M.1225/en (accessed April 20, 2015).

[5] R. V. Nee and R. Prasad, "*OFDM for Wireless Multimedia Communications*," Artech House Publishers, Boston, MA, 2000.

[6] P. Hoeher, S. Kaiser, and P. Robertson, "Two-dimensional Pilot-Symbol-Aided Channel Estimation by Wiener Filtering," *Proceedings of IEEE International Conference on Acoustics, Speech, and Signal Processing 1997 (ICASSP-97)*, vol. **3**, pp. 1845–1848, April 1997.

[7] R. Negi and J. Cioffi, "Pilot Tone Selection for Channel Estimation in a Mobile OFDM System," *IEEE Transactions on Consumer Electronics*, vol. **44**, no. 3, pp. 1122–1128, 1998.

[8] J. W. Cooley and J. Tukey, "An Algorithm for Machine Calculation of Complex Fourier Series," *Mathematics of Computation*, vol. **19**, no. 90, pp. 297–301, 1965.

[9] H. L. Groginsky and G. A. Works, "A Pipeline Fast Fourier Transform," *IEEE Transactions on Computers*, vol. **c-19**, no. 11, pp. 1015–1019, 1970.

[10] L. R. Rabiner and B. Gold, *Theory and Application of Digital Signal Processing*, Prentice-Hall, Englewood Cliffs, NJ, 1975.

[11] Y. T. Lin, P. Y. Tsai, and T. D. Chiueh, "Low-power Variable Length Fast Fourier Transform Processor," *IEE Proceedings Computers and Digital Techniques*, vol. **152**, no. 4, 2005.

8

Multiple Input Multiple Output

The Multiple Input Multiple Output (MIMO) techniques are an essential part of wireless communication systems because it significantly increases wireless communication system performances. The MIMO techniques enable us to obtain diversity gain, array gain, and multiplexing gain. Each different gain is related to different types of system performances. The diversity gain improves link reliability and transmission coverage by mitigating multipath fading. The array gain improves transmission coverage and QoS. The multiplexing gain increases spectral efficiency by transmitting independent signals via different antennas. In this chapter, we focus on space time codes and MIMO detection techniques. In addition, we look into the encoding and decoding process of a space time trellis code.

8.1   MIMO Antenna Design

The performance of MIMO systems depends on channel correlation caused by spatial correlation and antenna mutual coupling. In practical MIMO systems, each MIMO channel is related to another channel with different degrees. We call this spatial correlation. It depends on multipath channel environments. When one element of MIMO antennas transmits a signal, a power is radiated everywhere and a neighboring antenna element absorbs the power. This is undesirable situation. We call this antenna mutual coupling. It is caused by the interaction among transmit antennas [1]. This effect becomes a very serious problem if antenna spacing is very small. For example, a mobile terminal with many antennas has very small antenna spacing. Basically, a higher MIMO channel correlation means a lower channel capacity. If MIMO channels are fully correlated as an extreme case, the MIMO system is not different from a single antenna system. In MIMO systems, the channel transfer matrix should be of full rank in order to achieve the best performance. In other words, the transmitter should send signals without spatial correlation and the received signals should be uncorrelated. In order to

Wireless Communications Systems Design, First Edition. Haesik Kim.
© 2015 John Wiley & Sons, Ltd. Published 2015 by John Wiley & Sons, Ltd.

satisfy this requirement, antenna isolation (spacing or mutual coupling) is very important. Besides, we should keep many requirements in our mind such as antenna size, antenna radiation pattern, the number of antenna elements, antenna element configuration, antenna element polarization, *in situ* antenna performance, and compliance with regulation. Although the design rule of the MIMO antenna is not clearly defined, we should try to maximize the performance (high diversity order, high received SNR, high effective rank, etc.) and minimize the cost (small size, light weight, etc.). Sometimes, the complex mathematical analysis of the MIMO antenna prevents us to understand them intuitively. Thus, we describe simple MIMO antenna design requirements. In Refs [2] and [3], MIMO antenna design guidelines are described. Firstly, there are many types of antenna elements such as monopoles, patches, slots, and helices. In order to select a suitable antenna type, we should consider radiation pattern, gain, directivity, bandwidth, and so on. Among them, radiation pattern is very important. The radiation pattern should be omnidirectional because there is no location information about a receiver. If beamforming is considered, omnidirectional antenna is not suitable. Among many types of antennas, monopole or patch antennas are a suitable choice for a mobile terminal. Especially, a patch antenna is widely used in a smart phone due to many advantages such as easy implementation, various shapes, low weight, and cheap cost. Secondly, we consider antenna array configuration. The MIMO channel matrix depends on signal propagation as well as antenna array configuration. It is not easy to find an optimal antenna array configuration because it depends on specific channel propagation characteristics. Thirdly, we consider antenna mutual coupling caused by spacing, polarization, and antenna type. It degrades the capacity seriously. Thus, those antenna design parameters should be carefully chosen. One simple solution in a base station is that the antennas should be spaced enough. However, a mobile terminal has the limited device size. The number of MIMO antennas in a mobile terminal increases but the mobile terminal size is still fixed. Thus, the antenna spacing becomes narrow. It is a key research challenge.

8.2 Space Time Coding

Multipath fading was conventionally mitigated by temporal diversity, frequency diversity, and spatial diversity in a receiver. Since the advent of Space Time Codes (STCs), spatial diversity in a transmitter becomes an important part of wireless communication systems and it is widely used in the modern wireless communication systems. Especially, spatial diversity technique is more suitable for base stations to improve their link quality because the use of multiple antennas makes mobile stations larger and more expensive. An STC improves the reliability of a wireless link using multiple antennas. It performs jointly design of encoding, modulation, and diversity in space-time domain. It achieves both diversity gain and coding gain. There are two types of STCs: Space Time Block Codes (STBCs) and Space Time Trellis Codes (STTCs). STBCs transmit an orthogonal code and achieve full diversity gain with a simple maximum-likelihood decoding scheme. However, it does not provide coding gain. On the other hand, STTCs is the extended version of convolutional codes for multiple antennas and provides full diversity gain and coding gain like Trellis Coded Modulation (TCM). However, it requires a complex decoding scheme. The complexity exponentially increases as the data rate increases. In STCs, the knowledge of Channel State Information (CSI) is critical to the performance of STCs. According to the CSI knowledge of the receiver, STCs can be classified into coherent STCs, non-coherent STCs, and differential STCs. The receiver of coherent STCs can accurately estimate CSI using pilots.

The receiver of non-coherent STCs can estimate CSI statistically but it is not accurate. The receiver of differential STCs does not have any knowledge of CSI. Thus, a differential space time block code [4] uses differential modulation schemes which do not require channel estimation. Basically, a receiver can estimate CSI using pilots or statistics but a transmitter needs feedback information from a receiver. Thus, STCs can be classified into full CSI cases and no CSI cases. When a transmitter knows full CSI, the maximal capacity can be achieved by water-filling strategy [5]. However, if there is no CSI, it is not possible to allocate optimal powers to multiple antennas. A uniform power allocation is used.

The MIMO channel with N_t transmit antennas and N_r receive antennas is illustrated in Figure 8.1.

In the figure, h^t_{ji}, s^i_t, and r^j_t denote channel gain from transmit antenna i to receive antenna j, a transmitted symbol and a received symbol, respectively. The index t represents frame length in time domain. The received signal at each antenna can be expressed as follows:

$$r^j_t = \sum_{i=1}^{N_t} h^t_{ji} s^i_t + n^j_t \tag{8.1}$$

where n^j_t represents an independent complex Gaussian noise and $1 \le j \le N_r$. The MIMO channel is represented by H_t matrix as follows:

$$H_t = \begin{bmatrix} h^t_{11} & h^t_{12} & \cdots & h^t_{1N_t} \\ h^t_{21} & h^t_{22} & & h^t_{2N_t} \\ \vdots & & \ddots & \vdots \\ h^t_{N_r 1} & h^t_{N_r 2} & \cdots & h^t_{N_r N_t} \end{bmatrix} \tag{8.2}$$

and the received signals are expressed in the matrix form as follows:

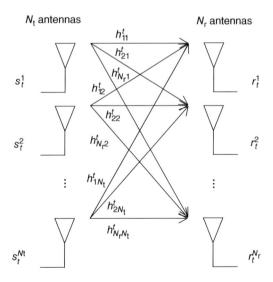

Figure 8.1   MIMO channel

$$
\begin{bmatrix} r_t^1 \\ r_t^2 \\ \vdots \\ r_t^{N_r} \end{bmatrix} = \begin{bmatrix} h_{11}^t & h_{12}^t & \cdots & h_{1N_t}^t \\ h_{21}^t & h_{22}^t & & h_{2N_t}^t \\ \vdots & & \ddots & \vdots \\ h_{N_r 1}^t & h_{N_r 2}^t & \cdots & h_{N_r N_t}^t \end{bmatrix} \begin{bmatrix} s_t^1 \\ s_t^2 \\ \vdots \\ s_t^{N_t} \end{bmatrix} + \begin{bmatrix} n_t^1 \\ n_t^2 \\ \vdots \\ n_t^{N_r} \end{bmatrix} \tag{8.3}
$$

The received signal vector is expressed as follows:

$$
r_t = H_t s_t + n_t. \tag{8.4}
$$

Every linear transformation can be expressed as a composition of rotation and scaling operations. The Singular Value Decomposition (SVD) [6] is commonly used to decompose the MIMO channel as follows:

$$
H_t = U_t \Lambda_t V_t^H \tag{8.5}
$$

where U_t and V_t represent the left and right eigenvector space, respectively. $(\ )^H$ denotes Hermitian operation. They have the following property:

$$
U_t^H U_t = I \text{ and } V_t V_t^H = I. \tag{8.6}
$$

In (8.5), Λ_t is a rectangular matrix with non-negative real numbers as diagonal elements and zero as off-diagonal elements. The diagonal elements of Λ_t are $\lambda_1 \geq \lambda_2 \geq \cdots \geq \lambda_{n_{min}}$ where $n_{min} = \min(N_t, N_r)$. This means that n_{min} parallel channels are created. We rewrite (8.4) as follows:

$$
r_t = U_t \Lambda_t V_t^H s_t + n_t \tag{8.7}
$$

$$
U_t^H r_t = U_t^H U_t \Lambda_t V_t^H s_t + U_t^H n_t \tag{8.8}
$$

$$
U_t^H r_t = \Lambda_t V_t^H s_t + U_t^H n_t \tag{8.9}
$$

where $U_t^H r_t$ is the projection of r_t on the left eigenvector space $U\left(\tilde{r}_t = U_t^H r_t\right)$ and $V_t^H s_t$ is the projection of s_t on the right eigenvector space $V\left(\tilde{s}_t = V_t^H s_t\right)$. We define $\tilde{n}_t = U_t^H n_t$ and rewrite (8.9) as follows:

$$
\tilde{r}_t = \Lambda_t \tilde{s}_t + \tilde{n}_t \tag{8.10}
$$

where $\tilde{n}_t$ is a complex Gaussian vector. Equation 8.10 is rewritten as n_{min} parallel channels as follows:

$$
\tilde{r}_t^k = \lambda_k \tilde{s}_t^k + \tilde{n}_t^k, k = 1, 2, \ldots, n_{min}. \tag{8.11}
$$

Figure 8.2 illustrates the MIMO channel conversion through SVD.

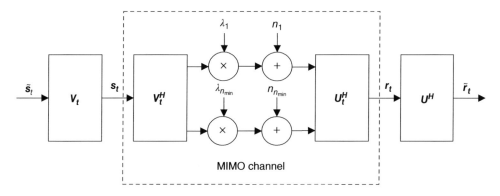

Figure 8.2   MIMO channel conversion through SVD

Now, we define a system model and design the STBC and STTC. As we can observe in Figure 8.3, the information source generates m binary symbols as follows:

$$c_t = \left(c_t^1, c_t^2, \ldots, c_t^m \right). \tag{8.12}$$

In the space time encoder, m binary symbols c_t are mapped into N_t modulation symbols from a signal set of $M = 2^m$ and the transmit vector is represented as follows:

$$s_t = \left(s_t^1, s_t^2, \ldots, s_t^{N_t} \right)^{\mathrm{T}}. \tag{8.13}$$

We assume the MIMO channel is memoryless and the MIMO system operates over a slowly-varying flat-fading MIMO channel. The transmit vector has L frame length at each antenna. The $N_t \times L$ space time codeword matrix is defined as follows:

$$S = [s_1, s_2, \ldots, s_L] = \begin{bmatrix} s_1^1 & s_2^1 & \cdots & s_L^1 \\ s_1^2 & s_2^2 & & s_L^2 \\ \vdots & & \ddots & \vdots \\ s_1^{N_t} & s_2^{N_t} & \cdots & s_L^{N_t} \end{bmatrix} \tag{8.14}$$

and the MIMO channel matrix is given as (8.2). The maximum likelihood decoding scheme and perfect CSI are assumed at the receiver. Thus, the receiver has the following decision metric:

$$\sum_{t=1}^{L} \sum_{j=1}^{N_r} \left| r_t^j - \sum_{i=1}^{N_t} h_{ji}^t s_t^i \right|^2. \tag{8.15}$$

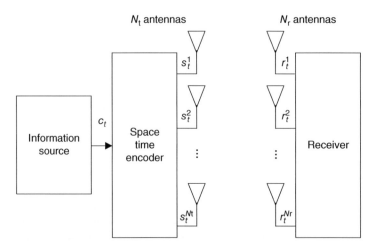

Figure 8.3   System model of space time coding

The ML decoder finds the codewords to minimize (8.15). We can calculate the Pairwise Error Probability (PEP) that the receiver decides erroneously as follows:

$$P(S,\hat{S}\,|\,H) = Q\!\left(\frac{d(S,\hat{S})}{2\sigma}\right) \le \exp\!\left(-\frac{d^2(S,\hat{S})E_s}{4N_0}\right) \tag{8.16}$$

where $\hat{S}$ is the estimated erroneous sequence as follows:

$$\hat{S} = \left[\hat{s}_1, \hat{s}_2, \ldots, \hat{s}_L\right] = \begin{bmatrix} \hat{s}_1^1 & \hat{s}_2^1 & \cdots & \hat{s}_L^1 \\ \hat{s}_1^2 & \hat{s}_2^2 & & \hat{s}_L^2 \\ \vdots & & \ddots & \vdots \\ \hat{s}_1^{N_t} & \hat{s}_2^{N_t} & \cdots & \hat{s}_L^{N_t} \end{bmatrix} \tag{8.17}$$

and $d^2(S,\hat{S})$ is the modified Euclidean distance as follows:

$$d^2(S,\hat{S}) = \sum_{t=1}^{L}\sum_{j=1}^{N_r}\left|\sum_{i=1}^{N_t} h_{ji}^t\left(\hat{s}_t^i - s_t^i\right)\right|^2 = \left\|H(\hat{S}-S)\right\|^2 \tag{8.18}$$

where H is channel response sequences at each time as follows:

$$H = [H_1, H_2, \ldots, H_L]. \tag{8.19}$$

In (8.16), $Q(\ )$ is the complementary error function and σ is standard deviation of Gaussian noise. We define the codeword difference matrix $B(S,\hat{S})$ and codeword distance matrix $A(S,\hat{S})$ as follows:

$$B(S,\hat{S}) = S - \hat{S} = \begin{bmatrix} s_1^1 - \hat{s}_1^1 & s_2^1 - \hat{s}_2^1 & \cdots & s_L^1 - \hat{s}_L^1 \\ s_1^2 - \hat{s}_1^2 & s_2^2 - \hat{s}_2^2 & & s_L^2 - \hat{s}_L^2 \\ \vdots & & \ddots & \vdots \\ s_1^{N_t} - \hat{s}_1^{N_t} & s_2^{N_t} - \hat{s}_2^{N_t} & \cdots & s_L^{N_t} - \hat{s}_L^{N_t} \end{bmatrix} \tag{8.20}$$

and

$$A(S,\hat{S}) = B(S,\hat{S})B^H(S,\hat{S}) \tag{8.21}$$

$$= \begin{bmatrix} \left(s_1^1 - \hat{s}_1^1\right)\overline{\left(s_1^1 - \hat{s}_1^1\right)} & \left(s_2^1 - \hat{s}_2^1\right)\overline{\left(s_2^1 - \hat{s}_2^1\right)} & \cdots & \left(s_L^1 - \hat{s}_L^1\right)\overline{\left(s_L^1 - \hat{s}_L^1\right)} \\ \left(s_1^2 - \hat{s}_1^2\right)\overline{\left(s_1^2 - \hat{s}_1^2\right)} & \left(s_2^2 - \hat{s}_2^2\right)\overline{\left(s_2^2 - \hat{s}_2^2\right)} & & \left(s_L^2 - \hat{s}_L^2\right)\overline{\left(s_L^2 - \hat{s}_L^2\right)} \\ \vdots & & \ddots & \vdots \\ \left(s_1^{N_t} - \hat{s}_1^{N_t}\right)\overline{\left(s_1^{N_t} - \hat{s}_1^{N_t}\right)} & \left(s_2^{N_t} - \hat{s}_2^{N_t}\right)\overline{\left(s_2^{N_t} - \hat{s}_2^{N_t}\right)} & \cdots & \left(s_L^{N_t} - \hat{s}_L^{N_t}\right)\overline{\left(s_L^{N_t} - \hat{s}_L^{N_t}\right)} \end{bmatrix} \tag{8.22}$$

where $\overline{()}$ is complex conjugate. After simple manipulations [7], the modified Euclidean distance (8.18) is rewritten as follows:

$$d^2(S,\hat{S}) = \sum_{j=1}^{N_r} h_j A(S,\hat{S}) h_j^H \tag{8.23}$$

where $h_j = [h_{j1}, h_{j2}, \ldots, h_{jN_t}]$ because a slow fading channel is assumed and channel gains at each frame are constant as follows:

$$h_{ji}^1 = h_{ji}^2 = \cdots = h_{ji}^L. \tag{8.24}$$

The matrix $A(S,\hat{S})$ is written as follows:

$$V \cdot A(S,\hat{S}) \cdot V^H = \Lambda \tag{8.25}$$

where Λ is a diagonal matrix as follows:

$$\Lambda = \begin{bmatrix} \lambda_1 & 0 & \cdots & 0 \\ 0 & \lambda_2 & & 0 \\ \vdots & & \ddots & \vdots \\ 0 & 0 & \cdots & \lambda_{N_t} \end{bmatrix}, \lambda_i \geq 0 \text{ is the eigenvalues of } A(S,\hat{S}) \tag{8.26}$$

and V is an orthonormal matrix ($V^H V = I$). The row vectors $[v_1, v_2, \ldots, v_{N_t}]$ of V are the eigenvectors of $A(S, \hat{S})$. Let $[\beta_{j1}, \beta_{j2}, \ldots, \beta_{jN_t}] = h_j V^H$, we express the term $h_j A(S, \hat{S}) h_j^H$ as follows:

$$h_j A(S, \hat{S}) h_j^H = h_j V^H \Lambda V h_j^H = \sum_{i=1}^{N_t} \lambda_i |\beta_{ji}|^2. \tag{8.27}$$

Thus, (8.23) is rewritten as follows:

$$d^2(S, \hat{S}) = \sum_{j=1}^{N_r} \sum_{i=1}^{N_t} \lambda_i |\beta_{ji}|^2 \tag{8.28}$$

and (8.16) is rewritten using (8.28) as follows:

$$P(S, \hat{S}|H) \leq \prod_{j=1}^{N_r} \exp\left(-\frac{E_s}{4N_0} \sum_{i=1}^{N_t} \lambda_i |\beta_{ji}|^2\right). \tag{8.29}$$

For Rayleigh fading [7], the upper bound of PEP is written as follows:

$$P(S, \hat{S}) \leq \left(\frac{1}{\prod_{i=0}^{N_t}(1 + (E_s/4N_0)\lambda_i)}\right)^{N_r}. \tag{8.30}$$

At a high SNR, we express the term $1 + (E_s/4N_0)\lambda_i$ as follows:

$$1 + \frac{E_s}{4N_0}\lambda_i \approx \frac{E_s}{4N_0}\lambda_i \tag{8.31}$$

and (8.30) is simplified as follows:

$$P(S, \hat{S}) \leq \left(\prod_{i=0}^{N_t}\frac{E_s}{4N_0}\lambda_i\right)^{-N_r} = \left(\prod_{i=1}^{r}\lambda_i\right)^{-N_r}\left(\frac{E_s}{4N_0}\right)^{-rN_r} = \left(\prod_{i=1}^{r}\lambda_i^{1/r}\right)^{-rN_r}\left(\frac{E_s}{4N_0}\right)^{-rN_r} \tag{8.32}$$

where r is the rank of $A(S, \hat{S})$. From (8.32), we can achieve diversity gain of rN_r and coding gain of $\prod_{i=1}^{r} \lambda_i^{1/r}$. In other words, diversity gain is the exponent of PEP upper bound and coding gain is independent of SNR in PEP upper bound. Thus, we design to maximize diversity gain through maximizing the rank of a codeword difference matrix and then maximize coding gain through maximizing the minimum product of the Euclidean distance (or $\prod_{i=1}^{r} \lambda_i$). Both diversity gain and coding gain affect BER versus SNR performance curve differently. The diversity gain changes the slope of BER curve and the coding gain shifts it horizontally. A larger diversity gain and a greater coding gain make it more a negative slope and a larger left shift, respectively. Now, we define three design criteria for the space time coding in slow Rayleigh fading. The first design criterion is *the rank criterion*. The maximum diversity $N_t N_r$ can be achieved if a codeword difference matrix $B(S,\hat{S})$ has full rank for any two codeword vector sequences S and $\hat{S}$. If it has the minimum rank r over two tuples of distinct codeword vector sequences, the diversity is rN_r. The second design criterion is about coding gain. Since the determinant of $A(S,\hat{S})$ is the product of the eigenvalues, we calculate the determinant and call it *the determinant criterion*. Assume rN_r is the target diversity gain. We should maximize the minimum determinant $\Pi_{j=1}^{r} \lambda_i$ of $A(S,\hat{S})$ along the pairs of distinct codewords with the minimum rank. Therefore, we can minimize the PEP. This determinant criterion is related to coding gain but does not calculate an accurate coding gain. Thus, this criterion should be considered as one design rule of the space time coding. In Ref. [8], one more design criterion was discussed. The third design criterion is *the trace criterion*. In order to maximize the minimum Euclidean distance among all possible codewords, the minimum trace $\Pi_{i=1}^{r} \lambda_i$ of $A(S,\hat{S})$ should be maximized.

In Chapter 5, a 2×1 simple transmit diversity scheme by Alamouti was discussed. He extended his scheme to $2\times N_r$ codes and achieved a diversity order of $2N_r$ where a diversity order simply means the number of independent signal paths between a transmitter and a receiver. After that, V. Tarokh et al. [9, 10] generalized the simple transmit diversity scheme using orthogonal designs. Thus, STBCs with any number of antennas can be constructed and they achieve a full diversity. A generalized STBC is defined by a $p\times N_t$ transmission matrix S_p which generates N_t parallel signal sequences with the length p and whose elements are linear combinations of the k modulated signals $(s_1, s_2, \ldots, s_k)$ and their conjugates $\left(s_1^*, s_2^*, \ldots, s_k^*\right)$. The data rate of the STBC is defined as k/p. In a STBC encoder, the k modulated signal is encoded by a $p\times N_t$ transmission matrix S_p as follows:

$$[s_1, s_2, \ldots, s_k] \rightarrow S_p = \begin{bmatrix} s_1^1 & s_2^1 & \cdots & s_p^1 \\ s_1^2 & s_2^2 & & s_p^2 \\ \vdots & & \ddots & \vdots \\ s_1^{N_t} & s_2^{N_t} & \cdots & s_p^{N_t} \end{bmatrix}. \tag{8.33}$$

In (8.33), S_p is constructed to satisfy

$$S_p \cdot S_p^H = c\left(|s_1|^2 + |s_2|^2 + \cdots + |s_k|^2\right) \cdot I_p \tag{8.34}$$

where c and I_p are constant and an identity matrix, respectively. The STBC can achieve a full diversity order of pN_r if S_p satisfies (8.34). When s_i is real and p is equal to N_t, then S_2, S_4, and S_8 are defined as follows [9]:

$$S_2 = \begin{bmatrix} s_1 & s_2 \\ -s_2 & s_1 \end{bmatrix}, \tag{8.35}$$

$$S_4 = \begin{bmatrix} s_1 & s_2 & s_3 & s_4 \\ -s_2 & s_1 & -s_4 & s_3 \\ -s_3 & s_4 & s_1 & -s_2 \\ -s_4 & -s_3 & s_2 & s_1 \end{bmatrix}, \tag{8.36}$$

and

$$S_8 = \begin{bmatrix} s_1 & s_2 & s_3 & s_4 & s_5 & s_6 & s_7 & s_8 \\ -s_2 & s_1 & s_4 & -s_3 & s_6 & -s_5 & -s_8 & s_7 \\ -s_3 & -s_4 & s_1 & s_2 & s_7 & s_8 & -s_5 & -s_6 \\ -s_4 & s_3 & -s_2 & s_1 & s_8 & -s_7 & s_6 & -s_5 \\ -s_5 & -s_6 & -s_7 & -s_8 & s_1 & s_2 & s_3 & s_4 \\ -s_6 & s_5 & -s_8 & s_7 & -s_2 & s_1 & -s_4 & s_3 \\ -s_7 & s_8 & s_5 & -s_6 & -s_3 & s_4 & s_1 & -s_2 \\ -s_8 & -s_7 & s_6 & s_5 & -s_4 & -s_3 & s_2 & s_1 \end{bmatrix}. \tag{8.37}$$

When s_i is complex, S_p construction is similar to the construction when it is real. The elements of S_p contain their conjugates and S_2^c and S_4^c are defined as follows [9]:

$$S_2^c = \begin{bmatrix} s_1 & s_2 \\ -s_2^* & s_1^* \end{bmatrix} \tag{8.38}$$

and

$$S_4^c = \begin{bmatrix} s_1 & s_2 & s_3 & s_4 \\ -s_2 & s_1 & -s_4 & s_3 \\ -s_3 & s_4 & s_1 & -s_2 \\ -s_4 & -s_3 & s_2 & s_1 \\ s_1^* & s_2^* & s_3^* & s_4^* \\ -s_2^* & s_1^* & -s_4^* & s_3^* \\ -s_3^* & s_4^* & s_1^* & -s_2^* \\ -s_4^* & -s_3^* & s_2^* & s_1^* \end{bmatrix}. \tag{8.39}$$

In the $p \times N_t$ transmission matrix S_p, each row represents a time slot and each column represents a transmit antenna. Thus, the code rate of $S_2, S_2^c, S_4,$ and S_8 is 1 and the code rate of S_4^c is 1/2.

Example 8.1   STBC diversity gain

Consider a 2×1 space time block code and any two distinct codewords S and $\hat{S}$ by S_2^c construction as follows:

$$S = \begin{bmatrix} s_1 & s_2 \\ -s_2^* & s_1^* \end{bmatrix} \quad \text{and} \quad \hat{S} = \begin{bmatrix} \hat{s}_1 & \hat{s}_2 \\ -\hat{s}_2^* & \hat{s}_1^* \end{bmatrix}$$

where $(s_1, s_2) \ne (\hat{s}_1, \hat{s}_2)$. Find the diversity gain.

Solution

From (8.20) and (8.21), the codeword difference matrix $B(S,\hat{S})$ is

$$B(S,\hat{S}) = \begin{bmatrix} s_1 - \hat{s}_1 & s_2 - \hat{s}_1 \\ -s_2^* + \hat{s}_2^* & s_1^* - \hat{s}_2^* \end{bmatrix}$$

and the codeword distance matrix $A(S,\hat{S})$ is

$$A(S,\hat{S}) = B(S,\hat{S})B^H(S,\hat{S}) = \begin{bmatrix} s_1 - \hat{s}_1 & s_2 - \hat{s}_2 \\ -s_2^* + \hat{s}_2^* & s_1^* - \hat{s}_1^* \end{bmatrix}\begin{bmatrix} s_1^* - \hat{s}_1^* & -s_2 + \hat{s}_2 \\ s_2^* - \hat{s}_2^* & s_1 - \hat{s}_1 \end{bmatrix}$$

$$= \begin{bmatrix} |s_1 - \hat{s}_1|^2 + |s_2 - \hat{s}_2|^2 & 0 \\ 0 & |s_1 - \hat{s}_1|^2 + |s_2 - \hat{s}_2|^2 \end{bmatrix}.$$

The determinant of $A(S,\hat{S})$ is

$$\det(A(S,\hat{S})) = \left(|s_1 - \hat{s}_1|^2 + |s_2 - \hat{s}_2|^2 \right)^2 \ne 0$$

and the rank of $A(S,\hat{S})$ is 2. Therefore, the diversity gain is $2 \, (= rN_r = 2 \cdot 1)$ from (8.32).   ∎

The STTCs are able to mitigate the effects of fading and provide us with a significant performance improvement. Now, we discuss the STTC design. The STTC encoder is similar to the TCM encoder. Figure 8.4 illustrates an STTC encoder.

As shown in Figure 8.4, the input sequence c_t is a block of information (or coded bits) at time t and is denoted by $\left(c_t^1, c_t^2, \ldots, c_t^m \right)$. The kth input sequence c_t^k goes through the kth shift register and is multiplied by the STTC encoder coefficient set g^k. It is defined as follows:

$$g^k = \left[\left(g_{0,1}^k, g_{0,2}^k, \ldots, g_{0,N_t}^k \right), \left(g_{1,1}^k, g_{1,2}^k, \ldots, g_{1,N_t}^k \right), \ldots, \left(g_{v_m,1}^k, g_{v_m,2}^k, \ldots, g_{v_m,N_t}^k \right) \right]. \qquad (8.40)$$

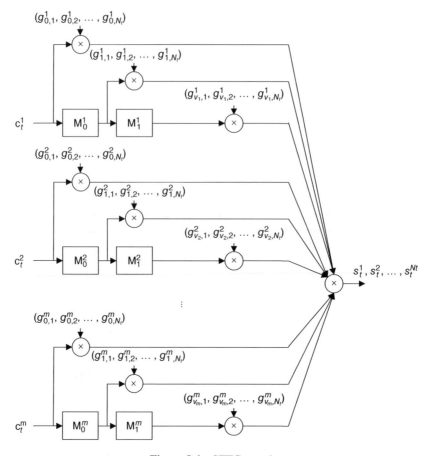

Figure 8.4   STTC encoder

Each $g_{l,i}^k$ represents an element of M-ary signal constellation set and v_m represents the memory order of the kth shift register. If QPSK modulation is considered, it has one of signal constellation set $\{0,1,2,3\}$. The STTC encoder maps them into an M-ary modulated symbol and is denoted by $\left(s_t^1, s_t^2, \ldots, s_t^{N_t}\right)$. The output of the STTC encoder is calculated as follows:

$$s_t^i = \sum_{k=1}^{m}\sum_{l=0}^{v_k} g_{l,i}^k c_{t-j}^k \bmod 2^m \tag{8.41}$$

where $i = 1, 2, \ldots, N_t$. The outputs of the multipliers are summed modulo 2^m. The modulated symbols s_t^i are transmitted in parallel through N_t transmit antennas. The total memory order of the STTC encoder is

$$v = \sum_{k=1}^{m} v_k \tag{8.42}$$

and v_k is defined as follows:

$$v_k = \frac{v+k-1}{m}.$$ (8.43)

The trellis state of the STTC encoder is 2^v. We assume that r_t^j is the received signal at the received antenna j at time t, the receiver obtains perfect CSI, and the branch metric is calculated as the squared Euclidean distance between the actual received signal and the hypothesized received signals as follows:

$$\sum_{j=1}^{N_r} \left| r_t^j - \sum_{i=1}^{N_t} h_{ji}^t s_t^i \right|^2.$$ (8.44)

The STTC decoder uses the Viterbi algorithm to select the path with the lowest path metric. In order to achieve full diversity, the STTC encoder requires a long memory and a large number of antennas. Thus, many states and branches of the Trellis diagram are needed. The complexity of the STTC decoder increases in proportion to them. This is one disadvantage of the STTC.

Example 8.2   STTC diversity gain

Consider a 2×1 space time trellis coded QPSK scheme with two transmit antennas with the following coefficient set of the generator matrix:

$$g^1 = [(0,2),(2,0)] \quad \text{and} \quad g^2 = [(0,1),(1,0)].$$

Find the diversity gain.

Solution

The STTC encoder can be implemented by a shift register, multiplier, and adder. Figure 8.5 illustrates an example of the STTC encoder with a constraint length of 2 and an input block length of 2. As we can observe from Figure 8.5, each output of the STTC encoder depends on both the current input block and the previous input block. This process is pretty much similar to conventional convolutional encoding. The STTC can be represented by the Trellis diagram as shown in Figure 8.6. Figure 8.7 illustrates QPSK signal constellation and mapping for the Trellis diagram. Similar to convolutional codes, each state and branch in the Trellis diagram represents by the memory state and output block, respectively.

The error event probability of Trellis codes is equivalent to the pairwise error probability. Thus, we consider two paths in the Trellis diagram as shown in Figure 8.8.

The dashed lines in Figure 8.8 diverge at time t_1 and remerge at time t_2. The first dashed line represents all zero input sequence (00, 00) and their output sequences are calculated using the following equation:

$$\left(s_t^1, s_t^2\right) = c_t^1(0,2) + c_{t-1}^1(2,0) + c_t^2(0,1) + c_{t-1}^2(1,0) \bmod 4.$$

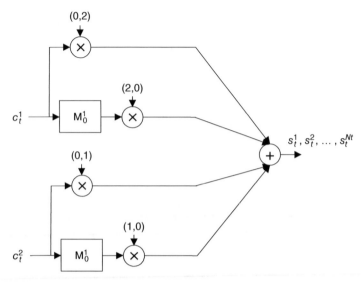

Figure 8.5 Example of STTC encoder

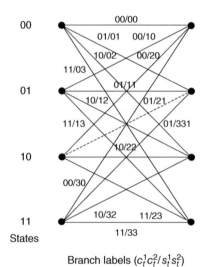

Branch labels $(c_t^1 c_t^2 / s_t^1 s_t^2)$

Figure 8.6 Trellis diagram for STTC with four states, QPSK, and two transmit antennas

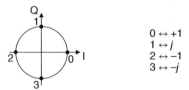

Figure 8.7 QPSK signal constellation and mapping

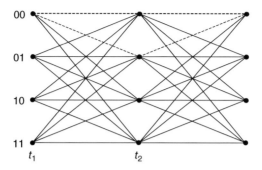

Figure 8.8 Path diverging at time t_1 and remerging at time t_2

The output at time t_1 is

$$\left(s_1^1, s_1^2\right) = c_1^1(0,2) + c_0^1(2,0) + c_1^2(0,1) + c_0^2(1,0) \bmod 4$$
$$= 0 \cdot (0,2) + 0 \cdot (2,0) + 0 \cdot (0,1) + 0 \cdot (1,0) \bmod 4$$
$$= (0,0) \leftrightarrow (+1,+1)$$

and the output at time t_2 is

$$\left(s_2^1, s_2^2\right) = c_2^1(0,2) + c_1^1(2,0) + c_2^2(0,1) + c_1^2(1,0) \bmod 4$$
$$= 0 \cdot (0,2) + 0 \cdot (2,0) + 0 \cdot (0,1) + 0 \cdot (1,0) \bmod 4$$
$$= (0,0) \leftrightarrow (+1,+1).$$

Thus, the output sequences are $(+1, +1, +1, +1)$. The other diverging dashed line represents the nonzero input sequence $(01, 00)$. Their output sequences are calculated as follows:
The output at time t_1 is

$$\left(s_1^1, s_1^2\right) = c_1^1(0,2) + c_0^1(2,0) + c_1^2(0,1) + c_0^2(1,0) \bmod 4$$
$$= 0 \cdot (0,2) + 0 \cdot (2,0) + 1 \cdot (0,1) + 0 \cdot (1,0) \bmod 4$$
$$= (0,1) \leftrightarrow (+1, j)$$

and the output at time t_2 is

$$\left(s_2^1, s_2^2\right) = c_2^1(0,2) + c_1^1(2,0) + c_2^2(0,1) + c_1^2(1,0) \bmod 4$$
$$= 0 \cdot (0,2) + 0 \cdot (2,0) + 0 \cdot (0,1) + 1 \cdot (1,0) \bmod 4$$
$$= (1,0) \leftrightarrow (j,+1).$$

Thus, the codeword difference matrix B is

$$B = \begin{bmatrix} 0 & 1-j \\ 1-j & 0 \end{bmatrix}$$

and the codeword distance matrix A is

$$A = BB^H = \begin{bmatrix} 0 & 1-j \\ 1-j & 0 \end{bmatrix}\begin{bmatrix} 0 & 1+j \\ 1+j & 0 \end{bmatrix} = \begin{bmatrix} 2 & 0 \\ 0 & 2 \end{bmatrix}.$$

Thus, the STTC has rank 2 and the diversity gain is 2 ($= rN_r = 2 \cdot 1$) from (8.32). This is full diversity. The determinant of the STTC is 4 ($= 2 \cdot 2 - 0 \cdot 0$). A good design of the STTC is to achieve full diversity and then increase coding gain for full diversity through maximizing the minimum determinant of the codeword distance matrix A. ∎

Summary 8.1   Space time coding design

$$P(S,\hat{S}) \le \left(\prod_{i=1}^{r} \lambda_i^{1/r}\right)^{-rN_r}\left(\frac{E_s}{4N_0}\right)^{-rN_r}$$

1. The rank criterion: The maximum diversity $N_t N_r$ can be achieved if a codeword difference matrix $B(S,\hat{S})$ has full rank for any two codeword vector sequences S and $\hat{S}$. If it has a minimum rank r over two tuples of distinct codeword vector sequences, the diversity is rN_r.
2. The determinant criterion: The second design criterion is about coding gain. Since the determinant of $A(S,\hat{S})$ is the product of the eigenvalues, we calculate the determinant. Assume rN_r is the target diversity gain. We should maximize the minimum determinant $\prod_{i=1}^{r} \lambda_i$ of $A(S,\hat{S})$ along the pairs of distinct codewords with minimum rank. Therefore, we minimize the PEP. This determinant criterion is related to coding gain but does not calculate an accurate coding gain.
3. The trace criterion: In order to maximize the minimum Euclidean distance among all possible codewords, the minimum trace $\prod_{i=1}^{r} \lambda_i$ of $A(S,\hat{S})$ should be maximized.

8.3   Example of STTC Encoding and Decoding

We consider a 2×2 space time trellis code with Gray coding QPSK and same coefficient set of the generator matrix as the Example 8.2. The input sequence [10, 01, 11, 01, 00] is encoded by the space time trellis encoder with the coefficient set of the generator matrix in Example 8.2. The output sequence is [02, 21, 13, 31, 10]. Figure 8.9 illustrates input/output sequences in the trellis diagram. We can observe a unique path.

In the trellis diagram, a branch label represents the input $(c_t^1 c_t^2)$ and output $(s_t^1 s_t^2)$ of the STTC encoder. $s_t^1 s_t^2$ means QPSK symbols transmitting via the transmit antenna 1 and 2. Figure 8.10 illustrates the STTC encoder and decoder.

As the decoder of the STTC, the Viterbi algorithm is used to perform maximum likelihood decoding. The branch metric of the trellis diagram is calculated using (8.44). For the 2×2 MIMO system with Gray coding QPSK, (8.44) is represented as follows:

$$\sum_{j=1}^{N_r}\left|r_t^j - \sum_{i=1}^{N_t}h_{ji}^t s_t^i\right|^2 = \left|r_t^{1I} - \left(h_{11}^t s_t^{1I} + h_{12}^t s_t^{2I}\right)\right|^2 + \left|r_t^{1Q} - \left(h_{11}^t s_t^{1Q} + h_{12}^t s_t^{2Q}\right)\right|^2 + \left|r_t^{2I} - \left(h_{21}^t s_t^{1I} + h_{22}^t s_t^{2I}\right)\right|^2$$

$$+ \left|r_t^{2Q} - \left(h_{21}^t s_t^{1Q} + h_{22}^t s_t^{2Q}\right)\right|^2$$

(8.45)

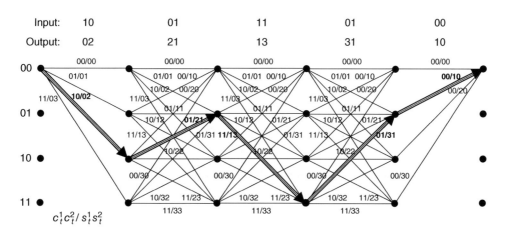

Figure 8.9   Encoding of the STTC

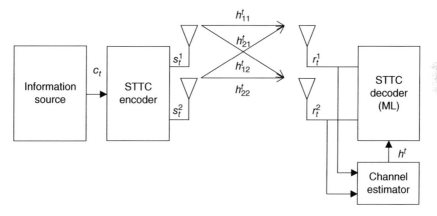

Figure 8.10   STTC transmitter and receiver

where r_t^{1I} and r_t^{2I} represent in-phase components of the received QPSK signal and r_t^{1Q} and r_t^{2Q} represent quadrature-phase components of the received QPSK signal. Figure 8.11 illustrates Gray coding QPSK signal constellation.

In the receiver, we assume the received symbols and the estimated 2×2 MIMO channel gains as shown in Table 8.1. Unlike the Viterbi algorithm, the STTC decoder needs to estimate MIMO channel gains.

As the first step of STTC decoding, the branch metrics $(\mathrm{BM}_t^{s_t^1 s_t^2})$ are calculated as follows:
At time 1,

$$
\begin{aligned}
\mathrm{BM}_1^{00} &= \left|r_1^{1I}-\left(h_{11}^1 s_1^{1I}+h_{12}^1 s_1^{2I}\right)\right|^2+\left|r_1^{1Q}-\left(h_{11}^1 s_1^{1Q}+h_{12}^1 s_1^{2Q}\right)\right|^2 \\
&\quad +\left|r_1^{2I}-\left(h_{21}^1 s_1^{1I}+h_{22}^1 s_1^{2I}\right)\right|^2+\left|r_1^{2Q}-\left(h_{21}^1 s_1^{1Q}+h_{22}^1 s_1^{2Q}\right)\right|^2 \\
&= \left|0.98-(1.28\cdot0.7+0.62\cdot0.7)\right|^2+\left|1.21-(1.28\cdot0.7+0.62\cdot0.7)\right|^2 \\
&\quad +\left|-0.87-(0.86\cdot0.7+1.16\cdot0.7)\right|^2+\left|-0.95-(0.86\cdot0.7+1.16\cdot0.7)\right|^2=10.94
\end{aligned}
$$

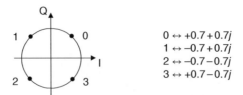

Figure 8.11   Gray coding QPSK signal constellation and mapping

Table 8.1   Transmitted symbols, channel gain, and received symbols in 2×2 MIMO

Time slot	Transmitted symbols $s_t^1 s_t^2 \leftrightarrow \left(s_t^{1I} + s_t^{1Q} j, s_t^{2I} + s_t^{2Q} j \right)$	MIMO channel gain $\begin{bmatrix} h_{11}^t & h_{12}^t \\ h_{21}^t & h_{22}^t \end{bmatrix}$	Received symbols $r_t^1 r_t^2 \leftrightarrow \left(r_t^{1I} + r_t^{1Q} j, r_t^{2I} + r_t^{2Q} j \right)$
1	$02 \leftrightarrow (+0.7 +0.7j, -0.7 -0.7j)$	$\begin{bmatrix} 1.28 & 0.62 \\ 0.86 & 1.16 \end{bmatrix}$	$(+0.98 +1.21j, -0.87 -0.95j)$
2	$21 \leftrightarrow (-0.7 -0.7j, -0.7 +0.7j)$	$\begin{bmatrix} 1.03 & 0.75 \\ 0.87 & 1.27 \end{bmatrix}$	$(+0.75 +1.2j, -1.59 -1.22j)$
3	$13 \leftrightarrow (-0.7 +0.7j, +0.7 -0.7j)$	$\begin{bmatrix} 0.83 & 0.52 \\ 0.88 & 0.31 \end{bmatrix}$	$(+0.93 +0.82j, -0.75 +0.12j)$
4	$31 \leftrightarrow (+0.7 -0.7j, -0.7 +0.7j)$	$\begin{bmatrix} 0.85 & 0.86 \\ 0.92 & 0.81 \end{bmatrix}$	$(+1.25 +0.77j, -0.86 -0.97j)$
5	$10 \leftrightarrow (-0.7 +0.7j, +0.7 +0.7j)$	$\begin{bmatrix} 0.72 & 0.79 \\ 0.71 & 0.51 \end{bmatrix}$	$(+0.81 +0.71j, -0.68 -0.79j)$

$$\begin{aligned}
\mathrm{BM}_1^{01} &= \left|0.98 - (1.28 \cdot 0.7 + 0.62 \cdot (-0.7))\right|^2 + \left|1.21 - (1.28 \cdot 0.7 + 0.62 \cdot 0.7)\right|^2 \\
&\quad + \left|-0.87 - (0.86 \cdot 0.7 + 1.16 \cdot (-0.7))\right|^2 \\
&\quad + \left|-0.95 - (0.86 \cdot 0.7 + 1.16 \cdot 0.7)\right|^2 = 6.31
\end{aligned}$$

$$\begin{aligned}
\mathrm{BM}_1^{02} &= \left|0.98 - (1.28 \cdot 0.7 + 0.62 \cdot (-0.7))\right|^2 + \left|1.21 - (1.28 \cdot 0.7 + 0.62 \cdot (-0.7))\right|^2 \\
&\quad + \left|-0.87 - (0.86 \cdot 0.7 + 1.16 \cdot (-0.7))\right|^2 \\
&\quad + \left|-0.95 - (0.86 \cdot 0.7 + 1.16 \cdot (-0.7))\right|^2 = 1.81
\end{aligned}$$

$$\begin{aligned}
\mathrm{BM}_1^{03} &= \left|0.98 - (1.28 \cdot 0.7 + 0.62 \cdot 0.7)\right|^2 + \left|1.21 - (1.28 \cdot 0.7 + 0.62 \cdot (-0.7))\right|^2 \\
&\quad + \left|-0.87 - (0.86 \cdot 0.7 + 1.16 \cdot 0.7)\right|^2 \\
&\quad + \left|-0.95 - (0.86 \cdot 0.7 + 1.16 \cdot (-0.7))\right|^2 = 6.54.
\end{aligned}$$

At time 2,

$$\text{BM}_2^{00} = \left| r_2^{1I} - \left(h_{11}^2 s_2^{1I} + h_{12}^2 s_2^{2I} \right) \right|^2 + \left| r_2^{1Q} - \left(h_{11}^2 s_2^{1Q} + h_{12}^2 s_2^{2Q} \right) \right|^2$$
$$+ \left| r_2^{2I} - \left(h_{21}^2 s_2^{1I} + h_{22}^2 s_2^{2I} \right) \right|^2 + \left| r_2^{2Q} - \left(h_{21}^2 s_2^{1Q} + h_{22}^2 s_2^{2Q} \right) \right|^2$$
$$= \left| 0.75 - (1.03 \cdot 0.7 + 0.75 \cdot 0.7) \right|^2 + \left| 1.2 - (1.03 \cdot 0.7 + 0.75 \cdot 0.7) \right|^2$$
$$+ \left| -1.59 - (0.87 \cdot 0.7 + 1.27 \cdot 0.7) \right|^2$$
$$+ \left| -1.22 - (0.87 \cdot 0.7 + 1.27 \cdot 0.7) \right|^2 = 17.17$$

$$\text{BM}_2^{01} = \left| 0.75 - (1.03 \cdot 0.7 + 0.75 \cdot (-0.7)) \right|^2 + \left| 1.2 - (1.03 \cdot 0.7 + 0.75 \cdot 0.7) \right|^2$$
$$+ \left| -1.59 - (0.87 \cdot 0.7 + 1.27 \cdot (-0.7)) \right|^2$$
$$+ \left| -1.22 - (0.87 \cdot 0.7 + 1.27 \cdot 0.7) \right|^2 = 9.41$$

$$\text{BM}_2^{02} = \left| 0.75 - (1.03 \cdot 0.7 + 0.75 \cdot (-0.7)) \right|^2 + \left| 1.2 - (1.03 \cdot 0.7 + 0.75 \cdot (-0.7)) \right|^2$$
$$+ \left| -1.59 - (0.87 \cdot 0.7 + 1.27 \cdot (-0.7)) \right|^2$$
$$+ \left| -1.22 - (0.87 \cdot 0.7 + 1.27 \cdot (-0.7)) \right|^2 = 3.91$$

$$\text{BM}_2^{03} = \left| 0.75 - (1.03 \cdot 0.7 + 0.75 \cdot 0.7) \right|^2 + \left| 1.2 - (1.03 \cdot 0.7 + 0.75 \cdot (-0.7)) \right|^2$$
$$+ \left| -1.59 - (0.87 \cdot 0.7 + 1.27 \cdot 0.7) \right|^2$$
$$+ \left| -1.22 - (0.87 \cdot 0.7 + 1.27 \cdot (-0.7)) \right|^2 = 11.67$$

$$\text{BM}_2^{10} = \left| 0.75 - (1.03 \cdot (-0.7) + 0.75 \cdot 0.7) \right|^2 + \left| 1.2 - (1.03 \cdot 0.7 + 0.75 \cdot 0.7) \right|^2$$
$$+ \left| -1.59 - (0.87 \cdot (-0.7) + 1.27 \cdot 0.7) \right|^2$$
$$+ \left| -1.22 - (0.87 \cdot 0.7 + 1.27 \cdot 0.7) \right|^2 = 11.78$$

$$\text{BM}_2^{11} = \left| 0.75 - (1.03 \cdot (-0.7) + 0.75 \cdot (-0.7)) \right|^2 + \left| 1.2 - (1.03 \cdot 0.7 + 0.75 \cdot 0.7) \right|^2$$
$$+ \left| -1.59 - (0.87 \cdot (-0.7) + 1.27 \cdot (-0.7)) \right|^2$$
$$+ \left| -1.22 - (0.87 \cdot 0.7 + 1.27 \cdot 0.7) \right|^2 = 11.38$$

$$\text{BM}_2^{12} = \left| 0.75 - (1.03 \cdot (-0.7) + 0.75 \cdot (-0.7)) \right|^2 + \left| 1.2 - (1.03 \cdot 0.7 + 0.75 \cdot (-0.7)) \right|^2$$
$$+ \left| -1.59 - (0.87 \cdot (-0.7) + 1.27 \cdot (-0.7)) \right|^2$$
$$+ \left| -1.22 - (0.87 \cdot 0.7 + 1.27 \cdot (-0.7)) \right|^2 = 5.88$$

$$\text{BM}_2^{13} = \left| 0.75 - (1.03 \cdot (-0.7) + 0.75 \cdot 0.7) \right|^2 + \left| 1.2 - (1.03 \cdot 0.7 + 0.75 \cdot (-0.7)) \right|^2$$
$$+ \left| -1.59 - (0.87 \cdot (-0.7) + 1.27 \cdot 0.7) \right|^2$$
$$+ \left| -1.22 - (0.87 \cdot 0.7 + 1.27 \cdot (-0.7)) \right|^2 = 6.28$$

$$BM_2^{20} = |0.75 - (1.03 \cdot (-0.7) + 0.75 \cdot 0.7)|^2 + |1.2 - (1.03 \cdot (-0.7) + 0.75 \cdot 0.7)|^2$$
$$+ |-1.59 - (0.87 \cdot (-0.7) + 1.27 \cdot 0.7)|^2$$
$$+ |-1.22 - (0.87 \cdot (-0.7) + 1.27 \cdot 0.7)|^2 = 8.59$$

$$BM_2^{21} = |0.75 - (1.03 \cdot (-0.7) + 0.75 \cdot (-0.7))|^2 + |1.2 - (1.03 \cdot (-0.7) + 0.75 \cdot 0.7)|^2$$
$$+ |-1.59 - (0.87 \cdot (-0.7) + 1.27 \cdot (-0.7))|^2$$
$$+ |-1.22 - (0.87 \cdot (-0.7) + 1.27 \cdot 0.7)|^2 = 8.19$$

$$BM_2^{22} = |0.75 - (1.03 \cdot (-0.7) + 0.75 \cdot (-0.7))|^2 + |1.2 - (1.03 \cdot (-0.7) + 0.75 \cdot (-0.7))|^2$$
$$+ |-1.59 - (0.87 \cdot (-0.7) + 1.27 \cdot (-0.7))|^2$$
$$+ |-1.22 - (0.87 \cdot (-0.7) + 1.27 \cdot (-0.7))|^2 = 10.05$$

$$BM_2^{23} = |0.75 - (1.03 \cdot (-0.7) + 0.75 \cdot 0.7)|^2 + |1.2 - (1.03 \cdot (-0.7) + 0.75 \cdot (-0.7))|^2$$
$$+ |-1.59 - (0.87 \cdot (-0.7) + 1.27 \cdot 0.7)|^2$$
$$+ |-1.22 - (0.87 \cdot (-0.7) + 1.27 \cdot (-0.7))|^2 = 10.45$$

$$BM_2^{30} = |0.75 - (1.03 \cdot 0.7 + 0.75 \cdot 0.7)|^2 + |1.2 - (1.03 \cdot (-0.7) + 0.75 \cdot 0.7)|^2$$
$$+ |-1.59 - (0.87 \cdot 0.7 + 1.27 \cdot 0.7)|^2$$
$$+ |-1.22 - (0.87 \cdot (-0.7) + 1.27 \cdot 0.7)|^2 = 13.98$$

$$BM_2^{31} = |0.75 - (1.03 \cdot 0.7 + 0.75 \cdot (-0.7))|^2 + |1.2 - (1.03 \cdot (-0.7) + 0.75 \cdot 0.7)|^2$$
$$+ |-1.59 - (0.87 \cdot 0.7 + 1.27 \cdot (-0.7))|^2$$
$$+ |-1.22 - (0.87 \cdot (-0.7) + 1.27 \cdot 0.7)|^2 = 6.22$$

$$BM_2^{32} = |0.75 - (1.03 \cdot 0.7 + 0.75 \cdot (-0.7))|^2 + |1.2 - (1.03 \cdot (-0.7) + 0.75 \cdot (-0.7))|^2$$
$$+ |-1.59 - (0.87 \cdot 0.7 + 1.27 \cdot (-0.7))|^2$$
$$+ |-1.22 - (0.87 \cdot (-0.7) + 1.27 \cdot (-0.7))|^2 = 8.08$$

$$BM_2^{33} = |0.75 - (1.03 \cdot 0.7 + 0.75 \cdot 0.7)|^2 + |1.2 - (1.03 \cdot (-0.7) + 0.75 \cdot (-0.7))|^2$$
$$+ |-1.59 - (0.87 \cdot 0.7 + 1.27 \cdot 0.7)|^2$$
$$+ |-1.22 - (0.87 \cdot (-0.7) + 1.27 \cdot (-0.7))|^2 = 15.84.$$

At time 3,

$$\text{BM}_3^{00} = \left| r_3^{1I} - \left(h_{11}^3 s_3^{1I} + h_{12}^3 s_3^{2I} \right) \right|^2 + \left| r_3^{1Q} - \left(h_{11}^3 s_3^{1Q} + h_{12}^3 s_3^{2Q} \right) \right|^2$$
$$+ \left| r_3^{2I} - \left(h_{21}^3 s_3^{1I} + h_{22}^3 s_3^{2I} \right) \right|^2 + \left| r_3^{2Q} - \left(h_{21}^3 s_3^{1Q} + h_{22}^3 s_3^{2Q} \right) \right|^2$$
$$= \left| 0.93 - (0.83 \cdot 0.7 + 0.52 \cdot 0.7) \right|^2 + \left| 0.82 - (0.83 \cdot 0.7 + 0.52 \cdot 0.7) \right|^2$$
$$+ \left| -0.75 - (0.88 \cdot 0.7 + 0.31 \cdot 0.7) \right|^2$$
$$+ \left| 0.12 - (0.88 \cdot 0.7 + 0.31 \cdot 0.7) \right|^2 = 3.03$$

$$\text{BM}_3^{01} = \left| 0.93 - (0.83 \cdot 0.7 + 0.52 \cdot (-0.7)) \right|^2 + \left| 0.82 - (0.83 \cdot 0.7 + 0.52 \cdot 0.7) \right|^2$$
$$+ \left| -0.75 - (0.88 \cdot 0.7 + 0.31 \cdot (-0.7)) \right|^2$$
$$+ \left| 0.12 - (0.88 \cdot 0.7 + 0.31 \cdot 0.7) \right|^2 = 2.35$$

$$\text{BM}_3^{02} = \left| 0.93 - (0.83 \cdot 0.7 + 0.52 \cdot (-0.7)) \right|^2 + \left| 0.82 - (0.83 \cdot 0.7 + 0.52 \cdot (-0.7)) \right|^2$$
$$+ \left| -0.75 - (0.88 \cdot 0.7 + 0.31 \cdot (-0.7)) \right|^2$$
$$+ \left| 0.12 - (0.88 \cdot 0.7 + 0.31 \cdot (-0.7)) \right|^2 = 2.27$$

$$\text{BM}_3^{03} = \left| 0.93 - (0.83 \cdot 0.7 + 0.52 \cdot 0.7) \right|^2 + \left| 0.82 - (0.83 \cdot 0.7 + 0.52 \cdot (-0.7)) \right|^2$$
$$+ \left| -0.75 - (0.88 \cdot 0.7 + 0.31 \cdot 0.7) \right|^2$$
$$+ \left| 0.12 - (0.88 \cdot 0.7 + 0.31 \cdot (-0.7)) \right|^2 = 2.95$$

$$\text{BM}_3^{10} = \left| 0.93 - (0.83 \cdot (-0.7) + 0.52 \cdot 0.7) \right|^2 + \left| 0.82 - (0.83 \cdot 0.7 + 0.52 \cdot 0.7) \right|^2$$
$$+ \left| -0.75 - (0.88 \cdot (-0.7) + 0.31 \cdot 0.7) \right|^2$$
$$+ \left| 0.12 - (0.88 \cdot 0.7 + 0.31 \cdot 0.7) \right|^2 = 1.96$$

$$\text{BM}_3^{11} = \left| 0.93 - (0.83 \cdot (-0.7) + 0.52 \cdot (-0.7)) \right|^2 + \left| 0.82 - (0.83 \cdot 0.7 + 0.52 \cdot 0.7) \right|^2$$
$$+ \left| -0.75 - (0.88 \cdot (-0.7) + 0.31 \cdot (-0.7)) \right|^2$$
$$+ \left| 0.12 - (0.88 \cdot 0.7 + 0.31 \cdot 0.7) \right|^2 = 4.05$$

$$\text{BM}_3^{12} = \left| 0.93 - (0.83 \cdot (-0.7) + 0.52 \cdot (-0.7)) \right|^2 + \left| 0.82 - (0.83 \cdot 0.7 + 0.52 \cdot (-0.7)) \right|^2$$
$$+ \left| -0.75 - (0.88 \cdot (-0.7) + 0.31 \cdot (-0.7)) \right|^2$$
$$+ \left| 0.12 - (0.88 \cdot 0.7 + 0.31 \cdot (-0.7)) \right|^2 = 3.96$$

$$BM_3^{13} = |0.93 - (0.83 \cdot (-0.7) + 0.52 \cdot 0.7)|^2 + |0.82 - (0.83 \cdot 0.7 + 0.52 \cdot (-0.7))|^2$$
$$+ |-0.75 - (0.88 \cdot (-0.7) + 0.31 \cdot 0.7)|^2$$
$$+ |0.12 - (0.88 \cdot 0.7 + 0.31 \cdot (-0.7))|^2 = 1.88$$

$$BM_3^{20} = |0.93 - (0.83 \cdot (-0.7) + 0.52 \cdot 0.7)|^2 + |0.82 - (0.83 \cdot (-0.7) + 0.52 \cdot 0.7)|^2$$
$$+ |-0.75 - (0.88 \cdot (-0.7) + 0.31 \cdot 0.7)|^2$$
$$+ |0.12 - (0.88 \cdot (-0.7) + 0.31 \cdot 0.7)|^2 = 2.78$$

$$BM_3^{21} = |0.93 - (0.83 \cdot (-0.7) + 0.52 \cdot (-0.7))|^2 + |0.82 - (0.83 \cdot (-0.7) + 0.52 \cdot 0.7)|^2$$
$$+ |-0.75 - (0.88 \cdot (-0.7) + 0.31 \cdot (-0.7))|^2$$
$$+ |0.12 - (0.88 \cdot (-0.7) + 0.31 \cdot 0.7)|^2 = 4.87$$

$$BM_3^{22} = |0.93 - (0.83 \cdot (-0.7) + 0.52 \cdot (-0.7))|^2 + |0.82 - (0.83 \cdot (-0.7) + 0.52 \cdot (-0.7))|^2$$
$$+ |-0.75 - (0.88 \cdot (-0.7) + 0.31 \cdot (-0.7))|^2$$
$$+ |0.12 - (0.88 \cdot (-0.7) + 0.31 \cdot (-0.7))|^2 = 7.55$$

$$BM_3^{23} = |0.93 - (0.83 \cdot (-0.7) + 0.52 \cdot 0.7)|^2 + |0.82 - (0.83 \cdot (-0.7) + 0.52 \cdot (-0.7))|^2$$
$$+ |-0.75 - (0.88 \cdot (-0.7) + 0.31 \cdot 0.7)|^2$$
$$+ |0.12 - (0.88 \cdot (-0.7) + 0.31 \cdot (-0.7))|^2 = 5.46$$

$$BM_3^{30} = |0.93 - (0.83 \cdot 0.7 + 0.52 \cdot 0.7)|^2 + |0.82 - (0.83 \cdot (-0.7) + 0.52 \cdot 0.7)|^2$$
$$+ |-0.75 - (0.88 \cdot 0.7 + 0.31 \cdot 0.7)|^2$$
$$+ |0.12 - (0.88 \cdot (-0.7) + 0.31 \cdot 0.7)|^2 = 3.85$$

$$BM_3^{31} = |0.93 - (0.83 \cdot 0.7 + 0.52 \cdot (-0.7))|^2 + |0.82 - (0.83 \cdot (-0.7) + 0.52 \cdot 0.7)|^2$$
$$+ |-0.75 - (0.88 \cdot 0.7 + 0.31 \cdot (-0.7))|^2$$
$$+ |0.12 - (0.88 \cdot (-0.7) + 0.31 \cdot 0.7)|^2 = 3.17$$

$$BM_3^{32} = |0.93 - (0.83 \cdot 0.7 + 0.52 \cdot (-0.7))|^2 + |0.82 - (0.83 \cdot (-0.7) + 0.52 \cdot (-0.7))|^2$$
$$+ |-0.75 - (0.88 \cdot 0.7 + 0.31 \cdot (-0.7))|^2$$
$$+ |0.12 - (0.88 \cdot (-0.7) + 0.31 \cdot (-0.7))|^2 = 5.85$$

$$BM_3^{33} = \left|0.93 - (0.83 \cdot 0.7 + 0.52 \cdot 0.7)\right|^2 + \left|0.82 - (0.83 \cdot (-0.7) + 0.52 \cdot (-0.7))\right|^2$$
$$+ \left|-0.75 - (0.88 \cdot 0.7 + 0.31 \cdot 0.7)\right|^2$$
$$+ \left|0.12 - (0.88 \cdot (-0.7) + 0.31 \cdot (-0.7))\right|^2 = 6.53.$$

At time 4,

$$BM_4^{00} = \left|r_4^{1I} - \left(h_{11}^4 s_4^{1I} + h_{12}^4 s_4^{2I}\right)\right|^2 + \left|r_4^{1Q} - \left(h_{11}^4 s_4^{1Q} + h_{12}^4 s_4^{2Q}\right)\right|^2$$
$$+ \left|r_4^{2I} - \left(h_{21}^4 s_4^{1I} + h_{22}^4 s_4^{2I}\right)\right|^2 + \left|r_4^{2Q} - \left(h_{21}^4 s_4^{1Q} + h_{22}^4 s_4^{2Q}\right)\right|^2$$
$$= \left|1.25 - (0.85 \cdot 0.7 + 0.86 \cdot 0.7)\right|^2 + \left|0.77 - (0.85 \cdot 0.7 + 0.86 \cdot 0.7)\right|^2$$
$$+ \left|-0.86 - (0.92 \cdot 0.7 + 0.81 \cdot 0.7)\right|^2$$
$$+ \left|-0.97 - (0.92 \cdot 0.7 + 0.81 \cdot 0.7)\right|^2 = 9.23$$

$$BM_4^{01} = \left|1.25 - (0.85 \cdot 0.7 + 0.86 \cdot (-0.7))\right|^2 + \left|0.77 - (0.85 \cdot 0.7 + 0.86 \cdot 0.7)\right|^2$$
$$+ \left|-0.86 - (0.92 \cdot 0.7 + 0.81 \cdot (-0.7))\right|^2$$
$$+ \left|-0.97 - (0.92 \cdot 0.7 + 0.81 \cdot 0.7)\right|^2 = 7.40$$

$$BM_4^{02} = \left|1.25 - (0.85 \cdot 0.7 + 0.86 \cdot (-0.7))\right|^2 + \left|0.77 - (0.85 \cdot 0.7 + 0.86 \cdot (-0.7))\right|^2$$
$$+ \left|-0.86 - (0.92 \cdot 0.7 + 0.81 \cdot (-0.7))\right|^2$$
$$+ \left|-0.97 - (0.92 \cdot 0.7 + 0.81 \cdot (-0.7))\right|^2 = 4.16$$

$$BM_4^{03} = \left|1.25 - (0.85 \cdot 0.7 + 0.86 \cdot 0.7)\right|^2 + \left|0.77 - (0.85 \cdot 0.7 + 0.86 \cdot (-0.7))\right|^2$$
$$+ \left|-0.86 - (0.92 \cdot 0.7 + 0.81 \cdot 0.7)\right|^2$$
$$+ \left|-0.97 - (0.92 \cdot 0.7 + 0.81 \cdot (-0.7))\right|^2 = 5.99$$

$$BM_4^{10} = \left|1.25 - (0.85 \cdot (-0.7) + 0.86 \cdot 0.7)\right|^2 + \left|0.77 - (0.85 \cdot 0.7 + 0.86 \cdot 0.7)\right|^2$$
$$+ \left|-0.86 - (0.92 \cdot (-0.7) + 0.81 \cdot 0.7)\right|^2$$
$$+ \left|-0.97 - (0.92 \cdot 0.7 + 0.81 \cdot 0.7)\right|^2 = 7.10$$

$$BM_4^{11} = \left|1.25 - (0.85 \cdot (-0.7) + 0.86 \cdot (-0.7))\right|^2 + \left|0.77 - (0.85 \cdot 0.7 + 0.86 \cdot 0.7)\right|^2$$
$$+ \left|-0.86 - (0.92 \cdot (-0.7) + 0.81 \cdot (-0.7))\right|^2$$
$$+ \left|-0.97 - (0.92 \cdot 0.7 + 0.81 \cdot 0.7)\right|^2 = 11.05$$

$$BM_4^{12} = \left|1.25 - (0.85 \cdot (-0.7) + 0.86 \cdot (-0.7))\right|^2 + \left|0.77 - (0.85 \cdot 0.7 + 0.86 \cdot (-0.7))\right|^2$$
$$+ \left|-0.86 - (0.92 \cdot (-0.7) + 0.81 \cdot (-0.7))\right|^2$$
$$+ \left|-0.97 - (0.92 \cdot 0.7 + 0.81 \cdot (-0.7))\right|^2 = 7.81$$

$$BM_4^{13} = \left|1.25 - (0.85 \cdot (-0.7) + 0.86 \cdot 0.7)\right|^2 + \left|0.77 - (0.85 \cdot 0.7 + 0.86 \cdot (-0.7))\right|^2$$
$$+ \left|-0.86 - (0.92 \cdot (-0.7) + 0.81 \cdot 0.7)\right|^2$$
$$+ \left|-0.97 - (0.92 \cdot 0.7 + 0.81 \cdot (-0.7))\right|^2 = 3.86$$

$$BM_4^{20} = \left|1.25 - (0.85 \cdot (-0.7) + 0.86 \cdot 0.7)\right|^2 + \left|0.77 - (0.85 \cdot (-0.7) + 0.86 \cdot 0.7)\right|^2$$
$$+ \left|-0.86 - (0.92 \cdot (-0.7) + 0.81 \cdot 0.7)\right|^2$$
$$+ \left|-0.97 - (0.92 \cdot (-0.7) + 0.81 \cdot 0.7)\right|^2 = 3.54$$

$$BM_4^{21} = \left|1.25 - (0.85 \cdot (-0.7) + 0.86 \cdot (-0.7))\right|^2 + \left|0.77 - (0.85 \cdot (-0.7) + 0.86 \cdot 0.7)\right|^2$$
$$+ \left|-0.86 - (0.92 \cdot (-0.7) + 0.81 \cdot (-0.7))\right|^2$$
$$+ \left|-0.97 - (0.92 \cdot (-0.7) + 0.81 \cdot 0.7)\right|^2 = 7.49$$

$$BM_4^{22} = \left|1.25 - (0.85 \cdot (-0.7) + 0.86 \cdot (-0.7))\right|^2 + \left|0.77 - (0.85 \cdot (-0.7) + 0.86 \cdot (-0.7))\right|^2$$
$$+ \left|-0.86 - (0.92 \cdot (-0.7) + 0.81 \cdot (-0.7))\right|^2$$
$$+ \left|-0.97 - (0.92 \cdot (-0.7) + 0.81 \cdot (-0.7))\right|^2 = 10.04$$

$$BM_4^{23} = \left|1.25 - (0.85 \cdot (-0.7) + 0.86 \cdot 0.7)\right|^2 + \left|0.77 - (0.85 \cdot (-0.7) + 0.86 \cdot (-0.7))\right|^2$$
$$+ \left|-0.86 - (0.92 \cdot (-0.7) + 0.81 \cdot 0.7)\right|^2$$
$$+ \left|-0.97 - (0.92 \cdot (-0.7) + 0.81 \cdot (-0.7))\right|^2 = 6.09$$

$$BM_4^{30} = \left|1.25 - (0.85 \cdot 0.7 + 0.86 \cdot 0.7)\right|^2 + \left|0.77 - (0.85 \cdot (-0.7) + 0.86 \cdot 0.7)\right|^2$$
$$+ \left|-0.86 - (0.92 \cdot 0.7 + 0.81 \cdot 0.7)\right|^2$$
$$+ \left|-0.97 - (0.92 \cdot (-0.7) + 0.81 \cdot 0.7)\right|^2 = 5.67$$

$$BM_4^{31} = \left|1.25 - (0.85 \cdot 0.7 + 0.86 \cdot (-0.7))\right|^2 + \left|0.77 - (0.85 \cdot (-0.7) + 0.86 \cdot 0.7)\right|^2$$
$$+ \left|-0.86 - (0.92 \cdot 0.7 + 0.81 \cdot (-0.7))\right|^2$$
$$+ \left|-0.97 - (0.92 \cdot (-0.7) + 0.81 \cdot 0.7)\right|^2 = 3.84$$

$$BM_4^{32} = \left|1.25 - (0.85 \cdot 0.7 + 0.86 \cdot (-0.7))\right|^2 + \left|0.77 - (0.85 \cdot (-0.7) + 0.86 \cdot (-0.7))\right|^2$$
$$+ \left|-0.86 - (0.92 \cdot 0.7 + 0.81 \cdot (-0.7))\right|^2$$
$$+ \left|-0.97 - (0.92 \cdot (-0.7) + 0.81 \cdot (-0.7))\right|^2 = 6.39$$

$$BM_4^{33} = \left|1.25 - (0.85 \cdot 0.7 + 0.86 \cdot 0.7)\right|^2 + \left|0.77 - (0.85 \cdot (-0.7) + 0.86 \cdot (-0.7))\right|^2$$
$$+ \left|-0.86 - (0.92 \cdot 0.7 + 0.81 \cdot 0.7)\right|^2$$
$$+ \left|-0.97 - (0.92 \cdot (-0.7) + 0.81 \cdot (-0.7))\right|^2 = 8.22.$$

At time 5,

$$BM_5^{00} = \left|r_5^{1I} - \left(h_{11}^5 s_5^{1I} + h_{12}^5 s_5^{2I}\right)\right|^2 + \left|r_5^{1Q} - \left(h_{11}^5 s_5^{1Q} + h_{12}^5 s_5^{2Q}\right)\right|^2$$
$$+ \left|r_5^{2I} - \left(h_{21}^5 s_5^{1I} + h_{22}^5 s_5^{2I}\right)\right|^2 + \left|r_5^{2Q} - \left(h_{21}^5 s_5^{1Q} + h_{22}^5 s_5^{2Q}\right)\right|^2$$
$$= \left|0.81 - (0.72 \cdot 0.7 + 0.79 \cdot 0.7)\right|^2 + \left|0.71 - (0.72 \cdot 0.7 + 0.79 \cdot 0.7)\right|^2$$
$$+ \left|-0.68 - (0.71 \cdot 0.7 + 0.51 \cdot 0.7)\right|^2$$
$$+ \left|-0.79 - (0.71 \cdot 0.7 + 0.51 \cdot 0.7)\right|^2 = 5.24$$

$$BM_5^{10} = \left|0.81 - (0.72 \cdot (-0.7) + 0.79 \cdot 0.7)\right|^2 + \left|0.71 - (0.72 \cdot 0.7 + 0.79 \cdot 0.7)\right|^2$$
$$+ \left|-0.68 - (0.71 \cdot (-0.7) + 0.51 \cdot 0.7)\right|^2$$
$$+ \left|-0.79 - (0.71 \cdot 0.7 + 0.51 \cdot 0.7)\right|^2 = 3.69$$

$$BM_5^{20} = \left|0.81 - (0.72 \cdot (-0.7) + 0.79 \cdot 0.7)\right|^2 + \left|0.71 - (0.72 \cdot (-0.7) + 0.79 \cdot 0.7)\right|^2$$
$$+ \left|-0.68 - (0.71 \cdot (-0.7) + 0.51 \cdot 0.7)\right|^2$$
$$+ \left|-0.79 - (0.71 \cdot (-0.7) + 0.51 \cdot 0.7)\right|^2 = 1.73$$

$$BM_5^{30} = \left|0.81 - (0.72 \cdot 0.7 + 0.79 \cdot 0.7)\right|^2 + \left|0.71 - (0.72 \cdot (-0.7) + 0.79 \cdot 0.7)\right|^2$$
$$+ \left|-0.68 - (0.71 \cdot 0.7) + 0.51 \cdot 0.7)\right|^2$$
$$+ \left|-0.79 - (0.71 \cdot (-0.7) + 0.51 \cdot 0.7)\right|^2 = 3.27$$

Figure 8.12 summarizes the branch metric calculations on the trellis diagram. As the next step, the path metrics $\left(PM_t^{state}\right)$ are calculated as follows:
At time 1,

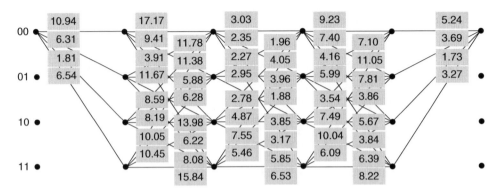

Figure 8.12 Branch metric calculations

$$PM_1^{00} = BM_1^{00} = 10.94$$
$$PM_1^{01} = BM_1^{01} = 6.31$$
$$PM_1^{10} = BM_1^{02} = 1.81$$
$$PM_1^{11} = BM_1^{03} = 6.54.$$

At time 2,

$$PM_2^{00} = Min\left(PM_1^{00} + BM_2^{00}, PM_1^{01} + BM_2^{10}, PM_1^{10} + BM_2^{20}, PM_1^{11} + BM_2^{30}\right)$$
$$= Min(10.94 + 17.17, 6.31 + 11.78, 1.81 + 8.59, 6.54 + 13.98)$$
$$= 10.4$$

$$PM_2^{01} = Min\left(PM_1^{00} + BM_2^{01}, PM_1^{01} + BM_2^{11}, PM_1^{10} + BM_2^{21}, PM_1^{11} + BM_2^{31}\right)$$
$$= Min(10.94 + 9.41, 6.31 + 11.38, 1.81 + 8.19, 6.54 + 6.22)$$
$$= 10$$

$$PM_2^{10} = Min\left(PM_1^{00} + BM_2^{02}, PM_1^{01} + BM_2^{12}, PM_1^{10} + BM_2^{22}, PM_1^{11} + BM_2^{32}\right)$$
$$= Min(10.94 + 3.91, 6.31 + 5.88, 1.81 + 10.05, 6.54 + 8.08)$$
$$= 11.86$$

$$PM_2^{11} = Min\left(PM_1^{00} + BM_2^{03}, PM_1^{01} + BM_2^{13}, PM_1^{10} + BM_2^{23}, PM_1^{11} + BM_2^{33}\right)$$
$$= Min(10.94 + 11.67, 6.31 + 6.28, 1.81 + 10.45, 6.54 + 15.84)$$
$$= 12.26.$$

At time 3,

$$PM_3^{00} = Min\left(PM_2^{00} + BM_3^{00}, PM_2^{01} + BM_3^{10}, PM_2^{10} + BM_3^{20}, PM_2^{11} + BM_3^{30}\right)$$
$$= Min(10.4 + 3.03, 10 + 1.96, 11.86 + 2.78, 12.26 + 3.85)$$
$$= 11.96$$

$$PM_3^{01} = Min\left(PM_2^{00} + BM_3^{01}, PM_2^{01} + BM_3^{11}, PM_2^{10} + BM_3^{21}, PM_2^{11} + BM_3^{31}\right)$$
$$= Min(10.4 + 2.35, 10 + 4.05, 11.86 + 4.87, 12.26 + 3.17)$$
$$= 12.75$$

$$PM_3^{10} = Min\left(PM_2^{00} + BM_3^{02}, PM_2^{01} + BM_3^{12}, PM_2^{10} + BM_3^{22}, PM_2^{11} + BM_3^{32}\right)$$
$$= Min(10.4 + 2.27, 10 + 3.96, 11.86 + 7.55, 12.26 + 5.85)$$
$$= 12.67$$

$$PM_3^{11} = Min\left(PM_2^{00} + BM_3^{03}, PM_2^{01} + BM_3^{13}, PM_2^{10} + BM_3^{23}, PM_2^{11} + BM_3^{33}\right)$$
$$= Min(10.4 + 2.95, 10 + 1.88, 11.86 + 5.46, 12.26 + 6.53)$$
$$= 11.88.$$

At time 4,

$$PM_4^{00} = Min\left(PM_3^{00} + BM_4^{00}, PM_3^{01} + BM_4^{10}, PM_3^{10} + BM_4^{20}, PM_3^{11} + BM_4^{30}\right)$$
$$= Min(11.96 + 9.23, 12.75 + 7.10, 12.67 + 3.54, 11.88 + 5.67)$$
$$= 16.2$$

$$PM_4^{01} = Min\left(PM_3^{00} + BM_4^{01}, PM_3^{01} + BM_4^{11}, PM_3^{10} + BM_4^{21}, PM_3^{11} + BM_4^{31}\right)$$
$$= Min(11.96 + 7.40, 12.75 + 11.05, 12.67 + 7.49, 11.88 + 3.84)$$
$$= 15.72$$

$$PM_4^{10} = Min\left(PM_3^{00} + BM_4^{02}, PM_3^{01} + BM_4^{12}, PM_3^{10} + BM_4^{22}, PM_3^{11} + BM_4^{32}\right)$$
$$= Min(11.96 + 4.16, 12.75 + 7.81, 12.67 + 10.04, 11.88 + 6.39)$$
$$= 16.12$$

$$PM_4^{11} = Min\left(PM_3^{00} + BM_4^{03}, PM_3^{01} + BM_4^{13}, PM_3^{10} + BM_4^{23}, PM_3^{11} + BM_4^{33}\right)$$
$$= Min(11.96 + 5.99, 12.75 + 3.86, 12.67 + 6.09, 11.88 + 8.22)$$
$$= 16.61.$$

At time 5,

$$PM_5^{00} = Min\left(PM_4^{00} + BM_5^{00}, PM_4^{01} + BM_5^{10}, PM_4^{10} + BM_5^{20}, PM_4^{11} + BM_5^{30}\right)$$
$$= Min(16.2 + 5.24, 15.72 + 3.69, 16.12 + 1.73, 16.61 + 3.27)$$
$$= 17.85.$$

Figure 8.13 summarizes the path metric calculations on the Trellis.

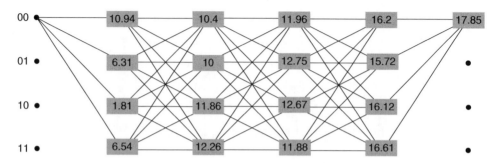

Figure 8.13   Path metric calculations

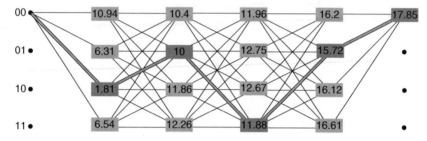

Figure 8.14   Traceback

Based on path metric calculations, the traceback is carried out and the most likely path is found. As we can observe one unique path in Figure 8.14, the decoded sequence of STTC decoding is [10 01 11 01 00]. This is same as the input sequence of STTC encoder. We can check whether the STTC is able to correct a symbol error. If there is no the smallest path metric at each time, namely we have more than two minimum path metrics at each time, one path are selected arbitrarily. Similar to the Viterbi algorithm, a normalization process is needed to avoid overflow if the packet length is long.

A hardware implementation of the STTC encoder and decoder is similar to the convolutional encoder and Viterbi decoder. The encoder of the STTC can be implemented by a shift register, modulo adder/multiplier, and memory. The decoder of the STTC is composed of branch metric calculation, path metric calculation, normalization, traceback, and channel estimation. Among the components of the STTC decoder, channel estimation is a different part from the Viterbi decoder and it is a critical part to decide STTC performance.

8.4   Spatial Multiplexing and MIMO Detection Algorithms

Space time coding we discussed in the previous section provides us with a better link reliability due to a better diversity and coding gain. It transmits same symbols multiple times via different antennas and time slots. Thus, it cannot achieve multiplexing gain. Multiplexing gain means the maximum number of symbols which can be transmitted simultaneously at the same

radio resource without additional power or bandwidth. A spatial multiplexing technique transmits multiple data streams via different transmit antennas in the same time and frequency. The
received signal includes the multiple data streams and each transmitted data stream is extracted
from the received signal. Thus, the spatial multiplexing technique can achieve a high multiplexing gain and increase a high channel capacity. However, the receiver complexity becomes
very high. In this section, we review several MIMO detection algorithms and focus on one
practical solution. Maximum likelihood (ML) detection is an optimal solution but the complexity exponentially grows according to the number of antennas. Thus, the application is
limited at less than 4 antennas. For other applications, sub-optimal algorithms such as Matched
filter (MF), Zero-forcing (ZF), Minimum Mean Square Error (MMSE), and Successive
Interference Cancellation (SIC) (or BLAST) are practically considered. Sphere Decoding
(SD) [11] and Fixed Sphere Decoding (FSD) [12] can provide near ML detection performance.
Thus, it attracts many MIMO system designers' attention. Figure 8.15 illustrates classification
of MIMO detection algorithms.

Consider a spatial multiplexing system with N_t transmit antennas and N_r receive antennas
and assume $N_r \geq N_t$. The received signal r ($N_r \times 1$ column vector) can be expressed as follows:

$$r = Hs + n \qquad (8.46)$$

where H, s, and n represent $N_r \times N_t$ MIMO channel matrix, $N_t \times 1$ transmit signal, and
$N_r \times 1$ white Gaussian noise vector, respectively. The element of r, r_j, is expressed as a
superposition of all elements of s. Figure 8.16 illustrates the MIMO system for spatial
multiplexing.

The maximum likelihood detection is to find the most likely $\hat{s}$ as follows:

$$\hat{s}_{\mathbf{ML}} = \arg \min_{s \in S} \left\| r - Hs^2 \right\| \qquad (8.47)$$

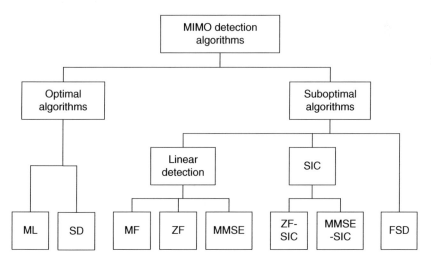

Figure 8.15   MIMO detection algorithms

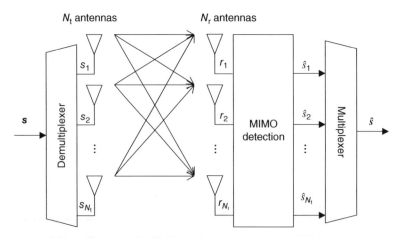

Figure 8.16   MIMO system for spatial multiplexing

where $\| \ \|$ denotes the norm of the matrix. The most likely $\hat{s}$ is an element of set S. Thus, it is simple solution to search all possible elements of set S and select one to satisfy (8.47). This is known as nondeterministic polynomial (NP) hard problem. The complexity increases exponentially according to the number of transmit antennas and the modulation order. Linear detection techniques such as MF, ZF, and MMSE use a MIMO channel inversion. The transmitted symbol estimation is calculated by the MIMO channel version, multiplication, and quantization. The MF detection technique is one of the simplest detection techniques as follows:

$$\hat{s}_{\mathrm{MF}} = \mathrm{Qtz}(\boldsymbol{H}^{H}\boldsymbol{r}) \tag{8.48}$$

where Qtz( ) represents quantization. The estimated symbols are obtained by multiplying the received symbols by Hermitian operation of the MIMO channel matrix. The ZF detection technique uses the pseudo-inverse of the MIMO channel matrix. When the MIMO channel matrix is square ($N_t = N_r$) and invertible, the estimated symbols by the ZF detection is expressed as follows:

$$\hat{s}_{\mathrm{ZF}} = \mathrm{Qtz}(\boldsymbol{H}^{-1}\boldsymbol{r}). \tag{8.49}$$

When the MIMO channel matrix is not square ($N_t \neq N_r$), it is expressed as follows:

$$\hat{s}_{\mathrm{ZF}} = \mathrm{Qtz}((\boldsymbol{H}^{H}\boldsymbol{H})^{-1}\boldsymbol{H}^{H}\boldsymbol{r}). \tag{8.50}$$

As we can observe from (8.49) and (8.50), the ZF detection technique forces amplitude of interferers to be zero with ignoring a noise effect. Thus, the MMSE detection technique considering a noise effect provides a better performance than the ZF detection technique. The MMSE detection technique minimizes the mean-squared error. It is expressed as follows:

$$\hat{s}_{\mathrm{MMSE}} = \mathrm{Qtz}((\boldsymbol{H}^{H}\boldsymbol{H} + N_{0}\boldsymbol{I})^{-1}\boldsymbol{H}^{H}\boldsymbol{r}). \tag{8.51}$$

where I and N_0 represent an identity matrix and a noise. As we can observe from (8.51), an accurate estimation of the noise is required. If the noise term N_0 is equal to zero, the MMSE detection technique is same as the ZF detection technique. In linear detections, an accurate estimation of the MIMO channel matrix is an essential part and these detections are useful at a high SNR. The SIC detection technique is located between ML detection and linear detection. It provides a better performance than linear detections. The SIC detection uses nulling and cancellation to extract the transmitted symbols from the received symbols. When one layer (a partial symbol sequence from one transmit antenna) is detected, an estimation of the transmitted layer is performed by subtracting from the detected layers in the previous time. The nulling and cancellation are carried out until all layers are detected. The SIC detection shows us good performance at a low SNR. One disadvantage is that the SIC suffers from error propagation. When a wrong decision is made in any layer, the wrong decision affects the other layers. Thus, the ordering technique is used to minimize the effect of error propagation. In the nulling and cancellation with the ordering technique, the first symbol with a high SNR is transmitted as the most reliable symbol. Then, the symbols with a lower SNR are transmitted. Both the linear detection and the SIC detection cannot achieve the ML detection performance even if they have a lower complexity than the ML detection. Thus, many MIMO system designers pay attention to the SD algorithm which can achieve near ML detection performance with a lower complexity than the exhaustive search. The basic idea of the SD algorithm is very simple. The ML detection searches all possible elements of the transmitted symbols but the SD algorithm searches a finite number of symbols within sphere radius $\mathcal{R}$ centered at the received symbol vector. The SD algorithm is expressed as follows:

$$\hat{s}_{\mathrm{SD}} = \arg \min_{s \in S} (r - Hs^2 < \mathcal{R}). \tag{8.52}$$

Consider a simple example about a two transmit antennas system with BPSK. The lattice points are shown in Figure 8.17.

In this figure, the candidate symbol $\hat{s}_i (i = 1, 2, 3,$ and $4)$ represents (BPSK symbol from antenna 1, BSPK symbol from antenna 2) $[+1, +1], [-1, +1], [-1, -1],$ and $[+1, -1]$, respectively.

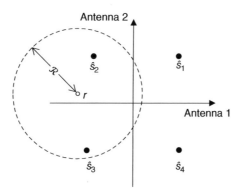

Figure 8.17   Lattice points for two transmit antenna and BPSK

The ML detection searches all possible lattice points ($\hat{s}_1$, $\hat{s}_2$, $\hat{s}_3$, and $\hat{s}_4$) but the SD algorithm searches only some lattice points ($\hat{s}_2$, $\hat{s}_3$) in a sphere radius $\mathcal{R}$ centered at the received symbol vector r. Thus, the computational complexity can be significantly reduced and the performance of the SD algorithm is almost same as the ML detection. The SD detection selects the lattice point $\hat{s}_2$ due to the closest one from the received symbol. This is same selection of the ML detection. As we can observe from (8.52), the sphere radius $\mathcal{R}$ is very important to performance and complexity. If the radius is too small, we may not have any candidate lattice points in the circle so that the performance may be worse. If the radius is too large, we may search too many candidate lattice points so that the complexity may increase very much. We reformulate (8.52) using QR decomposition and build a decision tree [13]. As we discussed in Chapter 5, the channel matrix H is expressed as follows:

$$H = QR \tag{8.53}$$

where Q and R denote an orthogonal matrix and an upper triangular matrix, respectively. Thus, (8.52) is rewritten as follows:

$$\hat{s}_{\mathrm{SD}} = \arg \min_{s \in S} \left(\left\| Q^H (r - Hs) \right\|^2 < \mathcal{R} \right) \tag{8.54}$$

$$= \arg \min_{s \in S} \left(\left\| \hat{r} - Rs \right\|^2 < \mathcal{R} \right) \tag{8.55}$$

where $\hat{r} = Q^H r$. An iterative solution is used in order to solve (8.55). When considering $s = [s_1, s_2, \ldots, s_{N_t}]^T$, (8.55) is expressed as a Partial Euclidean Distance (PED). Thus, we deal with the following equation:

$$T_i \left(s^{(i)} \right) = T_{i+1} \left(s^{(i+1)} \right) + \left| e_i \left(s^{(i)} \right) \right|^2 \tag{8.56}$$

$$e_i \left(s^{(i)} \right) = \hat{r}_i - \sum_{j=i}^{N_t} R_{ij} s_j \tag{8.57}$$

where $T_i(s^{(i)})$ represents the cumulative PED at level i from levels $i+1$ and $T_{N_t+i}(s^{(N_t+i)}) = 0$. $e_i(s^{(i)})$ represents the distance from level $i+1$ to level i. i represents each level of a tree and the total number of levels are same as the number of transmit antennas. Each node means signal constellation order. $s^{(i)} = [s^i, s^{i+1}, \ldots, s^{N_t}]$ is a sequence of symbols until level i. $s^{(1)}$ at level 1 means the total sequence of the symbol. Figure 8.18 illustrates an SD search tree with N_t transmit antennas and M-ary modulation.

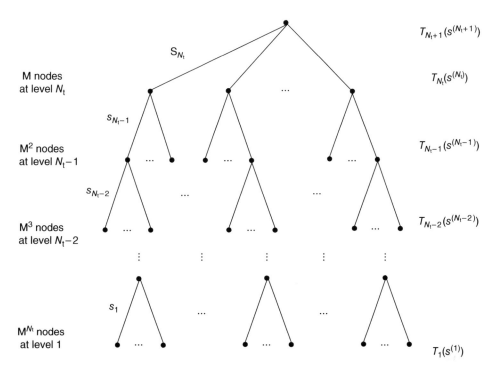

Figure 8.18   Tree structure for N_t transmit antenna and M-ary modulation

Example 8.3   SD search tree construction
Consider a three transmit antennas system with BPSK and draw the tree corresponding to the system.

Solution
The level of the tree is same as the number of transmit antennas. The tree has as many nodes as the modulation order. Thus, the levels and nodes of the tree are 3 and 2, respectively. Figure 8.19 illustrates the SD search tree for three transmit antenna and BPSK modulation.   ∎

Example 8.4   SD algorithm
Consider a 3×3 MIMO system with BPSK. In the receiver, we have the following received symbol $\hat{r}$ and upper triangular matrix $\boldsymbol{R}$:

$$\hat{r} = \begin{bmatrix} 0.1\,3.22 - 1.25 \end{bmatrix}^{T} \quad \text{and} \quad \boldsymbol{R} = \begin{bmatrix} 0.51 & -1.35 & -2.39 \\ 0 & 0.42 & -2.15 \\ 0 & 0 & 0.55 \end{bmatrix}$$

and assume the sphere radius $\mathcal{R}$ is 3. Estimate the transmitted symbols using the SD tree search.

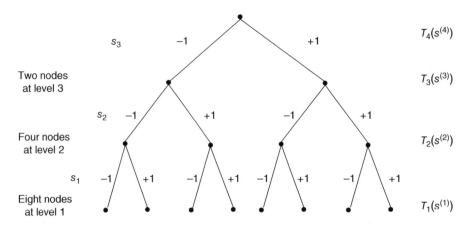

Figure 8.19   SD search tree for three transmit antenna and BPSK

Solution

As the first step, the SD search tree for three transmit antenna and BPSK is constructed as shown in Figure 8.19. From (8.56) and (8.57), we calculate the cumulative PED $T_i(s^{(i)})$ as follows:

At level 3,

$$e_3\left(s^{(3)}\right) = \hat{r}_3 - \sum_{j=3}^{3} R_{3j} s_j = -1.25 - 0.55 \cdot s_3$$

and

$$T_3\left(s^{(3)}\right) = T_4\left(s^{(4)}\right) + \left|e_3\left(s^{(3)}\right)\right|^2 = \left|-1.25 - 0.55 \cdot s_3\right|^2$$

$$= \begin{cases} 3.24 \text{ when } s_3 = +1 \\ 0.49 \text{ when } s_3 = -1 \end{cases}.$$

At level 2,

$$e_2\left(s^2\right) = \hat{r}_2 - \sum_{j=2}^{3} R_{2j} s_j = 3.22 - (0.42 \cdot s_2 - 2.15 \cdot s_3)$$

and

$$T_2\left(s^{(2)}\right) = T_3\left(s^{(3)}\right) + \left|e_2\left(s^{(2)}\right)\right|^2$$

$$= \left|-1.25 - 0.55 \cdot s_3\right|^2 + \left|3.22 - (0.42 \cdot s_2 - 2.15 \cdot s_3)\right|^2$$

$$= \begin{cases} 27.74 & \text{when } s_2 = +1 \text{ and } s_3 = +1 \\ 36.76 & \text{when } s_2 = -1 \text{ and } s_3 = +1 \\ 0.91 & \text{when } s_2 = +1 \text{ and } s_3 = -1 \\ 2.71 & \text{when } s_2 = -1 \text{ and } s_3 = -1 \end{cases}.$$

At level 1,

$$e_1\left(s^1\right)=\hat{r}_1-\sum_{j=1}^{3}R_{1j}s_j=0.1-(0.51\cdot s_1-1.35\cdot s_2-2.39\cdot s_3)$$

and

$$T_1\left(s^{(1)}\right)=T_2\left(s^{(2)}\right)+\left|e_1\left(s^{(1)}\right)\right|^2$$

$$=\left|-1.25-0.55\cdot s_3\right|^2+\left|3.22-(0.42\cdot s_2-2.15\cdot s_3)\right|^2+\left|0.1-(0.51\cdot s_1-1.35\cdot s_2-2.39\cdot s_3)\right|^2$$

$$=\begin{cases}38.83 & \text{when } s_1=+1, s_2=+1 \text{ and } s_3=+1 \\ 46.67 & \text{when } s_1=-1, s_2=+1 \text{ and } s_3=+1 \\ 37.16 & \text{when } s_1=+1, s_2=-1 \text{ and } s_3=+1 \\ 39.49 & \text{when } s_1=-1, s_2=-1 \text{ and } s_3=+1 \\ 3.02 & \text{when } s_1=+1, s_2=+1 \text{ and } s_3=-1 \\ 1.1 & \text{when } s_1=-1, s_2=+1 \text{ and } s_3=-1 \\ 19.93 & \text{when } s_1=+1, s_2=-1 \text{ and } s_3=-1 \\ 12.51 & \text{when } s_1=-1, s_2=-1 \text{ and } s_3=-1\end{cases}.$$

Figure 8.20 illustrates PED calculation in the SD search tree.

If one node PED exceeds the sphere radius $\mathcal{R}=3$, the SD detection algorithm does not search down from this node. Thus, the SD detection algorithm visits the solid branches and selects the white circle node only which is less than 3. Finally, we select the smallest node (the value is 1.1) at level 1 and estimate the transmitted symbol is [−1 +1 −1]. This estimation is same as the ML estimation but the SD algorithm visits only four nodes. Thus, the complexity is much less than the ML detection. ∎

The complexity of the SD detection algorithm is not fixed because the node visiting depends on channel condition. In addition, the SD algorithm is based on sequential search so that it is

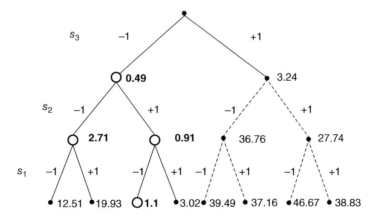

Figure 8.20   PED calculation in SD search tree

difficult to use pipelining techniques and system throughput is low. On the other hand, the complexity of FSD detection algorithm is fixed because it searches only a fixed number of nodes. The architecture is suitable for pipelining techniques. In the hardware implementation point of view, this is an important advantage. For example, one symbol from an antenna with a better channel condition or a higher SNR will be correctly detected. Thus, we do not need to search the other possible symbols. The FSD algorithm calculates a fixed number of transmitted symbols which are subset of all possible transmitted symbols. If the number of branches we consider in the FSD algorithm is $[s_1 \ s_2 \ s_3] = [1 \ 2 \ 1]$ (e.g., we decide $s_1 = -1$ and $s_3 = -1$ in advance.) in Example 8.3, the FSD search tree can be drawn as shown in Figure 8.21.

We calculate only two paths in the FSD search tree. The selection of the fixed number of the transmitted symbol affects complexity and performance. They are trade-off relationship. Figure 8.22 illustrates the FSD hardware architecture.

The FSD hardware architecture needs the received symbol r and the channel matrix H as the inputs where we assume H is perfectly known in the receiver. In the channel ordering block, the detection ordering is decided with satisfying the following relationship:

$$E(n_{N_t}) \geq E(n_{N_t-1}) \geq \cdots \geq E(n_1) \tag{8.58}$$

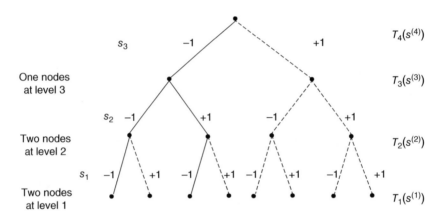

Figure 8.21 FSD search tree for $(s_1 = -1$ and $s_3 = -1)$

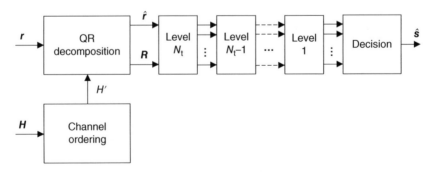

Figure 8.22 FSD hardware architecture

where $E(n_i)$ is the distribution of candidates (or the visited nodes at level i). The columns of the channel matrix H are ordered iteratively. For the received symbol with the lowest SNR, we select the maximum number of branches (same as modulation order) at the level. For the received symbol with a high SNR, we select smaller number of branches at the level. The QR decomposition block can be implemented by COordinate Rotation DIgital Computer (CORDIC) [14]. We obtain an upper triangular matrix R and $\hat{r}$ from QR decomposition. PEDs are calculated at each level block. In decision block, one node with the minimum PED is selected and the transmitted symbols $\hat{s}$ are estimated as the output of the FSD.

In this chapter, we mainly investigated two MIMO techniques: space time coding and spatial multiplexing. Space time coding shows us good performance at a low SNR and spatial multiplexing shows us good performance at a high SNR. Thus, adaptive MIMO technique combining both techniques would be a good solution. Figures 8.23 and 8.24 illustrate adaptive MIMO architecture and its performance, respectively.

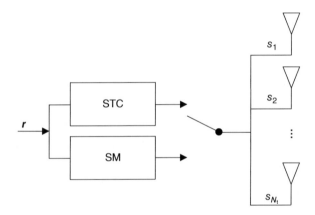

Figure 8.23   Adaptive MIMO transmitter

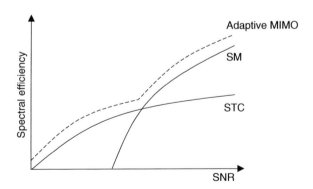

Figure 8.24   Adaptive MIMO performance

Summary 8.2   Sphere decoding algorithm

1. The SD algorithm can achieve near the ML detection performance with a lower complexity than the exhaustive search.
2. The SD algorithm searches a finite number of symbols within sphere radius R centered at the received symbol vector. The SD algorithm is expressed as follows:

$$\hat{s}_{SD} = \arg \min_{s \in S} \left(\|r - Hs\|^2 < \mathcal{R} \right)$$

$$= \arg \min_{s \in S} \left(\|\hat{r} - Rs\|^2 < \mathcal{R} \right)$$

$$= arg \min_{s \in S} \left(\|\hat{r} - Rs\|^2 < \mathcal{R} \right)$$

where $H = QR$ and $\hat{r} = Q^H r$.
3. In order to solve the above problem, an iterative solution is used as follows:

$$T_i \left(s^{(i)} \right) = T_{i+1} \left(s^{(i+1)} \right) + \left| e_i \left(s^{(i)} \right) \right|^2$$

$$e_i \left(s^{(i)} \right) = \hat{r}_i - \sum_{j=i}^{N_t} R_{ij} s_j$$

where $T_i(s^{(i)})$ represents the cumulative PED at level i from levels $i+1$ and $T_{N_t+i} \left(s^{(N_t+i)} \right) = 0$. $e_i(s^{(i)})$ represents the distance from level $i+1$ to level i. i represents each level of a tree and total number of levels are same as the number of transmit antennas. Each node means signal constellation order. $s^{(i)} = [s^i, s^{i+1}, \ldots, s^{N_t}]$ is a sequence of symbols until level i. $s^{(1)}$ at level 1 means the total sequence of the symbol.

8.5   Problems

8.1. Describe the MIMO antenna design considerations for indoor and outdoor applications.

8.2. SVD is fully parallelizing MIMO channel but each subchannel has different gain. On the other hands, QR decomposition has equal diagonal value but each subchannel is not fully parallelizing. Compare the BER performance of both SVD beamforming and QR decomposition beamforming.

8.3. Compare the diversity gain of both Alamouti scheme and V-BLAST for 2×2 and 4×4 MIMO channels.

8.4. Consider the space time code and any two distinct codeword by S_4^c (8.39). Find the diversity gain.

8.5. Describe the encoding and decoding process of 2×2 Alamouti scheme.

8.6. Consider the 4×4 MIMO system in Rayleigh fading channel. Compare the BER performance of ZF, MMSE, and SIC detection algorithm.

8.7. Adaptive MIMO performance is shown in Figure 8.24. Find the switching point from STC to SM.

8.8. Compare the MIMO schemes in LTE and WiMAX.

References

[1] R. Janaswamy "Effect of Element Mutual Coupling on the Capacity of Fixed Length Linear arrays," *IEEE Antennas and Wireless Propagation Letters*, vol. **1**, no. 1, pp. 157–160, 2002.

[2] M. A. Jensen and R. W. Wallace, "A Review of Antennas and Propagation for MIMO Wireless Communications," *IEEE Transactions on Antennas and Propagation*, vol. **52**, no. 11, 2004.

[3] S. W. Ellingson, "Antenna Design and Site Planning Considerations for MIMO," *Proceedings of IEEE the 62nd Vehicular Technology Conference (VTC Fall 2005)*, pp. 1718–1722, September 2005.

[4] V. Tarokh and H. Jafarkhani, "A Differential Detection Scheme for Transmit Diversity," *IEEE Journal on Selected Areas in Communications*, vol. **18**, no. 7, 2000.

[5] I. E. Telatar, "Capacity of Multi-Antenna Gaussian Channels," *European Transactions on Telecommunications and Related Technologies*, vol. **10**, no.6, pp.586–596, 1999.

[6] D. Tse and P. Viswanath, *Fundamentals of Wireless Communication*, Cambridge University Press, Cambridge, 2005.

[7] V. Tarokh, N. Seshadri, and A. R. Calderbank, "Space-Time Codes for High Data Rate Wireless Communication: Performance Criterion and Code Construction," *IEEE Transactions on Information Theory*, vol. **44**, no. 2, 1998.

[8] D. M. Ionescu, "New Results on Space-Time Code Design Criteria," *Proceedings of IEEE Wireless Communications and Networking Conference (WCNC'99)*, vol. **2**, pp. 684–687, 1999.

[9] V. Tarokh, H. Jafarkhani, and A. R. Calderbank, "Space-time Block Coding from Orthogonal Designs," *IEEE Transactions on Information Theory*, vol. **45**, pp. 1456–1467, 1999.

[10] V. Tarokh, H. Jafarkhani, and A. R. Calderbank, "Space-time Block Doing for Wireless Communications: Performance Results," *IEEE Journal on Selected Areas in Communications*, vol. **17**, no. 3, pp. 451–460, 1999.

[11] E. Viterbo and J. Boutros, "A Universal Lattice Code Decoder for Fading Channels," *IEEE Transactions on Information Theory*, vol. **45**, no. 5, pp. 1639–1642, 1999.

[12] L. G. Barbero and J. S. Thompson, "Fixing the Complexity of the Sphere Decoder for MIMO Detection," *IEEE Transactions on Wireless Communications*, vol. **7**, no. 6, pp. 2131–2142, 2008.

[13] B. Hassibi and H. Vikalo, "On the Sphere-Decoding Algorithm. Part I, Expected Complexity," *IEEE Transactions on Signal Processing*, vol. **53**, no. 8, pp. 2806–2818, 2005.

[14] J. E. Volder, "The CORDIC Trigonometric Computing Techniques," *IRE Transactions on Electronic Computers*, vol. **EC-8**, no. 3, pp. 330–334, 1959.

9

Channel Estimation and Equalization

As we discussed in Chapter 3, a wireless channel is no longer a black box. Wireless communication systems can obtain Channel State Information (CSI) such a fading effect and power delay. The mathematical models of the wireless channels enable us to understand how a received signal is affected by specific environments. Channel estimation is used to obtain CSI and has become a critical part of wireless communication systems. Many communication techniques use this information and improve system performances. Especially, the performance of equalizers and MIMO techniques depends on accuracy of channel estimation.

9.1   Channel Estimation

Channel estimation is simply defined as characterizing a mathematically modeled channel. The mathematical channel model is characterized by long-term CSI (or statistical CSI) and short-term CSI (or instantaneous CSI). Long-term CSI simply means statistical information such as channel statistical distribution and average channel gain. Short-term CSI simply means channel impulse response. Channel estimation algorithms typically find *channel impulse response* by time domain channel estimation (before DFT processing in OFDM systems) or *channel frequency response* by frequency domain channel estimation (after DFT processing in OFDM systems). While short-term CSI can be accurately estimated in a slow fading channel, it is difficult to obtain accurate short-term CSI in a fast fading channel. Thus, an adaptive channel estimation algorithm is used for a rapidly time varying channel. An adaptive channel estimation algorithm changes its parameters according to time varying environments. Figure 9.1 illustrates general channel estimation. The design criteria of channel estimators are to minimize a Mean Squared Error (MSE) and computational complexity.

Wireless Communications Systems Design, First Edition. Haesik Kim.
© 2015 John Wiley & Sons, Ltd. Published 2015 by John Wiley & Sons, Ltd.

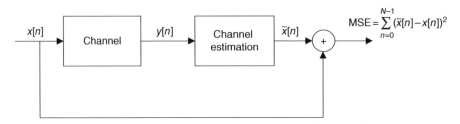

Figure 9.1   General channel estimation

The channel knowledge by channel estimation is very useful to mitigate channel impairments. An equalizer uses this knowledge and reduces inter-symbol interferences. A correlator of a rake receiver uses this knowledge and makes successful despreading. Especially, the capacity of MIMO systems depends on CSI in both a transmitter and a receiver.

Channel estimation can be classified into pilot-assisted channel estimation, blind channel estimation, and Decision-Directed Channel Estimation (DDCE). *Pilot-assisted channel estimation* is the most popular method that a transmitter sends a known signal, where pilot means the reference signal used by both a transmitter and a receiver. It can be applied to any wireless communication systems and its computational complexity is very low. However, the main disadvantage is the reduction of transmission rate because non-data symbols (pilots) are inserted. Thus, one design challenge of pilot-assisted channel estimation is to minimize the number of pilots and estimate the channel accurately. A pilot assignment method can be mainly classified into block type and comb type as we discussed in Chapter 5. A block-type pilot assignment method is suitable for a slow fading channel because the channel varies slowly over a number of OFDM symbols. On the other hands, a comb-type pilot assignment method is suitable for a fast fading channel because pilots are uniformly distributed over a number of OFDM symbols. Interpolation in frequency domain is required to find a channel response of non-pilot (data) symbols and it is more sensitive to frequency selective channel. *Blind channel estimation* does not require non-data symbols (pilots) and uses inherent information in the received symbols. Although blind channel estimation does not waist a bandwidth, its computational complexity and latency are significantly higher than pilot-assisted channel estimation. For example, about 100 symbols are required to obtain 1 channel coefficient. Thus, blind channel estimation is rarely used in practical wireless communication systems. *DDCE* uses both pilot symbols and detected data symbols. It updates the values of channel estimation. Thus, DDCE has a better performance than pilot-assisted channel estimation. Figure 9.2 illustrates simple DDCE structure.

The estimated symbol $\hat{S}_n$ in frequency domain is obtained by the previous estimated channel transfer function $\tilde{H}_{n-1}$ and the received symbol R_n in the equalizer (zero-forcing (ZF) algorithm is assumed) as follows:

$$\hat{S}_n = \frac{R_n}{\tilde{H}_{n-1}}. \tag{9.1}$$

This estimated symbol is demodulated to $\tilde{S}_n$. The channel transfer function is updated by this demodulated symbol $\tilde{S}_n$ and the previous channel transfer function $\tilde{H}_{n-1}$ as follows:

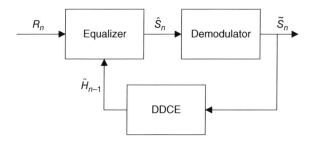

Figure 9.2   Decision-directed channel estimation

$$\hat{H}_n = \frac{R_n}{\tilde{S}_n} \tag{9.2}$$

and

$$\tilde{H}_n = \alpha \hat{H}_n + (1 - \alpha) \tilde{H}_{n-1} \tag{9.3}$$

where α is the update factor depending on the quality of the decision symbol. Since DDCE uses both pilot symbols and detected data symbols, fewer pilot symbols are required than pilot-assisted channel estimation.

In this chapter, we focus on pilot-assisted channel estimation in OFDM systems. There are two main design issues for channel estimation in OFDM systems. The first one is pilot design and the second one is channel estimator design with both low complexity and accuracy. The first design issue can be described as where pilot symbols are allocated and how many pilot symbols are used in time and frequency domain. Each question depends on pilot patterns, interval between pilot symbols, and redundancies of OFDM systems. We should select a suitable pilot design pattern among block type, comb type, and lattice type according to system requirements. Pilot symbol spacing is decided by (5.67) and (5.68). If a short pilot interval is designed, channel estimation can be accurate at frequency selective channel and time-varying channel but spectral efficiency would be low due to redundancy increase. The pilot design was already discussed in Chapter 7 and LS and MMSE channel estimation was introduced in Chapter 5. Thus, we design a channel estimator and equalizer for SISO/MIMO-OFDM systems in this chapter.

Channel estimation is carried out for OFDM systems. First of all, we consider a simple signal model as follows:

$$Y = XH + n \tag{9.4}$$

where $Y \, (\in C^{N_{\mathrm{IDFT}}})$ is a received OFDM symbol vector, $X (\in C^{N_{\mathrm{IDFT}} \times N_{\mathrm{IDFT}}})$ is a diagonal transmitted OFDM symbol matrix including transmitted data, pilots, and null signals, $H \big(H (\in C^{N_{\mathrm{IDFT}}}) \big)$ is a channel frequency response vector we should estimate, and $n (\in C^{N_{\mathrm{IDFT}}})$ is a white Gaussian noise vector. The channel frequency response H can be expressed as the channel impulse response $h \big(H = \mathrm{DFT}(h) \big)$ and (9.4) is rewritten as follows:

$$Y = XFh + n \tag{9.5}$$

where $h(\in C^{N_{\text{IDFT}}})$ is the channel impulse response and $F(\in C^{N_{\text{IDFT}} \times N_{\text{IDFT}}})$ is the DFT matrix as follows:

$$F = \begin{bmatrix} f_{1,1} & \cdots & f_{1,N_{\text{IDFT}}} \\ \vdots & \ddots & \vdots \\ f_{N_{\text{IDFT}},1} & \cdots & f_{N_{\text{IDFT}},N_{\text{IDFT}}} \end{bmatrix} \tag{9.6}$$

where N_{IDFT} is the IDFT size. Pilot-assisted channel estimation uses a part of subcarriers as a pilot. An interpolation technique should be performed to obtain a channel estimation value at each data subcarrier. In order to simplify the system model and reduce the complexity of the channel estimator, we consider two assumptions [1]. The first assumption is that we consider a finite length impulse response and deal with only significant channel tap. That is, when the channel impulse response h has the maximum delay at tap $[0, L-1]$, the first L columns of F are considered. The second assumption is that only rows of F corresponding to the position of pilots are considered for the transmitted matrix X because the pilots are distributed in the OFDM symbol. Thus, we have the following equation:

$$Y_{\text{p}} = X_{\text{p}} T_{\text{p}} h_{\text{p}} + n_{\text{p}} \tag{9.7}$$

where $Y_{\text{p}}(\in C^{N_{\text{p}} \times L})$, $h_{\text{p}}(\in C^{L \times L})$, and $n_{\text{p}}(\in C^{N_{\text{p}} \times L})$ are the received pilot symbol matrix, channel impulse response matrix, and white Gaussian noise matrix associated with the transmitted pilot symbol, respectively. X_{p} is the diagonal pilot symbol matrix and T_{p} is the DFT matrix corresponding to the transmitted pilot symbol as follows:

$$X_{\text{p}} = \begin{bmatrix} X_{\text{p}}(0) & \cdots & 0 \\ \vdots & \ddots & \vdots \\ 0 & \cdots & X_{\text{p}}(N_{\text{p}}-1) \end{bmatrix} \tag{9.8}$$

and

$$T_{\text{p}} = \frac{1}{\sqrt{N_{\text{DFT}}}} \begin{bmatrix} 1 & e^{-j\frac{2\pi}{N_{\text{DFT}}}(i_{\text{p}})(1)} & \cdots & e^{-j\frac{2\pi}{N_{\text{DFT}}}i_{\text{p}}(L-1)} \\ 1 & e^{-j\frac{2\pi}{N_{\text{DFT}}}(i_{\text{p}}+1)(1)} & \cdots & e^{-j\frac{2\pi}{N_{\text{DFT}}}(i_{\text{p}}+1)(L-1)} \\ \vdots & \vdots & \ddots & \vdots \\ 1 & e^{-j\frac{2\pi}{N_{\text{DFT}}}(i_{\text{p}}+N_{\text{p}}-1)(1)} & \cdots & e^{-j\frac{2\pi}{N_{\text{DFT}}}(i_{\text{p}}+N_{\text{p}}-1)(L-1)} \end{bmatrix} \tag{9.9}$$

where N_{DFT} is the DFT size, N_{p} is the number of pilot subcarriers, and i_{p} is the pilot index. The channel impulse response matrix is reduced to an $L \times L$ matrix. This result significantly affects LS and MMSE estimation.

The block-type pilot-assisted channel estimation method can be performed under the assumption of a slow fading channel. The pilots are inserted in all subcarriers. Thus, interpolation is not needed. However, the comb-type pilot-assisted channel estimation method can be developed under the assumption of channel change in OFDM symbol sequences. Thus, interpolation is needed to estimate a whole subcarrier. As we discussed in Chapter 5, the receiver estimates the channel condition based on LS and MMSE algorithms. When considering the comb-type pilot channel estimation method, the pilots are uniformly inserted and the kth subcarrier in the OFDM symbol can be expressed as follows:

$$X(k) = X(mD+d) = \begin{cases} X_p(m), & \text{when } d = 0 \\ \text{Data}, & \text{when } d = 1, 2, D-1 \end{cases} \qquad (9.10)$$

where m and D are a pilot subcarrier ($0 \le m \le N_p-1$) and N_{DFT}/N_p ($0 \le d < D$), respectively. LS channel estimation determining the channel frequency response is calculated as follows:

$$\boldsymbol{H}_{p,LS} = \left[H_p(0)\, H_p(1) \cdots H_p(N_p-1) \right] = \left[\frac{Y_p(0)}{X_p(0)}\, \frac{Y_p(1)}{X_p(1)} \cdots \frac{Y_p(N_p-1)}{X_p(N_p-1)} \right]. \qquad (9.11)$$

This is estimated for only pilot subcarriers. Thus, channel estimation results should be interpolated over a whole subcarrier including data subcarriers. One easiest way is to connect two pilots with straight lines. We call it piecewise constant interpolation (or nearest neighbor interpolation). It allocates same pilot values to near data subcarriers. Another simple interpolation is linear interpolation. It allocates the values on the straight line between two pilots. Linear interpolation has better performance than piecewise constant interpolation [2]. When we have two pilots (e.g., $(H(a), a)$ and $(H(b), b)$), the interpolant $(H(c), c)$ satisfies the following equation:

$$\frac{H(c)-H(a)}{c-a} = \frac{H(b)-H(a)}{b-a} \qquad (9.12)$$

and the interpolant at the data subcarrier is represented as follows:

$$H(c) = H(a)+(H(b)-H(a))\frac{c-a}{b-a}. \qquad (9.13)$$

Thus, we express channel estimation values $(H_e(k))$ at data subcarriers as follows [3]:

$$H_e(k) = \frac{d}{D}\left(H_p(m+1)-H_p(m)\right)+H_p(m) \qquad (9.14)$$

where k is the data subcarrier ($mD \le k < (m+1)D$). Linear interpolation is simple but does not produce accurate interpolants. Thus, polynomial interpolation is widely used. It is the generalized form of linear interpolation. The order of polynomial interpolation is defined as the number of given data (or pilots) minus one. The first-order interpolation (e.g., when we consider only two pilots) is linear interpolation. Basically, a high-order interpolation generates wild oscillations. The practical limitation is the fifth-order interpolation. Thus, the second-order polynomial interpolation is good choice according to performance and complexity. It shows us better performance than linear interpolation. The channel estimation values are decided by linear combination of the weighted pilot values. We express channel estimation values ($H_e(k)$) at data subcarriers as follows [3]:

$$H_e(k) = \alpha_1 H_p(m-1) + \alpha_0 H_p(m) + \alpha_{-1} H_p(m+1) \tag{9.15}$$

where

$$\alpha_1 = \frac{\beta(\beta-1)}{2}, \alpha_0 = -(\beta-1)(\beta+1), \alpha_{-1} = \frac{\beta(\beta+1)}{2} \text{ and } \beta = \frac{d}{N_{\text{DFT}}}. \tag{9.16}$$

Another practical interpolation is low-pass interpolation (or upsampling). It has simple two-step process. The first step is to insert zeros into the channel estimation position (data subcarriers) and the second step is to put the sequence into low-pass filter (LPF; FIR filter). Figure 9.3 illustrates one example of low-pass interpolation by a factor of 2. As we can observe, the inserted zeros by upsampling increase the data rate but there is no effect to sample distribution. After passing through LPF, the inserted zeros are replaced by interpolants.

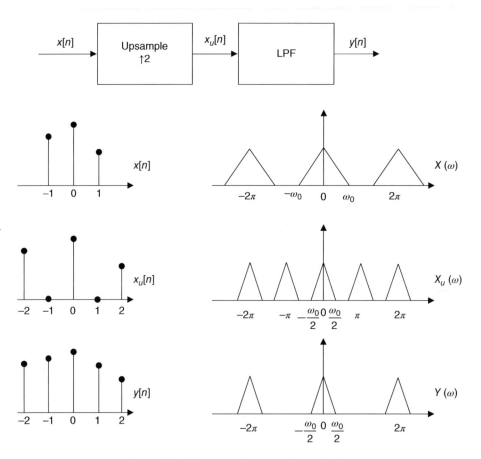

Figure 9.3 Example of low-pass interpolation

Example 9.1   Interpolation
Consider the following system parameters:

Parameters	Values
N_{DFT}	16
N_p	4
Pilot position k	[0 4 7 11]
Transmitted pilots $[X_p(0)\ X_p(1)\ X_p(2)\ X_p(3)]$	[1 1 1 1]
Received pilots $[Y_p(0)\ Y_p(1)\ Y_p(2)\ Y_p(3)]$	[0.2 2.5 2.8 1.2]

Find the channel estimation values at data subcarriers using piecewise constant interpolation, linear interpolation, and second-order polynomial interpolation.

Solution
From (9.11), LS channel estimation determining the channel frequency response is calculated at pilot positions as follows:

$$\left[H_p(0)H_p(1)\ H_p(2)H_p(3)\right]=\left[0.2\ 2.5\ 2.8\ 1.2\right]$$

and Figure 9.4 illustrates channel response at pilot positions.

The piecewise constant interpolation method allocates same pilot values to near data subcarriers. Thus, we can easily obtain channel estimation values as shown in Figure 9.5.

The linear interpolation method allocates channel estimation values on the straight line between two pilots. From (9.14), the channel estimation values $H_e(k)$ between $H_p(0)$ and $H_p(1)$ are calculated as follows:

$$D=N_{DFT}/N_p=16/4=4.$$

$$mD\leq k<(m+1)D \text{ when } m=0\rightarrow 0\leq k<4$$

$$0\leq d<D\rightarrow 0\leq d<4$$

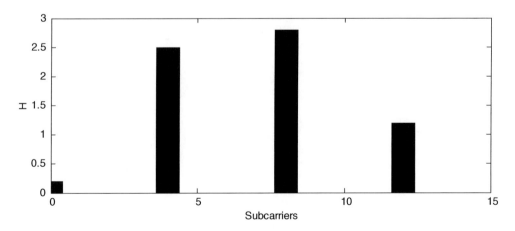

Figure 9.4   Channel response at pilot position

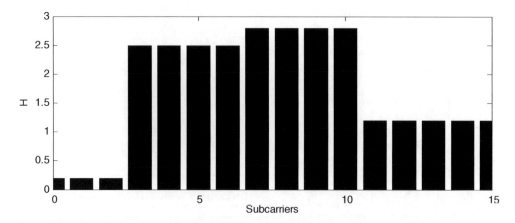

Figure 9.5   Piecewise constant interpolation

$$H_e(k) = \frac{d}{4}\big(H_p(1) - H_p(0)\big) + H_p(0)$$

$$\big[H_e(0)\,H_e(1)H_e(2)H_e(3)\big] = [0.2\ 0.78\ 1.35\ 1.92]$$

In the same way, the remaining channel estimation values $H_e(k)$ are calculated as follows:
$mD \leq k < (m+1)D$ when $m = 1 \rightarrow 4 \leq k < 8$

$$H_e(k) = \frac{d}{4}\big(H_p(2) - H_p(1)\big) + H_p(1)$$

$$\big[H_e(4)\,H_e(5)\,H_e(6)\,H_e(7)\big] = [2.5\ 2.58\ 2.65\ 2.73]$$

$mD \leq k < (m+1)D$ when $m = 2 \rightarrow 8 \leq k < 12$

$$H_e(k) = \frac{d}{4}\big(H_p(3) - H_p(2)\big) + H_p(2)$$

$$\big[H_e(8)\,H_e(9)\,H_e(10)\,H_e(11)\big] = [2.8\ 2.4\ 2\ 1.6]$$

Figure 9.6 illustrates linear interpolation.
   The second-order polynomial interpolation method allocates the channel estimation values using linear combination of the weighted pilot values. From (9.15) and (9.16), the channel estimation values $H_e(k)$ between $H_p(1)$ and $H_p(2)$ are calculated as follows:

$$mD \leq k < (m+1)D \text{ when } m = 1 \rightarrow 4 \leq k < 8$$

$$0 \leq d < D \rightarrow 0 \leq d < 4$$

$$\beta = \frac{d}{16}$$

$$\alpha_1 = \frac{\beta(\beta - 1)}{2}, \quad \alpha_0 = -(\beta - 1)(\beta + 1), \quad \alpha_{-1} = \frac{\beta(\beta + 1)}{2}$$

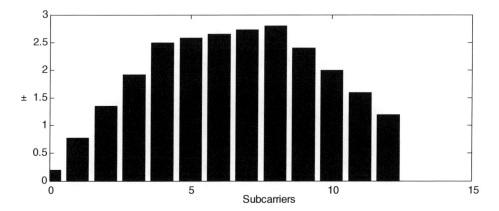

Figure 9.6   Linear interpolation

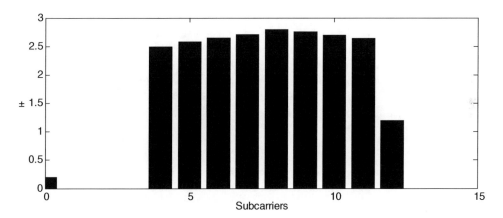

Figure 9.7   Second-order polynomial interpolation

$$H_e(k) = \alpha_1 H_p(0) + \alpha_0 H_p(1) + \alpha_{-1} H_p(2)$$

$$\left[H_e(4) H_e(5) H_e(6) H_e(7) \right] = [2.5\ 2.58\ 2.65\ 2.71]$$

In the same way, the remaining channel estimation values $H_e(k)$ are calculated as follows:
$mD \leq k < (m+1)D$ when $m = 2 \rightarrow 8 \leq k < 12$

$$H_e(k) = \alpha_1 H_p(1) + \alpha_0 H_p(2) + \alpha_{-1} H_p(3)$$
$$[H_e(8) H_e(9) H_e(10) H_e(11)] = [2.8\ 2.76\ 2.7\ 2.64]$$

Figure 9.7 illustrates the result of the second-order polynomial interpolation method.

In the linear interpolation method and the second-order polynomial interpolation method, several interpolants were not estimated because the pilots are not enough to cover all subcarriers.   ∎

Example 9.2   Low-pass interpolation
Consider the following discrete signal:

$$x[n] = \sin(2\pi30n) + \cos(2\pi50n) + \sin(2\pi10n)$$

and its discrete signal is illustrated as shown in Figure 9.8.
Perform the interpolation using upsampling by a factor of 2 and 4.

Solution
Applying a LPF after upsampling by 2 and 4, we obtain low-pass interpolations as shown in Figures 9.9 and 9.10.   ∎

LS channel estimation is simple but the performance is not good. In block-type pilot assignment, the MMSE estimator has about 10–15 dB better performance than the LS estimator for the same MSE value [1]. The MMSE estimator calculates the channel impulse response (CIR) which minimizes the means squared error between the exact CIRs and the estimated CIRs. One disadvantage of MMSE estimators is a high complexity. It increases exponentially according to observation samples. Thus, a linear MMSE (LMMSE) estimator is widely used in the OFDM system. In Ref. [4], low rank approximation based on the DFT is proposed and the complexity is reduced by singular value decomposition. The LMMSE estimator uses the frequency correlation of the channel. It can be calculated as follows [1]:

$$H_{p,\text{LMMSE}} = R_{HH_p}\left(R_{H_pH_p} + \sigma_n^2\left(XX^H\right)^{-1}\right)^{-1} H_{p,\text{LS}} \tag{9.17}$$

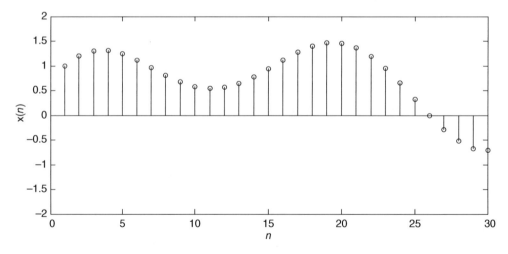

Figure 9.8   Signal $x[n]$

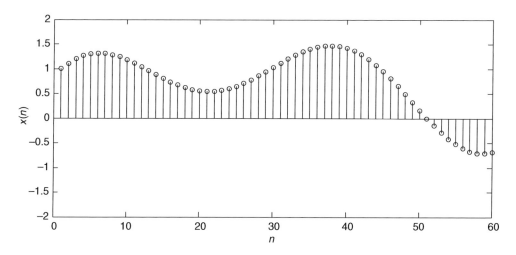

Figure 9.9   Low-pass interpolation by a factor of 2

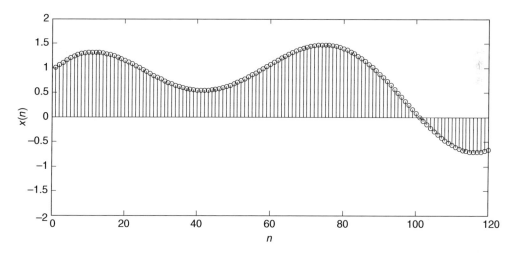

Figure 9.10   Low-pass interpolation by a factor of 4

where $\boldsymbol{R}_{HH_p}$ is the cross-correlation matrix between all subcarriers and pilots, $\boldsymbol{R}_{H_p H_p}$ is the auto-correlation matrix between pilots, and σ_n^2 is the noise variance. The correlation matrices are defined as follows:

$$\boldsymbol{R}_{HH_p} = E\left[\boldsymbol{H}\boldsymbol{H}_p^H\right] \tag{9.18}$$

$$\boldsymbol{R}_{H_p H_p} = E\left[\boldsymbol{H}_p\boldsymbol{H}_p^H\right]. \tag{9.19}$$

As we can observe from (9.17), the LMMSE calculation is very complex. We need simpler equation. If the LMMSE equation does not depend on the transmitted symbols, the calculation can be significantly reduced. Thus, the simplified LMMSE estimator is expressed as follows [4]:

$$H_{p,\text{LMMSE}} = R_{HH_p}\left(R_{H_pH_p} + \frac{\beta}{\text{SNR}}I_{N_p}\right)^{-1}H_{p,\text{LS}} \tag{9.20}$$

where

$$\beta = E\left[\left|X_k\right|^2\right]E\left[\frac{1}{\left|X_k\right|^2}\right] \tag{9.21}$$

and SNR is average signal to noise radio at subcarriers. β is constant depending on signal constellation. For example, $\beta = 1$ for QPSK and $\beta = 17/9$ for 16 QAM [5]. The simplification comes from SNR estimation simplification. Thus, this equation would not provide accurate channel estimation when it is difficult to obtain SNR levels in the whole OFDM subcarrier [6]. When a block-type pilot assignment method is considered, the LMMSE estimator is rewritten as follows:

$$H_{\text{LMMSE}} = WH_{\text{LS}} = R_{H_pH_p}\left(R_{H_pH_p} + \frac{\beta}{\text{SNR}}I_{N_{\text{DFT}}}\right)^{-1}H_{\text{LS}}. \tag{9.22}$$

In a practical system, the channel autocorrelation and SNR can be set to known factors at the receiver in advance. The channel autocorrelation calculation and SNR estimation affect the performance. The channel correlation matrices are derived in Ref. [4]. Thus, the term W is calculated only once. Equation 9.22 can be simplified in order to reduce multiplications. The correlation term can be decomposed by singular value decomposition (SVD) as follows:

$$R_{H_pH_p} = U\Lambda U^H \tag{9.23}$$

where U and Λ are a unitary matrix and a diagonal matrix containing the singular values, $\lambda_0, \lambda_1, \ldots, \lambda_{N_{\text{DFT}}-1}$, respectively. When m rank approximation is applied, (9.22) is rewritten as follows:

$$H_{m,\text{LMMSE}} = U\Delta_m U^H H_{\text{LS}}. \tag{9.24}$$

where Δ_m is a diagonal matrix containing the following entries:

$$\delta_i = \frac{\lambda_i}{\lambda_i + (\beta/\text{SNR})}, \quad \text{where } i = 0,1,\ldots,N_{\text{DFT}} - 1. \tag{9.25}$$

(9.24) is much simpler than (9.22) because the number of singular value m is much smaller than N_{DFT}.

Example 9.3   Comparison of LS and LMMSE
Consider the following system parameters:

Parameters	Values
Pilot assignment type	Block
N_{DFT}	8
SNR	10 dB
Modulation	16 QAM

Transmitted pilots

$$X_p = \left[X_p(0) \, X_p(1) \, X_p(2) \, X_p(3) \, X_p(4) \, X_p(5) \, X_p(6) \, X_p(7) \right]$$
$$= [-0.95 + 0.32i - 0.95 + 0.95i \quad -0.95 - 0.32i - 0.95 - 0.95i \quad 0.95 + 0.95i$$
$$-0.32 + 0.32i \, 0.32 + 0.95i \quad -0.95 + 0.32i]$$

Received pilots

$$Y_p = \left[Y_p(0) \, Y_p(1) \, Y_p(2) \, Y_p(3) \, Y_p(4) \, Y_p(5) \, Y_p(6) \, Y_p(7) \right]$$
$$= [-0.51 - 0.54i \quad -0.4 + 0.1i - 0.58 + 0.67i \quad 0.14 + 0.82i \quad 0.29 - 1.21i$$
$$-0.34 - 0.86i - 0.81 + 1.3i \quad -0.19 + 0.56i]$$

Channel autocorrelation function $R_{H_p H_p} =$

$$
\begin{bmatrix}
0.04 & 0.06+0.07i & -0.09+0.21i & -0.12-0.08i & -0.08+0.12i & -0.14-0.28i & 0.28+0.06i & -0.1+0.1i \\
0.06-0.07i & 0.2 & 0.23+0.48i & -0.33+0.1i & 0.09+0.33i & -0.7-0.18i & 0.52-0.5i & 0.02+0.32i \\
-0.1-0.2i & 0.23-0.48i & 1.4 & -0.15+0.89i & 0.87+0.15i & -1.2+1.45i & -0.37-1.68i & 0.79+0.32i \\
-0.12+0.08i & -0.33-0.1i & -0.15-0.89i & 0.58 & 0.57i & 1.05+0.62i & -1.03+0.42i & 0.12-0.54i \\
-0.08-0.12i & 0.09-0.33i & 0.87-0.15i & 0.57i & 0.57 & -0.61+1.04i & -0.41-1.02i & 0.53+0.12i \\
-0.14+0.28i & -0.7+0.18i & -1.21-1.45i & 1.05-0.62i & -0.6-1.04i & 2.57 & -1.43+1.85i & -0.35-1.1i \\
0.28-0.1i & 0.52+0.4i & -0.37+1.68i & -0.41+1.02i & -0.41+1.02i & -1.43-1.85i & 2.13 & -0.6+0.86i \\
-0.1-0.1i & 0.02-0.32i & 0.79-0.33i & 0.53-0.12i & 0.53-0.12i & -0.35+1.1i & -0.6-0.86i & 0.52
\end{bmatrix}
$$

Find the channel estimation values using LS and LMMSE.

Solution

From (9.11), LS channel estimation values can be calculated as follows:

$$H_{LS} = \left[H_p(0) \, H_p(1) \cdots H_p(N_p - 1) \right] = \left[\frac{Y_p(0)}{X_p(0)} \, \frac{Y_p(1)}{X_p(1)} \cdots \frac{Y_p(N_p - 1)}{X_p(N_p - 1)} \right]$$

$$= \left[\frac{-0.51 - 0.54i}{-0.95 + 0.32i} \, \frac{-0.4 + 0.1i}{-0.95 + 0.95i} \, \frac{-0.58 + 0.67i}{-0.95 - 0.32i} \, \frac{0.14 + 0.82i}{-0.95 - 0.95i} \right.$$

$$\left. \frac{0.29 - 1.21i}{0.95 + 0.95i} \, \frac{-0.34 - 0.86i}{-0.32 + 0.32i} \, \frac{-0.81 + 1.3i}{0.32 + 0.95i} \, \frac{-0.19 + 0.56i}{0.95 + 0.32i} \right]$$

$$= [0.31 + 0.67i \, 0.26 + 0.16i \, 0.34 - 0.82i - 0.5 - 0.36i$$
$$-0.48 - 0.79i - 0.81 + 1.88i \, 0.97 + 1.18i \, 0.59i].$$

From (9.22), LMMSE channel estimation can be found as follows:

$$H_{LMMSE} = W H_{LS} = R_{H_p H_p} \left(R_{H_p H_p} + \frac{\beta}{SNR} I_{N_{DFT}} \right)^{-1} H_{LS}$$

$$= R_{H_p H_p} \left(R_{H_p H_p} + 0.19 I_{N_{DFT}} \right)^{-1} H_{LS}$$

$$[0.15 + 0.09i \, 0.37 - 0.12i \, 0.14 - 1.02i - 0.66 + 0.02i$$
$$-0.02 - 0.65i - 1.18 + 0.75i \, 1.2 + 0.43i - 0.16 - 0.6i]$$

■

When a receiver has additional statistical information such as channel autocorrelation matrix and average SNR, the LMMSE estimator shows us good performance. Especially, in a low SNR scenario, it brings much better performance than the LS estimator. On the other hand, additional statistical information produces additional latency and complexity.

Summary 9.1   Channel estimation

1. Channel estimation is simply defined as characterizing a mathematically modeled channel. Channel estimation algorithms typically find channel impulse response or channel frequency response.
2. Channel estimation can be classified into pilot-assisted channel estimation, blind channel estimation, and decision DDCE. Pilot-assisted channel estimation where pilot means the reference signal used by both a transmitter and a receiver is the most popular method in which a transmitter sends a known signal. Blind channel estimation does not require non-data symbols (pilots) and uses inherent information in the received symbol. DDCE uses both pilot symbols and detected data symbols and updates the values of channel estimation.
3. LS channel estimation determining the channel frequency response is calculated as follows:

$$H_{p,LS} = [H_p(0)H_p(1)\cdots H_p(N_p-1)] = \left[\frac{Y_p(0)}{X_p(0)} \frac{Y_p(1)}{X_p(1)} \cdots \frac{Y_p(N_p-1)}{X_p(N_p-1)} \right].$$

4. LMMSE estimation is calculated as follows:

$$H_{\text{LMMSE}} = WH_{\text{LS}} = R_{H_pH_p} \left(R_{H_pH_p} + \frac{\beta}{\text{SNR}} I_{N_{\text{DFT}}} \right)^{-1} H_{\text{LS}}.$$

Summary 9.2   Interpolation

1. Piecewise constant interpolation (or nearest neighbor interpolation) allocates same pilot value to near data subcarriers.
2. Linear interpolation allocates the values on the straight line between two pilots as follows:

$$H(c) = H(a) + \left(H(b) - H(a)\right)\frac{c-a}{b-a}$$

where $(H(a),a)$ and $(H(b),b)$ are two pilots and $(H(c),c)$ is interpolant.

3. The second-order polynomial interpolation method is expressed as follows:

$$H_e(k) = \alpha_1 H_p(m-1) + \alpha_0 H_p(m) + \alpha_{-1} H_p(m+1)$$

$$\alpha_1 = \frac{\beta(\beta-1)}{2}, \ \alpha_0 = -(\beta-1)(\beta+1), \ \alpha_{-1} = \frac{\beta(\beta+1)}{2} \text{ and } \beta = \frac{d}{N_{DFT}}.$$

where $H_e(k)$ is channel estimation value and $H_p(m)$ is pilot.

4. The low-pass interpolation (or upsampling) method has simple two steps. The first step is to insert zeros into the channel estimation position (data subcarriers) and the second step is to put the sequence into LPF (FIR filter).

9.2 Channel Estimation for MIMO–OFDM System

MIMO–OFDM systems send symbols via multiple transmit antennas and receive the superposition of the symbols from multiple receive antennas. The superposed symbols should be extracted and the inter-antenna interference should be eliminated. Thus, channel estimation for multiple antenna systems cannot be simply transformed from single antenna systems. Channel estimation for MIMO–OFDM systems is more complicated than single antenna systems but the basic principle is similar to the single antenna systems.

One simple pilot assignment for MIMO–OFDM systems is the grid-type pilot assignment in both frequency domain and time domain as shown in Figure 9.11.

One transmit antenna sends its pilot at a given subcarrier while the other antenna remains silent (null subcarrier) [7]. This pilot assignment method protects each pilot from signals of the other antennas. The inter-antenna interference is easily eliminated. However, the disadvantage is to reduce spectral efficiency due to null subcarriers. In addition, the null subcarriers increase the Peak to Average Power Ratio (PAPR) [8]. It is essential that a transmitter has a

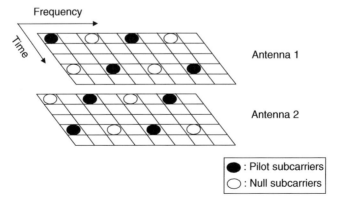

Figure 9.11   Pilots assignment for MIMO OFDM system

power amplifier which provides a suitable power for transmission. The power amplifier is very sensitive to operational area which requires a linear area. A high PAPR causes a nonlinear operation problem so that it causes signal distortion. Therefore, the PAPR should be well managed in the OFDM system. In order to estimate the MIMO channel for a space time block code, the channel can be assumed stationary during L OFDM symbols transmission. Thus, the MIMO channel matrix is represented as follows:

$$
\boldsymbol{H}_{ji}^{t,c_k} =
\begin{bmatrix}
h_{00}^{t,c_k} & h_{01}^{t,c_k} & \cdots & h_{0N_t-1}^{t,c_k} \\
h_{10}^{t,c_k} & h_{11}^{t,c_k} & & h_{1N_t-1}^{t,c_k} \\
\vdots & & \ddots & \vdots \\
h_{N_r-10}^{t,c_k} & h_{N_r-11}^{t,c_k} & \cdots & h_{N_r-1N_t-1}^{t,c_k}
\end{bmatrix}
\tag{9.26}
$$

and

$$
\boldsymbol{H}_{ji}^{0,c_k} = \boldsymbol{H}_{ji}^{1,c_k} = \cdots = \boldsymbol{H}_{ji}^{L-1,c_k}
\tag{9.27}
$$

where N_t, N_r, and c_k are the total transmit antennas, the total receive antennas, and the kth subcarrier, respectively. The pilot signals at subcarrier c_k are transmitted at the ith transmit antenna as follows:

$$
\boldsymbol{s}_{t,c_k}^{i} = \left[s_{t,c_k}^{i} \; s_{t+1,c_k}^{i} \cdots s_{t+L-1,c_k}^{i} \right]^{\mathrm{T}}
\tag{9.28}
$$

and the $L \times L$ pilot matrix is represented as follows:

$$
\boldsymbol{S}_{c_k} = \left[\boldsymbol{s}_{t,c_k}^{0} \; \boldsymbol{s}_{t,c_k}^{1} \cdots \boldsymbol{s}_{t,c_k}^{L-1} \right].
\tag{9.29}
$$

The received pilot signals at the jth received antenna are represented as follows:

$$
\boldsymbol{r}_{t,c_k}^{j} = \left[r_{t,c_k}^{j} \; r_{t+1,c_k}^{j} \cdots r_{t+L-1,c_k}^{j} \right]^{\mathrm{T}}
\tag{9.30}
$$

and the corresponding MIMO channel is represented as follows:

$$
\boldsymbol{H}_{j}^{t,c_k} = \left[h_{j0}^{t,c_k} \; h_{j1}^{t,c_k} \cdots h_{jL-1}^{t,c_k} \right]^{\mathrm{T}}.
\tag{9.31}
$$

Thus, the received signal vector at the jth receive antenna is expressed as follows:

$$
\boldsymbol{r}_{t,c_k}^{j} = \boldsymbol{S}_{c_k} \boldsymbol{H}_{j}^{t,c_k} + \boldsymbol{n}_{j}^{t}
\tag{9.32}
$$

where $\boldsymbol{n}_{j}^{t}$ is a Gaussian noise vector. If the pilot signal matrix $\boldsymbol{S}_{c_k}$ is a unitary matrix, the matrix $\boldsymbol{S}_{c_k}$ is invertible. The channel frequency response is obtained using LS estimation as follows:

$$
\hat{\boldsymbol{H}}_{j}^{t,c_k} = \boldsymbol{S}_{c_k}^{-1} \boldsymbol{r}_{t,c_k}^{j}.
\tag{9.33}
$$

The MIMO channel estimation strategy does not fit in the non-stationary channel. Thus, the superposed pilot based channel estimation strategy and the optimal pilot assignment method are investigated in Ref. [9]. They do not use null subcarriers and a prior information. It provides higher spectral efficiency than when null subcarriers are employed. The disadvantage is that the pilot design condition is tricky. Thus, in practical systems such as WiMAX and LTE, the grid-type pilot assignment method using null subcarriers is adopted as shown in Figure 9.11.

9.3   Equalization

As we briefly discussed in Chapter 5, the purpose of equalizers is to find the inverse function of the channel response and eliminate an Inter-Symbol Interference (ISI) caused by the dispersive nature of the channel. The equalizers can be classified into several different ways. One criterion of the classifications is whether they are adaptive equalizers or non-adaptive equalizers. Adaptive equalizers are used when the channel is time varying. The filter coefficients of the equalizers are changed according to time-varying channel. Non-adaptive equalizers are used for timing invariant channels and designed by the inverse function of the channel response. Another criterion of the classifications is whether they are linear or non-linear equalizers. The linear equalizers are based on the tap delayed equalization. They do not have a feedback path. The output is linear combination of the inputs. On the other hands, the non-linear equalizers have a feedback path and are used when a receiver copes with a large ISI. In OFDM systems, it can be classified in Time Domain eQualization (TEQ) or Frequency Domain eQualization (FEQ). The TEQ deals with the time domain symbol before carrying out DFT/FFT of the receiver. The FEQ is simple and less complex than the TEQ. Once channel estimation is finished, the FEQ is carried out in order to compensate for signal distortion at each subcarrier. A one-tap FEQ is widely used in the OFDM system because a frequency selective channel is regarded as a flat fading channel at each subcarrier. However, when we consider a fast fading channel, the channel state is changed in one OFDM symbol period and an Inter-Carrier Interference (ICI) occurs. Thus, the OFDM receiver should remove the ICI. FEQ was invented in 1970s but was not widely recognized. After multicarrier systems appeared, FEQ was recognized because it well matches with OFDM systems and provides us with good performance in a frequency selective fading channel. Another important equalizer is the Turbo equalizer which is inspired from Turbo codes [10]. Basically, the received signals suffer from an ISI and noise. An equalizer suppresses the ISI and an error control code corrects or detects a corrupted codeword bit. Conventional approach is to suppress the interference from the received signals and then decode the received codewords. However, the Turbo equalizer performs equalization and decoding together. In other words, it estimates, equalizes, decodes, and re-encodes, iteratively. During the iteration, accurate estimation results can be obtained and simultaneously it provides us with more precise equalization.

In a time-invariant channel (for example, when the channel impulse response is constant in one OFDM symbol period), the received signal $Y_{k,i}$ is expressed as follows:

$$Y_{k,i} = H_{k,i} X_{k,i} + N_{k,i} \tag{9.34}$$

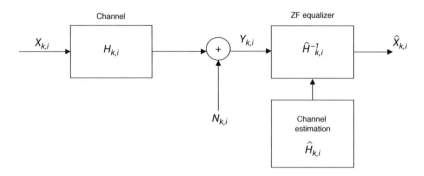

Figure 9.12   Zero forcing equalizer

where $H_{k,i}$, $X_{k,i}$, and $N_{k,i}$ are the channel frequency response, the transmitted signal, and the Gaussian noise at the kth subcarrier and the ith OFDM symbol, respectively. The one tap FEQ is carried out as follows:

$$\hat{X}_{k,i} = G_{k,i} Y_{k,i} \tag{9.35}$$

where $G_{k,i}$ is the equalizer coefficient. The ZF equalizer simply applies the inverse of the channel frequency response to the received OFDM symbol. It changes a frequency selective fading at a whole OFDM symbol to a flat fading at one subcarrier. The ZF equalizer coefficient is expressed as follows:

$$G_{k,i} = \hat{H}_{k,i}^{-1} \tag{9.36}$$

where $\hat{H}_{k,i}$ is the channel response estimation. Figure 9.12 illustrates the ZF equalizer.

Since this approach ignores the Gaussian noise of (9.34), it may amplify a noise in the subcarriers. In particular, the equalizer performance is poor at a low SNR. The MMSE equalizer takes the Gaussian noise into account and achieves a better performance. It tries to minimize the variance of the difference (or the mean square error) between the output of the equalizer and the transmitted symbol. It is expressed as follows [11]:

$$G_{k,i} = \frac{\hat{H}_{k,i}^*}{\left|\hat{H}_{k,i}\right|^2 + 1/\mathrm{SNR}}. \tag{9.37}$$

Figure 9.13 illustrates the MMSE equalizer.

Adaptive equalizers compare the output of the equalizer with the pilot signals and then calculate an error signal. Thus, the equalizer coefficients are adjusted by the error signal $\left(E_{k,i} = \left|\hat{X}_{k,i} - X_{k,i}\right|\right)$ as follows:

$$G_{k,i+1} = G_{k,i} - w_{k,i} E_{k,i} \tag{9.38}$$

where $w_{k,i+1}$ is the weight factor.

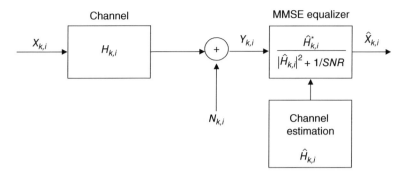

Figure 9.13 MMSE equalizer

Example 9.4 ZF equalizer

Consider the system parameters in Example 9.3 and the received signals as follows:

Received signals

$$Y = [Y(0)Y(1)Y(2)Y(3)Y(4)Y(5)Y(6)Y(7)]$$
$$= [0.21 - 0.74i \quad -0.3 + 0.3i \quad -0.76 - 0.12i \quad 0.16 + 0.22i$$
$$-0.43 + 1.81i 1.21 + 0.49i \quad 0.53 - 1.1i \quad 0.72 - 0.29i]$$

Find the equalized signals using the ZF equalizer and LS estimation.

Solution
In Example 9.3, LS channel estimation values are calculated as follows:

$$H_{LS} = [0.31 + 0.67i \; 0.26 + 0.16i \; 0.34 - 0.82i - 0.5 - 0.36i$$
$$-0.48 - 0.79i - 0.81 + 1.88i \; 0.97 + 1.18i 0.59i].$$

From (9.35) and (9.36), the equalized signals can be obtained as follows:

$$\hat{X}_{k,i} = \left[\frac{0.21 - 0.74i}{0.31 + 0.67i} \; \frac{-0.3 + 0.3i}{0.26 + 0.16i} \; \frac{-0.76 - 0.12i}{0.34 - 0.82i} \; \frac{0.16 + 0.22i}{-0.5 - 0.36i} \right.$$
$$\left. \frac{-0.43 + 1.81i}{-0.48 - 0.79i} \; \frac{1.21 + 0.49i}{-0.81 + 1.88i} \; \frac{0.53 - 1.1i}{0.97 + 1.18i} \; \frac{0.72 - 0.29i}{0.59i} \right]$$

$$= [-0.79 - 0.68i - 0.32 + 1.35i - 0.2 - 0.84i - 0.42 - 0.14i$$
$$-1.43 - 1.41i - 0.01 - 0.64i - 0.34 - 0.73i - 0.49 - 1.22i]. \quad ■$$

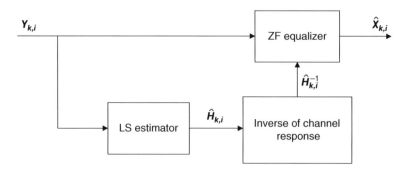

Figure 9.14   LS estimator and ZF equalizer

9.4   Hardware Implementation of Channel Estimation and Equalizer for OFDM System

The channel estimator and the equalizer are designed according to the system requirements such as pilot assignment, antenna array, system complexity, latency, and power. Especially, the latency is strict design parameter because channel estimation and interpolation must be finished and the equalizer must be ready when the data OFDM symbol is arrived at the receiver. In this section, we briefly investigate a simple LS estimator and ZF equalizer design in the OFDM system with the block-type pilot assignment and single antenna. Figure 9.14 illustrates a simple LS estimator and ZF equalizer block diagram. When the received pilot OFDM symbol $Y_{k,i}$ is arrived, the LS estimator calculate (9.11) with the stored pilot OFDM symbol. In this calculation, we deal with a complex number and division. Thus, Euler's theorem $\left(A + Bi = Re^{j\theta}\right)$ can be used for simple calculation. The conversion can be obtained by the CORDIC algorithm. This conversion is also useful in the inverse of the channel response block. The ZF equalizer collects the data OFDM symbol and the inverse of the channel response and then calculates simple multiplications. Sometime, a phase offset is compensated in this process. Once the pilot subcarriers are detected, they are compared with the stored pilot subcarriers and the phase difference is calculated. In the ZF equalizer, it is compensated.

9.5   Problems

9.1. Compare the performances of time domain channel estimation and frequency domain channel estimation for an OFDM system.

9.2. Compare the MSEs of pilot-assisted channel estimation, blind channel estimation, and DDCE.

9.3. Describe the pilot assignment of a 4×4 MIMO OFDM system in the LTE standard.

9.4. Compare the Lagrangian interpolation with the Newton interpolation.

9.5. Define the error that can occur in polynomial interpolation and describe the method to minimize the error.

9.6. Consider the following system model: $y(k) = Hx(k) + z(k)$ where $y(k)$, H, $x(k)$, and $z(k)$ are the received signal vector, channel matrix, transmitted signal vector, and white Gaussian noise vector, respectively. Find the optimal precoding matrix W such that $\hat{x}(k) = Wy(k)$.

9.7. Channel estimation errors result in system performance degradation. Define the channel estimation error in an OFDM system and describe its effect to system performance.

References

[1] J. Beek, O. Edfors, M. Sandell, S. Wilson, and P. Börjesson, "On Channel Estimation in OFDM Systems," *Proceedings of IEEE the 45th Vehicular Technology Conference (VTC'95)*, Chicago, IL, vol. 2, pp. 815–819, July 25–28, 1995.

[2] L. J. Cimini, "Analysis and Simulation of a Digital Mobile Channel Using Orthogonal Frequency Division Multiplexing," *IEEE Transactions on Communications*, vol. 33, no. 7, pp. 665–675, 1985.

[3] M. Hsieh and C. Wei, "Channel Estimation for OFDM Systems Based on Comb-Type Pilot Arrangement in Frequency Selective Fading Channels," *IEEE Transactions on Consumer Electronics*, vol. 44, no. 1, p. 217, 1998.

[4] O. Edfors, M. Sandell, J. J. van de Beek, S. K. Wilson, and P. O. Borjesson, "OFDM Channel Estimation by Singular Value Decomposition," *IEEE Transactions on Communications*, vol. 46, no. 7, pp. 931–939, 1998.

[5] A. Khlifi and R. Bouallegue, "Performance Analysis of LS and LMMSE Channel Estimation Techniques for LTE Downlink Systems," *International Journal of Wireless and Mobile Network*, vol. 3, no. 5, pp. 141–149, 2011.

[6] M. Rim, J. Ahn, and Y. Kim, "Decision-Directed Channel Estimation for M-QAM Modulated OFDM Systems," *Proceedings of IEEE the 55th Vehicular Technology Conference (VTC spring '02)*, vol. 4, pp. 1742–1746, Birmingham, AL, USA, May 6–9, 2002.

[7] W. G. Jeon, K. H. Paik, and Y. S. Cho, "An Efficient Channel Estimation Technique for OFDM systems with Transmitter Diversity," *Proceedings of IEEE the 11th Personal, Indoor and Mobile Radio Communication Conference (PIMRC 2000)*, vol. 2, pp. 1246–1250, London, UK, September 18–21, 2000.

[8] A. Dowler and A. Nix, "Performance Evaluation of Channel Estimation Techniques in a Multiple Antenna OFDM System," *Proceedings of IEEE Vehicular Technology Conference (VTC'03)*, vol. 2, pp. 1214–1218, Orlando, FL, USA, October 6–9, 2003.

[9] Y. Li, "Simplified Channel Estimation for OFDM Systems with Multiple Transmit Antennas," *IEEE Transactions on Wireless Communications*, vol. 1, no. 1, pp. 67–75, 2002.

[10] C. Douillard, M. Jézéquel, and C. Berrou, "Iterative Correction of Intersymbol-Interference: Turbo-Equalization," *European Transactions on Telecommunications*, vol. 6, no. 5, pp. 507–511, 1995.

[11] T. Chiueh and P. Tsai, *OFDM Baseband Receiver Design for Wireless Communications*, John Wiley & Sons, Inc., Hoboken, NJ, 2007.

10

Synchronization

In wireless communications, a transmitter put data in radio resources such as time and frequency and a receiver should be able to pull data out. The synchronization performs this role in the receiver of wireless communication systems. Synchronization techniques are an indispensable part of the wireless communication receiver like a conductor of orchestras unifies performers and forms an ensemble. The receiver needs to detect a packet and find an accurate starting position. In addition, it should correct several synchronization errors such as timing offsets (caused by channel delay), carrier frequency offsets (caused by Doppler effect and local oscillator mismatch between transmitter and receiver), phase noises (caused by phase rotation of the channel), and sampling frequency offsets (caused by mismatch between DAC and ADC). In particular, OFDM systems are very sensitive and vulnerable to these synchronization errors. It is a tricky part of the wireless communication system design. In this chapter, we will look into synchronization errors and their effects, review synchronization techniques, and design ML synchronization block in the OFDM system.

10.1 Fundamental Synchronization Techniques for OFDM System

Synchronization techniques can be classified into data-aided methods, decision-directed methods, and non-data-aided methods. The *data-aided methods* use a reference signal (training symbols, preambles, pilots, etc.) which is known in the receiver side. It provides us with the best synchronization performance among them even if it suffers from the leakage of a bandwidth or data transmission capacity due to overhead signals. Thus, the reference signals for synchronization should be designed to provide high reliability and spend a few radio resources. The *decision-directed methods* use the detected symbols as the reference signals. They are sensitive to detection errors. The *non-data-aided methods* (or blind methods) do not depend on a known signal and their complexity is high. It can be used for symbol synchronization of

Wireless Communications Systems Design, First Edition. Haesik Kim.
© 2015 John Wiley & Sons, Ltd. Published 2015 by John Wiley & Sons, Ltd.

the OFDM system. For example, they make use of autocorrelation between the Cyclic Prefix (CP) and the end of the OFDM symbol. Among the three methods, the data aided methods are widely used in the OFDM systems because of their accuracy.

Many synchronization techniques are based on two important algorithms: autocorrelation and crosscorrelation. As we discussed in Chapters 2 and 4, autocorrelation is based on the similarity between one signal and its time lag version, whereas crosscorrelation is based on the similarity between a transmitted signal and a stored signal in the receiver. In Ref. [1], the non-data-aided method using autocorrelation is introduced. This method exploits the cyclic prefix of the OFDM system. Basically, the cyclic prefix is designed to avoid an inter-symbol interference (ISI). Thus, it is difficult to obtain the ISI free cyclic prefix. In order to use it as a reference signal, more number of OFDM symbols are required and they should be averaged. The autocorrelation, $R_r(m)$, using the CP is defined as follows:

$$R_r(m) = \sum_{k=m}^{m+v-1} r(k)r^*(k+N) + E \qquad (10.1)$$

where N, v, and E are the DFT/IDFT size, the CP length, and the normalized energy, respectively. $r(k)$ is the received OFDM symbol. This method estimates the timing offset using the peak of the autocorrelation function and then calculates the frequency offset using the phase shift between the cyclic prefix and the subcarriers in the end of the OFDM symbol. In the wireless LAN (IEEE802.11a), short and long training symbols are included in the frame. Thus, the data-aided method is possible. For the purpose of coarse time synchronization, the autocorrelation of the periodic short training symbols is performed as follows:

$$R_r(v_p) = \sum_{k=0}^{v_p-1} r(k)r^*(k+v_p) \qquad (10.2)$$

where v_p is the periodicity factor. In IEEE802.11a, the value of v_p is 16 (the short training symbol) or 64 (the long training symbol). The data-aided method provides us with more accurate and faster synchronization than the non-data-aided method. The most popular OFDM synchronization is the crosscorrelation using a preamble [2]. It correlates a stored preamble in the receiver with a received symbol. The crosscorrelation $R_{sr}(v_p)$ is defined as follows:

$$R_{sr}(v_p) = \sum_{k=0}^{v_p-1} s(k)r^*(k+v_p) \qquad (10.3)$$

where $s(k)$ is the stored preamble. The stored preamble is not affected by noise, fading, and nonlinearity but the received symbol includes them. Thus, it fits well into a low SNR situation. In OFDM systems, coarse timing synchronization and fine timing synchronization are usually based on autocorrelation and crosscorrelation, respectively.

Example 10.1   Autocorrelation
Consider two signals with different length as shown in Figure 10.1
    Compare two autocorrelation functions.

Solution
The received signal $r_1(t)$ and the time lag received signal $r_1(t+v_p)$ with 10 samples are fed into autocorrelation function as shown in (10.1). Figure 10.2 illustrates the autocorrelation of $r_1(t)$.

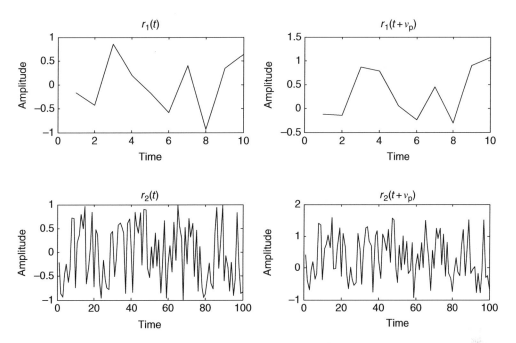

Figure 10.1 Two received signals ($r_1(t)$ and $r_2(t)$) and two time lag received signals ($r_1(t+\nu_p)$ and $r_2(t+\nu_p)$)

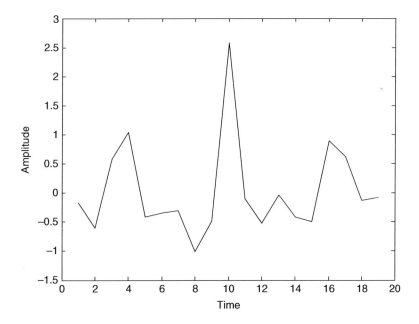

Figure 10.2 Autocorrelation of $r_1(t)$

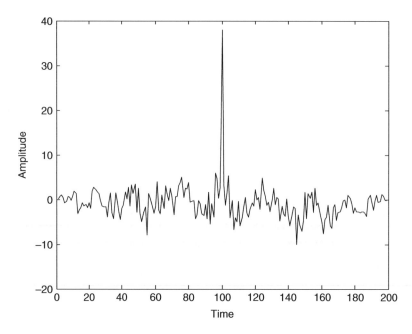

Figure 10.3   Autocorrelation of $r_2(t)$

Likewise, the received signal $r_2(t)$ and the time lag received signal $r_2(t+v_p)$ with 100 samples are fed into autocorrelation function as shown in (10.1). Figure 10.3 illustrates the autocorrelation of $r_2(t)$.

As we can see from both figures, the autocorrelation function with many samples produces a higher peak. The signal detection probability complies with the number of samples. In particular, a false alarm probability increases when a noise level is high. Thus, we need to observe enough samples to detect a signal accurately.   ∎

Example 10.2   Crosscorrelation
Consider the stored signal and two signals with different noise level. In the Figure 10.4, $s(t)$ is the stored preamble in the receiver. $r_1(t)$ is the received signal with a high noise. $r_2(t)$ is received signal with a low noise.

Find two crosscorrelation functions and compare them.

Solution
From (10.3), we can calculate the crosscorrelation of the stored signal $s(t)$ and the received signal $r_1(t)$ with a high noise. Figure 10.5 illustrates the crosscorrelation of them.

Likewise, we can calculate the crosscorrelation of the stored signal $s(t)$ and the received signal $r_2(t)$ with a low noise. Figure 10.6 illustrates the crosscorrelation of them.

We can barely see a peak at Figure 10.5. The performance of the crosscorrelation at a high noise level is poor. On the other hand, Figure 10.6 clearly shows us a high peak.   ∎

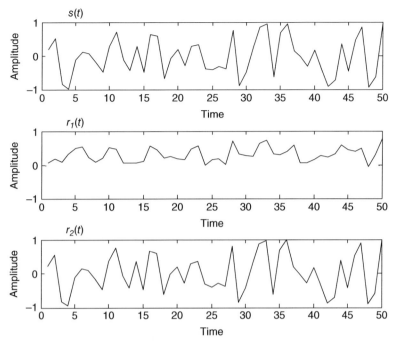

Figure 10.4 Stored signal $s(t)$ and two received signals $r_1(t)$ and $r_2(t)$

10.2 Synchronization Errors

The OFDM technique is suitable for a high-speed wireless communication system. However, one important condition is that the orthogonality among subcarriers should be maintained. In a practical wireless communication system, it suffers from some imperfections such as nonlinear distortions, thermal noises, and synchronization errors (time and frequency offsets, sampling frequency offsets, and phase offsets). The orthogonality may not be preserved and the system performance may be significantly degraded. Thus, it is very important to compensate for those imperfections. There are many literatures dealing with the effect of time and frequency offsets [3, 4]. In this section, synchronization errors are discussed and the effect of these offsets is investigated.

An OFDMA packet is composed of multiple OFDM symbols. The receiver should find the start position of the OFDM symbols. The transmitted symbols may reach the receiver with different delays. When the channel maximum excess delay is shorter than the guard interval, we can consider several scenarios as shown in Figure 10.7.

The DFT window in scenario 2 is located in safe position so that the signal is not contaminated by the previous signal and there is no ISI. The only phase shift occurs. The received signal in frequency domain is expressed as follows:

$$Y_{k,l} = X_{k,l} H_k e^{-j2\pi T_d k / NT_s} + W_{k,l} \tag{10.4}$$

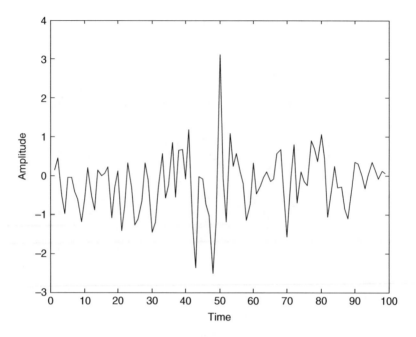

Figure 10.5　Crosscorrelation of $s(t)$ and $r_1(t)$

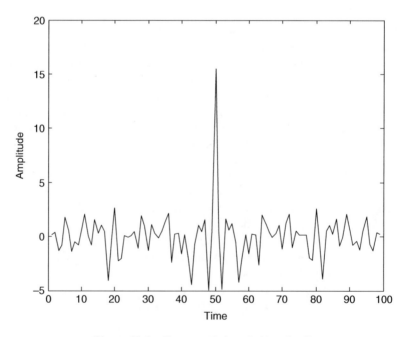

Figure 10.6　Crosscorrelation of $s(t)$ and $r_2(t)$

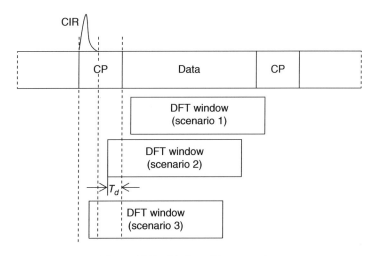

Figure 10.7   Timing offset scenarios

where $Y_{k,l}$, $X_{k,l}$, H_k, and $W_{k,l}$ are the frequency domain received signal, the frequency domain transmitted signal, the channel frequency response, and the Gaussian noise at the kth subcarrier and the lth symbol, respectively. T_d and T_s are the symbol timing offset (the delay to the correct DFT window starting position) and the subcarrier signal interval, respectively. In this case, the subcarrier spacing f_s is $1/NT_s$ and the signal bandwidth is approximately $1/T_s$. On the other hand, the DFT windows in scenario 1 and 3 are located in unsafe position so that ISI occurs and both magnitude and phase are distorted. The received signal in frequency domain is expressed as follows:

$$Y_{k,l} = \frac{N - |T_d|/T_s}{N} X_{k,l} H_k e^{-j2\pi T_d k/NT_s} + \text{ISI} + W_{k,l} \qquad (10.5)$$

where the term $(N - (|T_d|/T_s))/N$ means a slight magnitude attenuation because the DFT windows collect smaller samples of the original data. The other sample collected by the DFT window is expressed as the ISI term.

The OFDM system is vulnerable to *Carrier Frequency Offset* (CFO). The CFO induces the loss of orthogonality among the subcarriers in the OFDM system. The CFO is caused by the Doppler effect and the local oscillator mismatch between a transmitter and a receiver. The wireless communication system designer should consider the CFO and define the OFDM system parameters. The CFO is one of the most common channel impairments in the OFDM system. An accurate estimation of the CFO is a critical part of the synchronization block design in the OFDM system. For example, if we consider that the maximum carrier frequency offset is ±40 ppm, the carrier frequency is 2 GHz, and the subcarrier spacing is 100 kHz, the maximum allowed frequency offset is ±80 kHz. Figure 10.8 illustrates the maximum frequency offset and subcarrier spacing.

The normalized carrier frequency offset ε_{CFO} is defined as the ratio of CFO Δf to subcarrier spacing f_s. It is composed of three parts as follows:

$$\varepsilon_{\text{CFO}} = \frac{\Delta f}{f_s} = \varepsilon_{\text{I}} + \varepsilon_{\text{f}} + \varepsilon_{\text{r}} \qquad (10.6)$$

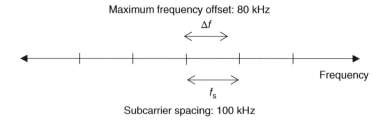

Figure 10.8   Maximum frequency offset and subcarrier spacing

where ε_i, ε_f, and ε_r are the integer frequency offset, the fractional frequency offset, and the residual frequency offset, respectively. The integer frequency offset ($\varepsilon_\text{i} \geq 1$) literally shifts an index and produces a phase shift. However, it does not affect the orthogonality. The fractional frequency offset ($-0.5 \leq \varepsilon_\text{f} < 0.5$) produces magnitude attenuation, phase rotation, and inter-carrier interference (ICI). It destroys the orthogonality of the OFDM symbol. The residual frequency offset is less than 5% of the subcarrier spacing [5]. Thus, it does not significantly affect ICI. However, the phase rotation is affected and frequency offset estimation is not perfect. There is always some residual frequency offset. The compensation of the residual frequency offset is carried out separately. The received signal with CFO in frequency domain is expressed as follows:

$$Y_{k,l} = X_{k-\varepsilon_\text{i},l}H_{k-\varepsilon_\text{i}}\frac{\sin(\pi\varepsilon_\text{f})}{N\sin(\pi\varepsilon_\text{f}/N)}e^{j2\pi\frac{l(N+N_\text{g})+N_\text{g}}{N}(\varepsilon_\text{i}+\varepsilon_\text{f})}e^{j\pi\frac{N-1}{N}\varepsilon_\text{f}} +$$

$$\sum_{i=0,i\neq k-\varepsilon_\text{i}}^{N-1} X_{i,l}H_i\frac{\sin(\pi(\varepsilon_\text{i}+\varepsilon_\text{f}+i-k))}{N\sin(\pi(\varepsilon_\text{i}+\varepsilon_\text{f}+i-k)/N)}e^{j2\pi\frac{l(N+N_\text{g})+N_\text{g}}{N}(\varepsilon_\text{i}+\varepsilon_\text{f})}e^{j\pi\frac{N-1}{N}(\varepsilon_\text{i}+\varepsilon_\text{f}+i-k)} + W_{k,l} \qquad (10.7)$$

where N_g is the guard interval. The first term represents the received signal including magnitude attenuation and phase rotation, the second term represents the ICI, and the third term is the Gaussian noise. The ICI affects the orthogonality as well as degrades SNR. The SNR degradation [6] is expressed as follows:

$$L_\text{SNR,CFO} \approx \frac{10}{3\ln 10}(\pi\varepsilon_\text{f})^2\frac{E_\text{s}}{N_0} \qquad (10.8)$$

where E_s/N_0 is the symbol to noise ratio. Figure 10.9 illustrates the inter-carrier interference by frequency offset. We can observe the frequency offset Δf at the center subcarrier in the figure. The ICI occurs due to the loss of orthogonality.

The *phase noise* is caused by many reasons such as carrier phase offset, channel impulse response, residual time, and frequency offset. Although time and frequency synchronizations are finished, the residual time and frequency offsets still exist and they produce the phase noise. The phase noise caused by the frequency offset is constant and time-variant while the phase noise caused by the time offset is not constant and time-invariant. The local oscillator difference between a transmitter and a receiver produces the phase noise as well

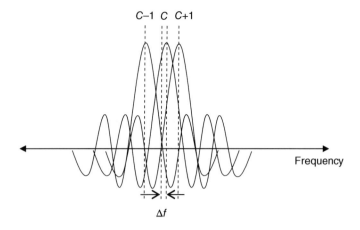

Figure 10.9   Inter-carrier interference and frequency offset

as the frequency offset. The received signal with the carrier phase offset in frequency domain is expressed as follows:

$$Y_{k,l} = X_{k,l}H_k e^{j\varphi_0}\left(\frac{1}{N}\sum_{n=0}^{N-1}e^{j\phi_{n,l}}\right) + \sum_{i=0,i\neq k}^{N-1}X_{i,l}H_i e^{j\varphi_0}\left(\frac{1}{N}\sum_{n=0}^{N-1}e^{j\varphi_{n,l}}e^{j2\pi n(k-i)/N}\right) + W_{k,l} \qquad (10.9)$$

where φ_0 is the constant phase offset and $\varphi_{n,l}$ is the phase offset at the nth sample in the lth symbol. The SNR degradation caused by the phase noise [7] is expressed as follows:

$$L_{\text{SNR,PO}} \approx 10\log\left(1+\sigma^2\frac{E_\text{s}}{N_0}\right) \qquad (10.10)$$

where the phase noise variance is small ($\sigma^2 \ll 1$).

The system clock should be derived from same oscillators but a mismatch between a transmitter oscillator and a receiver oscillator exists. The received signal is sampled at an interval of $(1+\delta)T_\text{s}$ where δ is *Sample Frequency Offset* (SFO). The received signal with the SFO in frequency domain is expressed as follows:

$$Y_{k,l} = X_{k,l}H_k \frac{\sin(\pi\delta k)}{N\sin\left(\dfrac{\pi\delta k}{N}\right)}e^{j2\pi\frac{l(N+N_g)+N_g}{N}\delta k}e^{j\pi\frac{N-1}{N}\delta k}$$

$$+ \sum_{i=0,i\neq k}^{N-1}X_{i,l}H_i \frac{\sin(\pi((1+\delta)i-k))}{N\sin\left(\dfrac{\pi((1+\delta)i-k)}{N}\right)}e^{j2\pi\frac{l(N+N_g)+N_g}{N}\delta i}e^{j\pi\frac{N-1}{N}((1+\delta)i-k)} + W_{k,l}. \qquad (10.11)$$

As we can observe from (10.11), the effect of the SFO is similar to the effect of the CFO. The SFO affects the symbol timing shift and the SNR loss. The symbol timing shift produces the phase rotation. The SNR degradation [8] caused by the SFO is expressed as follows:

$$L_{\mathrm{SNR,SFO}} \approx 10\log\left(1+\frac{1}{3}(\pi\delta k)^2\frac{E_{\mathrm{s}}}{N_0}\right). \tag{10.12}$$

As we can observe from (10.12), the SNR degradation increases in accordance with the subcarrier index k and the SFO δ. It can cause the loss of the orthogonality among the subcarriers.

Example 10.3   Carrier frequency offset
Compare the effect of carrier frequency offsets (100 Hz, 500 Hz, 1 kHz, and 5 kHz) when considering 16 QAM modulation, -60 dBc/Hz phase noise, and 100 kHz sampling frequency.

Solution
Their effects to the baseband received signal can be simply modeled as follows:

$$r_{\mathrm{wCFO}}(t) = e^{j2\pi\left(f_{\mathrm{c}}^{\mathrm{t}}-f_{\mathrm{c}}^{\mathrm{r}}\right)t}r_{\mathrm{woCFO}}(t)$$

where $r_{\mathrm{wCFO}}(t)$, $r_{\mathrm{woCFO}}(t)$, $f_{\mathrm{c}}^{\mathrm{t}}$, and $f_{\mathrm{c}}^{\mathrm{r}}$ represent the time domain received signal with CFO, the time domain received signal without CFO, the transmitter carrier frequency, and the receiver carrier frequency, respectively. As we can observe from the above equation, the effect of the frequency offset produces phase rotation in signal constellation. The phase rotation gets more severe as the mismatch between the transmitter carrier frequency and the receiver carrier frequency (e.g., the frequency offset) gets bigger. From simple computer simulation, we can obtain the phase rotated signal constellations as shown in Figure 10.10.   ∎

10.3   Synchronization Techniques for OFDM System

The purpose of the OFDM synchronization block is to detect an OFDM symbol, find the time and frequency offset, and maintain the orthogonality of each subcarrier. There are many algorithms to estimate the time and frequency offset. In order to build the synchronization block, the OFDM system designer should consider transmission type, system performance, latency, and complexity and then find the most suitable algorithm. For example, when comparing wireless LAN (IEEE802.11a) with Digital Video Broadcasting-Terrestrial (DVB-T), they have different transmission type. The IEEE802.11a is based on the burst transmission. Its receiver should detect the signal as quickly as it receives the preamble. Thus, the preamble is periodic and suitable for fast synchronization. On the other hand, the DVB-T is based on the continuous transmission and synchronization time is not critical. Averaging some OFDM symbols can be used while adaptive compensation techniques are needed because of the time-varying channel effect.

In the synchronization process, the first step is to detect OFDM symbols in the received signal and find the beginning of the OFDM symbol. As we discussed in Section 10.1, coarse timing synchronization based on autocorrelation acquires rough timing, and then fine timing synchronization based on crosscorrelation improves the timing accuracy. One of the simplest coarse time synchronizations is autocorrelation using the repeated symbols as shown in (10.1).

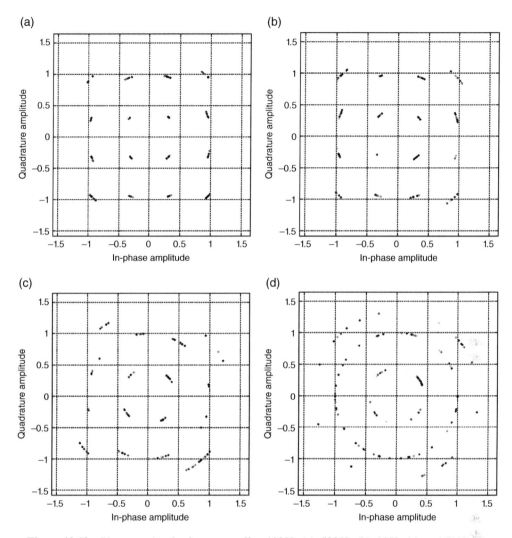

Figure 10.10   Phase rotation for frequency offset 100 Hz (a), 500 Hz (b), 1 kHz (c), and 5 kHz (d)

The generalized form with the periodicity factor ν and the separation between two intervals L is expressed as follows:

$$R_r(m) = \left| \sum_{k=m}^{m+\nu-1} r(k) r^*(k+L) \right|. \tag{10.13}$$

The time index with the maximum autocorrelation value is as follows:

$$\hat{m} = \arg\max_m R_r(m). \tag{10.14}$$

The correlator output in different symbols has a wide range of values and fluctuates because it is not normalized. Thus, it is difficult to set a threshold level and detect the maximum value.

In addition, the correlator output does not drop enough when the correlator window moves away from the preamble. The correlation performance decreases as SNR decreases.

In Ref. [2], T. M. Schmidl and D. C. Cox proposed the power normalized metric which takes the input signal power into account. Thus, the range of the output is decreased so that it is easy to set the threshold level and detect the maximum value. This is the most popular method for the OFDM system synchronization, which is close to the Cramer-Rao bound. They constructed the preamble with two identical segments which has $N/2$ length and contains a Pseudo-Noise (PN) sequence on the odd frequencies and a null symbol on the even frequencies. Due to the DFT property, we have the following autocorrelation:

$$R_r(m) = \left| \sum_{k=m}^{m+N/2-1} r(k)r^*(k+N/2) \right| \tag{10.15}$$

and the received energy of the second half symbol is defined as follows:

$$E_r(m) = \sum_{k=m}^{m+N/2-1} \left| r(k+N/2) \right|^2. \tag{10.16}$$

Thus, the symbol detection metric is defined as follows:

$$M(m) = \frac{R_r(m)^2}{E_r(m)^2}. \tag{10.17}$$

The time index with the maximum autocorrelation value is as follows:

$$\hat{m} = \arg\max_m M(m) \tag{10.18}$$

where the time index means the end of the preamble. This method has a low complexity and the autocorrelation can be implemented with iterative formula. However, it is difficult to find the beginning of the OFDM symbol. A plateau effect happens in the autocorrelation output. The cyclic prefix is the copy of the last several subcarriers. In Figure 10.11, the DFT windows in scenarios 1 and 2 produce same correlation result.

Thus, many algorithms tried to overcome this obstacle. One of them is Minn's algorithm [9]. He has modified the preamble structure consisting of several identical segments. A typical preamble has [A A −A −A] structure where A is the preamble segment with $N/4$ length. Thus, the autocorrelation and the received energy are defined as follows:

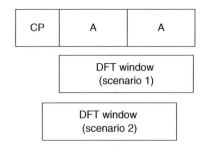

Figure 10.11   Inter-carrier interference and frequency offset

$$R_r(m) = \sum_{l=0}^{1} \sum_{k=m}^{m+N/4-1} r\left(\frac{N}{2}l+k\right) r^*\left(\frac{N}{2}l+k+\frac{N}{4}\right) \qquad (10.19)$$

and

$$E_r(m) = \sum_{l=0}^{1} \sum_{k=m}^{m+N/4-1} \left| r\left(\frac{N}{2}l+k+\frac{N}{4}\right) \right|^2. \qquad (10.20)$$

This method provides a sharper roll-off than Schmidl and Cox's power normalized metric. However, it shows us a lower performance under the multipath channel because it uses fewer samples.

In Ref. [10], Maximum Likelihood (ML) signal detection for OFDM synchronization is proposed. The timing offset is estimated as follows:

$$R_r(m) = 2\left| \sum_{k=m}^{m+\nu-1} r(k)r^*(k+L) \right| - \rho \sum_{k=m}^{m+\nu-1} \left(|r(k)|^2 + |r(k+L)|^2 \right) \qquad (10.21)$$

where ρ is the weighting factor depending on the SNR and is defined as follows:

$$\rho = \frac{\text{SNR}}{1+\text{SNR}}. \qquad (10.22)$$

Likewise, the time index with the maximum value is found as follows:

$$\hat{m} = \arg\max_m R_r(m). \qquad (10.23)$$

As we can observe from (10.21) and (10.22), the ML signal detection needs to estimate the SNR value. The SNR can be easily estimated in the continuous transmission such as DVB-T while it is difficult to estimate an accurate SNR in the first preamble. This SNR estimation error increases a false alarm probability. In addition, the ML signal detection has a high complexity due to this SNR calculation.

Example 10.4   ML signal detection
Consider the following system parameters and assume we can obtain ISI free cyclic prefix, accurate SNR estimation, and non-data-aided synchronization. Compare the ML signal detection performances.

Requirements	Values
IDFT/DFT size	1024
CP lengths	16 and 32
Frequency offset	0.5
Channel	AWGN
The number of OFDM symbols	6
SNR	5, 10, 15, 20 dB

Solution

In (10.21) and (10.22), the periodicity factor ν is equal to the CP length and the separation between two intervals L is the DFT/IDFT size. Computer simulations are carried out. In Figures 10.12 and 10.13, we can observe six high peaks. A high peak level increases as the SNR level increases. Thus, we can detect the OFDM symbol accurately when the SNR is high. However, it is difficult to find a high peak in Figure 10.12a and b. A false alarm probability will increase at a low SNR and few samples (a short CP length). On the other hand, we can observe clear high peaks from Figure 10.12c to 10.13d. In Figure 10.13d (the longest CP length and the highest SNR), we can find the highest peak.   ∎

Although coarse time synchronization is finished, a remaining symbol timing error may still exist. In order to achieve further accuracy, fine time synchronization is necessary. As we briefly discussed in Section 10.1, the crosscorrelation is used for fine time synchronization. It is defined in (10.3). The generalized form with the preamble length K is rewritten as follows:

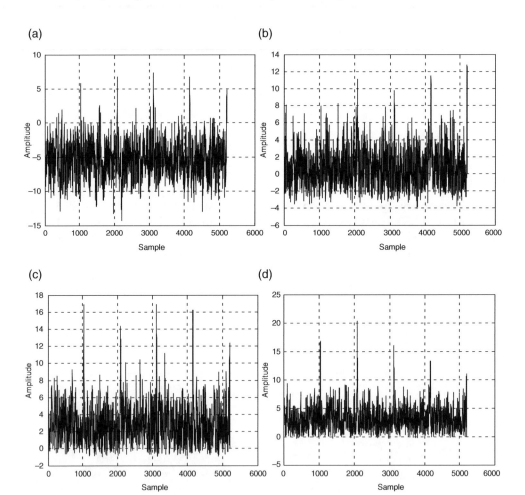

Figure 10.12   ML signal detection with 16 CP length at 5 (a), 10 (b), 15 (c), and 20 dB (d)

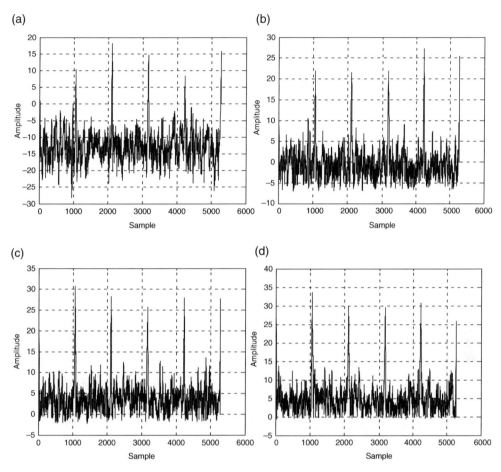

Figure 10.13 ML signal detection with 32 CP length at 5 (a), 10 (b), 15 (c), and 20 dB (d)

$$R_{sr}(m) = \left| \sum_{k=0}^{K-1} s(k)r^*(k+m) \right|. \tag{10.24}$$

The received signal $r(k)$ is correlated with the stored preamble $s(k)$ which is not affected by channel impairments and nonlinearity. Likewise, the time index with the maximum value is found as follows:

$$\hat{m} = \arg\max_m R_{sr}(m). \tag{10.25}$$

In a practical wireless communication system, frequency offset estimation is performed before fine time synchronization starts because it cannot manage a signal including frequency offset.

Fractional CFO estimation can be accomplished simultaneously when the symbol boundary is detected. The maximum likelihood CFO estimator [10] is expressed as follows:

$$\Delta\hat{f} = \frac{1}{2\pi LT_s} \tan^{-1} \left| \sum_{k=m}^{m+\nu-1} r(k)r^*(k+L) \right| \tag{10.26}$$

where the periodicity factor ν is the block size and the separation L means the distance between two identical blocks. When we consider synchronization using the CP, ν and L are the CP length and the DFT size, respectively. When we consider the preamble [A A] structure where A is the preamble segment with $N/2$ length, both ν and L are $N/2$. The acquisition range of the above formula is $[-1/2LT_s, 1/2LT_s]$.

Example 10.5   ML CFO estimation
Consider the following system parameters and assume we can obtain ISI free cyclic prefix, accurate SNR estimation, and non-data-aided synchronization. Find the ML CFO estimation.

Requirements	Values
IDFT/DFT size	1024
CP lengths	128
Frequency offset	0.3
Channel	AWGN
The number of OFDM symbols	6
SNR	20 dB

Solution
In (10.26), the periodicity factor ν is equal to the CP length and the separation between two intervals L is the DFT/IDFT size (N). The ML CFO can be estimated simultaneously when ML signal detection is performed. From computer simulation, we can obtain Figure 10.14.

As we can observe from Figure 10.14, the receiver detects 6 OFDM symbols and estimates the fractional frequency offset (~0.3) at the symbol detection positions.   ∎

Fractional CFO estimation is performed in time domain while an integer CFO is typically estimated in frequency domain. In Ref. [2], two pilot blocks with differential encoding at identical subcarrier positions are used. In Ref. [11], one pilot block with differential encoding among adjacent subcarriers is exploited. The first scheme has a better performance than the second scheme. However, the system overhead of the first scheme is higher. One simple integer CFO estimator uses frequency domain autocorrelation. It brings good estimation because the integer part of CFO causes a frequency shift of the received signal in frequency domain. Let $Y[l, k]$ be the frequency domain received symbol at the kth subcarrier of the lth OFDM symbol. Two consecutive OFDM symbols are correlated as follows:

$$\Delta \hat{f}(g) = \sum_{k=0}^{K-1} Y[l, \alpha_k + g] Y^*[l-1, \alpha_k + g] \qquad (10.27)$$

where $g = 0, \pm 1, \pm 2, \ldots$ means the possible integer subcarrier index and α_k is the kth pilot subcarrier. The integer CFO can be estimated by finding the integer subcarrier index g with the largest value as follows:

$$\hat{g} = \arg\max_g \left| \Delta \hat{f}(g) \right|. \qquad (10.28)$$

This method works well when the pilot assignment is comb-type and the channel is slow fading. Although the fractional CFO and the integer CFO are estimated and compensated, there still is a residual CFO. The residual CFO affects the phase shift but causes a minor SNR degradation. In a time-varying channel, it should be continuously tracked and compensated. Figure 10.15 illustrates synchronization block diagram in the OFDM system.

(a)

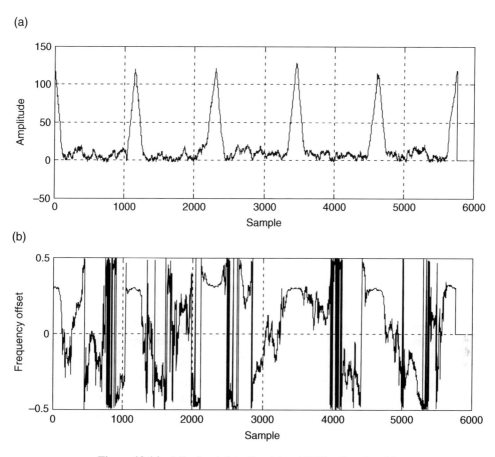

(b)

Figure 10.14   ML signal detection (a) and CFO estimation (b)

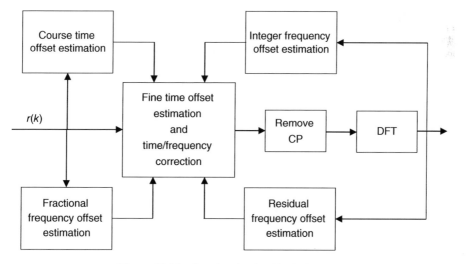

Figure 10.15   Synchronization block diagram

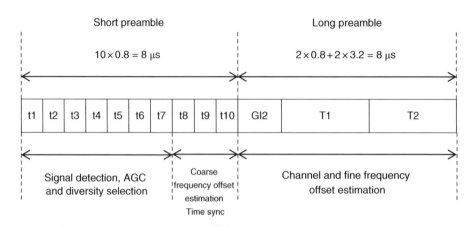

Figure 10.16   Preamble structure of IEEE802.11a

When we have a preamble and use the data-aided synchronization algorithm, a preamble structure should be designed in consideration of frame synchronization, Automatic Gain Control (AGC), carrier frequency offset estimation, symbol timing synchronization, channel estimation, and fine frequency offset estimation. For example, the preamble of IEEE802.11a contains 10 identical short training symbols, 2 identical long training symbols, and guard interval. Each symbol is used for each synchronization process as shown in Figure 10.16.

The overall synchronization procedure of IEEE802.11a is as follows: Firstly, the receiver observes a wireless channel and detects a frame using a correlation technique within $5.6\,\mu s$ ($=7\times0.8\,\mu s$). In this stage, the frame detection algorithm should be used to avoid false alarms. Secondly, once the frame is detected, the receiver estimates a symbol timing offset and coarse frequency offset within $2.4\,\mu s$ ($=3\times0.8\,\mu s$). In this stage, the receiver compensates the offsets and differentiates preamble symbols from data symbols. Thirdly, channel estimation/ equalization and fine frequency offset estimation/compensation are carried out within $8\,\mu s$.

Summary 10.1   Symbol timing synchronization

1. The purpose of the OFDM synchronization is to detect OFDM symbols, find the time and frequency offset, and maintain the orthogonality of each subcarrier.
2. The coarse timing synchronization based on autocorrelation acquires a rough timing. The generalized form of the periodicity factor v, the separation between two intervals L, and the received signal $r(k)$ is expressed as follows:

$$R_r(m) = \left| \sum_{k=m}^{m+v-1} r(k)r^*(k+L) \right|$$

and the time index with the maximum autocorrelation value is found as follows:

$$\hat{m} = \arg\max_m R_r(m).$$

3. The fine time synchronization is necessary in order to achieve further accuracy. The generalized form with the preamble length K is rewritten as follows:

$$R_{sr}(m) = \left| \sum_{k=0}^{K-1} s(k) r^*(k+m) \right|.$$

4. The received signal $r(k)$ is correlated with the stored preamble $s(k)$ which is not affected by channel impairments and nonlinearity. Likewise, the time index with the maximum value is found as follows:

$$\hat{m} = \arg\max_{m} R_{sr}(m).$$

Summary 10.2   Frequency synchronization

1. Fractional CFO estimation can be accomplished simultaneously when the symbol boundary is detected. The maximum likelihood CFO estimation is expressed as follows:

$$\Delta \hat{f} = \frac{1}{2\pi L T_s} \tan^{-1} \left| \sum_{k=m}^{m+v-1} r(k) r^*(k+L) \right|$$

where the periodicity factor v is the block size and the separation L means the distance between two identical blocks.

2. One of the simplest integer CFO estimators uses frequency domain autocorrelation. Let $Y[l, k]$ be the frequency domain received symbol at the kth subcarrier of the lth OFDM symbol. Two consecutive OFDM symbols are correlated as follows:

$$\Delta \hat{f}(g) = \sum_{k=0}^{K-1} Y[l, \alpha_k + g] Y^*[l-1, \alpha_k + g]$$

where $g = 0, \pm1, \pm2, \ldots$ means the possible integer subcarrier index and α_k is the kth pilot subcarrier. The integer CFO can be estimated by finding the integer subcarrier index g with the largest value as follows:

$$\hat{g} = \arg\max_{g} \left| \Delta \hat{f}(g) \right|.$$

This method works well when the pilot assignment is comb-type and the channel is slow fading.

10.4   Hardware Implementation of OFDM Synchronization

We briefly investigate the hardware design issues of the OFDM synchronization block in this section. There are many issues to consider when implementing the OFDM synchronization block and choosing its architecture. An OFDM synchronization block requires a real

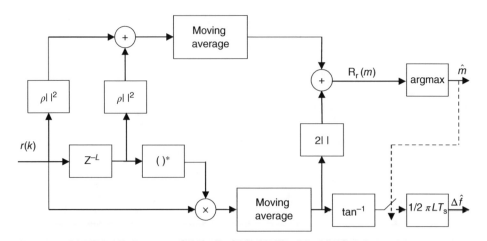

Figure 10.17   Block diagram of ML timing/frequency offset estimator

time implementation and a large size memory. Thus, the DSP or FPGA implementation may not be suitable. The ASIC design would be the best solution. Since many wireless communication systems support a high data rate service and they are mobile devices, acquisition speed and power consumption as well as the precision of the estimation are very important parameters. Basically, the synchronization block takes responsibility for estimating and compensating the symbol timing offset and carrier frequency offset. If a burst transmission mode is considered, a frame detection block is included in the symbol timing synchronization.

From (10.21), (10.23), and (10.26), the architecture of the simple ML timing/frequency offset estimator [1] can be designed as shown in Figure 10.17.

In this architecture, the weighting factor ρ depends on the SNR estimation. The synchronization block has a big computational burden in a burst transmission mode. The received signal is fed into autocorrelation and then the argmax block looks for a timing index from the maximum value. Simultaneously, the frequency offset is found at the timing index.

10.5   Problems

10.1. Plot the autocorrelation function of the rectangular signal.

10.2. Describe the properties of autrocorrelation function and crosscorrelation function.

10.3. Plot the crosscorrelation function of the following two signals:

$$r(t) = \cos(2\pi t)$$
$$s(t) = \cos(2\pi t) + n(t)$$

where $n(t)$ is the Gaussian noise.

10.4. Plot the crosscorrelation function between two triangular signals of different widths.

10.5. Consider the OFDM system with the following parameters:

Maximum carrier frequency offset	±30 ppm
Carrier frequency	5 GHz
Subcarrier spacing	200 kHz

10.6. Find the maximum allowed frequency offset.

10.7. Describe the SNR degradation caused by CFO, phase noise, and SFO in both a high mobility environment and a low mobility environment.

10.8. There are two types of CFO compensations: The first approach is to use a frequency domain interpolator and the other approach is to use a phase locked loop in time domain. Compare two techniques.

10.9. The preamble structure of wireless LAN (IEEE802.11.a) is different from the preamble structure of UWB (MB-OFDM). Compare their detection probability.

10.10. In LTE standard, there are two types of downlink synchronization signals: Primary Synchronization Signal (PSS) and Secondary Synchronization Signal (SSS). Describe the downlink synchronization process in LTE standard.

References

[1] J. J. van de Beek, M. Sandell, and P. O. Borjesson, "ML Estimation of Time and Frequency Offset in OFDM Systems," *IEEE Transactions on Signal Processing*, vol. **45**, no. 7, p. 1800, 1997.

[2] T. M. Schmidl and D. C. Cox, "Robust Frequency and Timing Synchronization for OFDM," *IEEE Transactions on Communications*, vol. **45**, no. 12, pp. 1613–1621, 1997.

[3] H. Zhou, A. V. Malipati, and Y. G. Huang, "Synchronization Issues in OFDM Systems," *Proceedings of IEEE Asia Pacific Conference on Circuits and Systems (APCCAS 2006)*, pp. 988–991, Singapore, December 4–7, 2006.

[4] B. Ai, Z. Yang, C. Pan, J. Ge, Y. Wang, and Z. Lu, "On the Synchronization Techniques for Wireless OFDM Systems," *IEEE Transaction on Broadcasting*, vol. **52**, no. 2, pp. 236–244, 2006.

[5] L. Chen, Q. Yang, K. Xue, and J. Shi, "A Residual Frequency Offset Compensation Scheme for OFDM System Base on MLE," *Journal of Information & Computational Science*, vol. **8**, no. 4, pp. 697–707, 2011.

[6] T. Pollet, M. Van Bladel, and M. Moeneclaey, "BER Sensitivity of OFDM Systems to Carrier Frequency Offset and Wiener Phase Noise," *IEEE Transactions on Communications*, vol. **43**, no. 2, pp. 192–193, 1995.

[7] M. Moeneclaey, "The Effect of Synchronization Errors on the Performance of Orthogonal Frequency Division Multiplexed (OFDM) Systems," *Proceedings of COST 254 (Emergent Techniques for Communication Terminals)*, Toulouse, France, July 7–9, 1997.

[8] T. Pollet, P. Spruyt, and M. Moeneclaey, "The BER Performance of OFDM Systems using Non-Synchronized Sampling," *Proceedings of IEEE Global Telecommunications Conference (Globecom'94)*, vol. **1**, pp. 253–257, San Francisco, CA, USA, November 28–December 2, 1994.

[9] H. Minn, V. K. Bhargava, and K. B. Letaief, "A Robust Timing and Frequency Synchronization for OFDM Systems," *IEEE Transactions on Wireless Communications*, vol. **2**, no. 4, pp. 822–839, 2003.

[10] M. Sandell, J. J. van de Beek, and P. O. Borjesson, "Timing and Frequency Synchronization in OFDM Systems Using the Cyclic Prefix," *Proceedings of the International Symposium on Synchronization*, pp. 16–19, Essen, Germany, December 14–15, 1995.

[11] Y. S. Lim and J. H. Lee, "An Efficient Carrier Frequency Offset Estimation Scheme for an OFDM System," *Proceedings of IEEE the 52nd Vehicular Technology Conference (VTC 2000)*, vol. **5**, pp. 2453–2458, Boston, MA, USA, September 24–28, 2000.

Part III

Wireless Communications Systems Design

11

Radio Planning

In Part III, we take a look at a big picture of wireless communication systems and use a top down approach. Many different interests are entangled in the wireless communications system design. The designed wireless communication system should satisfy various interested parties such as vendors, mobile operators, regulators, and mobile users. An officer of a regulatory agency takes a look at the wireless communication system in terms of an efficient spectrum usage and violation of the regulation for specific frequency bands. A mobile operator is interested in business opportunities, service quality, and the cost of the infrastructure. A mobile phone vendor or a network equipment vendor looks into the cost effective design of the wireless communication system. An end user is interested in good service quality and stylish mobile phones. For example, the designed wireless communication system copes with many channel impairments. The transmit power should be enough to cover users located in a cell edge. The battery life of a mobile phone should be long enough. It should support enough data rates at any place. The packet loss rate should be acceptable at any place. An end user is interested in whether or not it can use a network with enough data rates and seamless connection. Thus, we should consider all conditions and satisfy their requirements. In order to build wireless communication systems, we consider many system requirements such as spectrum, system capacity, service area, radio resource allocation, cost, and QoS. It is not easy to find an optimal solution for them but we should find a good trade-off. In Part III, we mainly deal with wireless communication network planning and wireless communication system integration. In this chapter, we focus on radio planning and link budget analysis.

11.1 Radio Planning and Link Budget Analysis

Since we have a limited radio resource, radio planning is an essential part of a wireless network design. The radio planning process deals with spectrum allocation, power assignment, base station location, configuration of a wireless communication system, and so on.

Wireless Communications Systems Design, First Edition. Haesik Kim.
© 2015 John Wiley & Sons, Ltd. Published 2015 by John Wiley & Sons, Ltd.

The goal of radio planning is to provide sufficient coverage, capacity, and service quality. Basically, the radio planning process is composed of three steps: *dimensioning, detailed planning*, and *optimization*. The process of radio planning is iteratively carried out. The first step is dimensioning. It provides the initial estimation of the number and placement of base stations to meet the network requirements. We firstly collect the required system coverage, capacity, and service quality and then take into consideration population, geographical area (dense urban area, urban area, and rural area), frequency, propagation model, and so on. Thus, we perform link budget analysis, traffic analysis, coverage estimation, and capacity evaluation. We initially decide how many base stations are needed, where to deploy them, and how to configure base station parameters such as antenna type, transmit power, height, and frequency allocation. Based on the results of the dimensioning step, the detailed planning step obtains further detailed information such as site selection and acquisition, coverage calculation, capacity planning, and base station configuration. In this step, many software tools are employed to predict coverage, capacity, interference, and so on. A wireless network designer creates databases including geographical and statistical information and decides base station location and configuration in detail. In the optimization step, we use measurements from the wireless networks and adjust the settings of the base stations to achieve a better coverage and capacity. Basically, the radio and network planning is very complex and there is no optimal solution. The only way we can do is to find a better setting iteratively. Figure 11.1 illustrates the summary of the radio and network planning process.

The dimensioning step starts from the link budget calculation. It can be defined as the signal strength gain and loss calculation on the path from a transmitter to a receiver. The purpose of the link budget calculation is to estimate the Maximum Allowable Path Loss (MAPL). It determines the cell size along with the geographical area and area coverage

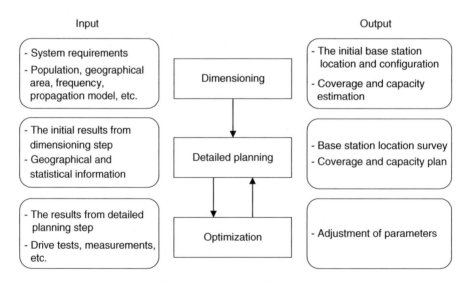

Figure 11.1   Radio and network planning process

probability. It estimates the required numbers of bases stations. The MAPL is calculated as follows:

$$\text{MAPL} = \text{EIRP} - P_{\text{Rx}} - M_{\text{F}} - M_{\text{ITF}} \tag{11.1}$$

where the Equivalent Isotropic Radiated Power (EIRP) is the RF power that would be emitted by a theoretical isotropic antenna and P_{Rx} is the received power. The EIRP considers all gains and losses in the transmitter and produces the peak power density observed in the direction of maximum antenna gain. It is represented as follows:

$$\text{EIRP} = P_{\text{Tx}} + G_{\text{Tx}} - L_{\text{Tx}} \tag{11.2}$$

where P_{Tx}, G_{Tx}, and L_{Tx} are the transmit power, the antenna gain, and the transmitter losses, respectively. The transmit power is determined by each regulation [1–4]. Tables 11.1 and 11.2 show the maximum transmit power levels for different Mobile Station (MS) (or User Equipment (UE)) classes in GSM and WCDMA.

Depending on MS (or UE) scenario, the maximum transmit powers are classified because it should not interfere with other MSs (or UEs). For LTE system, Tables 11.3 and 11.4 show the maximum transmit powers for UE and Base Station (BS), respectively.

In Table 11.3, the power class 2 and 3 are considered for indoor deployment and both indoor and outdoor deployment, respectively. The regulation basically allows about ±2 dBm tolerance

Table 11.1 Maximum transmit powers for each MS class in GSM [1]

MS classes	MS output power (dBm)
Class 1	43
Class 2	39
Class 3	37
Class 4	33

Table 11.2 Maximum transmit powers for each UE class in WCDMA [2]

UE classes	UE output power (dBm)
Class 1	33
Class 2	27
Class 3	24
Class 4	21

Table 11.3 Maximum transmit powers for each UE class in LTE [3]

CPE band	Class 2 (dBm)	Class 3 (dBm)
CPE_13	27	23

Table 11.4   Base station rated output power in LTE [4]

BS class	Rated output power (PRAT)
Wide area BS	There is no upper limit for the rated output power of the Wide Area Base Station
Medium range BS	≤ +38 dBm
Local area BS	≤ +24 dBm
Home BS	≤ +20 dBm (for one transmit antenna port) ≤ +17 dBm (for two transmit antenna ports) ≤ +14 dBm (for four transmit antenna ports) < +11 dBm (for eight transmit antenna ports)

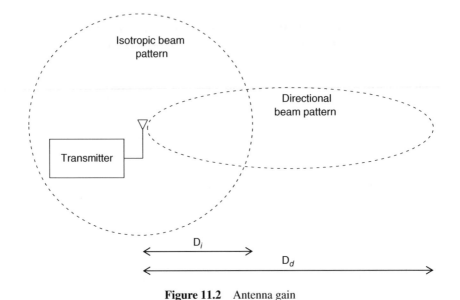

Figure 11.2   Antenna gain

of the maximum transmit power from each value. The antenna gain G_{Tx} is expressed in decibels-isotropic (dBi). Since the antenna is a passive device, it does not amplify a signal. However, a base station does not use omnidirectional antennas but directional (sectiorized) antennas. Thus, it can produce a relative gain due to directivity. Antenna gain is defined as the ratio of the power produced by the antenna in a given direction to the power produced by a loss free isotropic antenna at the same field strength and distance. Figure 11.2 illustrates antenna gain definition. It can be simply expressed as follows:

$$G_{Tx} = \varepsilon \frac{D_d}{D_i} \tag{11.3}$$

where ε is the efficiency of antenna.

There are several transmitter losses L_{Tx}. The cable and connector loss is one of the main transmission losses. The value depends on the length, material, temperature, frequency, and so on.

The cable manufacturer basically provides the attenuation value. The cable and connector loss of MS is zero because it does not use an external antenna. Instead, MS should take into account a body loss because a handset is close to a user's head. Basically, a human head absorbs transmission energy and change polarization. The value of the body loss is about 3–5 dB. If a mobile user is located in a car or a building, an extra loss (car and building penetration loss) should be considered [5]. The received power is represented as follows:

$$P_{Rx} = S_{Rx} - G_{Rx} + L_{Rx} \qquad (11.4)$$

where S_{Rx}, G_{Rx}, and L_{Rx} are the receiver sensitivity, the receiver antenna gain, and receiver losses, respectively. The receiver sensitivity is the minimum received signal power which can successfully demodulate and decode the received signals. It is always expressed as a negative value and the typical value is in the range of -70 to $-110\,$dBm. It is independent of a transmitter and depends on the receiver noise figure and the received signal-to-interference-plus-noise ratio (SINR). It can be represented as follows:

$$S_{Rx} = N_T + SINR + F_{Rx} + M_i \qquad (11.5)$$

where SINR, N_T, F_{Rx}, and M_i are the SINR, the thermal noise power per subcarrier, the receiver noise figure, and the implementation margin, respectively. The SINR can be defined as follows:

$$SINR = \frac{\text{Average received power}}{\text{Interference} + \text{Noise}} \geq \text{Required SINR} \qquad (11.6)$$

The required SINR is defined as the power ratio of the desired signal to total interferences and noise. The required SINR is one of the key parameters in wireless communication systems. It depends on modulation and coding schemes and channel models. Basically, a higher required SINR is needed when a high-order modulation and a low coding rate are used. The thermal noise N_T is calculated as follows:

$$N_T = kTB \qquad (11.7)$$

where k, T, and B are Boltzmann constant ($1.38 \times 10^{-23}\,$J/K), the temperature (K), and the bandwidth, respectively. The receiver noise figure is defined as follows:

$$F_{Rx} = \frac{SNR_{in}}{SNR_{out}} \qquad (11.8)$$

where SNR_{in} and SNR_{out} are the input SNR and the output SNR in a RF device, respectively. It means how much thermal noises are added to the received signal in a RF device. Basically, it is expressed in decibels. (11.8) can be rewritten as follows:

$$NF_{Rx}(dB) = 10 \log F_{Rx} = SNR_{in}(dB) - SNR_{out}(dB). \qquad (11.9)$$

If the receiver is composed of multiple cascaded RF devices, the total noise figure can be represented with Friis's formula as follows:

$$F_{\mathrm{Rx}} = F_1 + \frac{F_2 - 1}{G_1} + \frac{F_3 - 1}{G_1 G_2} + \cdots + \frac{F_n - 1}{G_1 G_2 \cdots G_{n-1}} \qquad (11.10)$$

where F_i and G_i are the noise figure and the power gain of the ith device, respectively. The power gain is given by the manufacturer if it is an active device. However, if it is a passive device, there is an inverse relationship between the power gain and the transmission loss.

Basically, a shadowing and multipath fading effect occurs in wireless channels. Thus, a wireless communication and network designer should keep sufficient system gain to accommodate the fading effect for maintaining good service quality. Higher the fading margin M_F provides the more reliable link quality. It depends on carrier frequency, distance between MS and BS, required reliability, and so on. As a rule of thumb, we need to have at least 15 dB as a fading margin.

In cellular systems, the interferences from own cells and neighboring cells always occur. A wireless communication and network designer should allow a network to secure an interference margin. This margin is related to cell load and coverage. Generally, a smaller interference margin is needed for coverage limited scenarios and a larger interference margin is needed for capacity limited scenarios. As cell load increases, addition interferences are generated from own cell and neighboring cells and the coverage shrinks. Namely, a larger interference margin is suggested when more loading is allowed in the network and smaller coverage is caused. The interference margin depends on cell load and G-factor. It can be defined as follows:

$$M_{\mathrm{ITF}}(\mathrm{dB}) = \mathrm{SNR}(\mathrm{dB}) - \mathrm{SINR}(\mathrm{dB}). \qquad (11.11)$$

The typical values of the interference margin are 1–3 dB and the corresponding cell loading is 20–50% [6]. Figure 11.3 illustrates signal strength change from a transmitter to a receiver.

We now obtained the MAPL and can calculate the cell radius using a propagation model such as the free space model, the ray tracing model, and the empirical model. The empirical model is widely used in the network planning phase. For example, we have Hata's formula [7] for an urban area as follows:

$$\mathrm{MAPL} = 69.55 + 26.16 \log_{10} f_c - 13.82 \log_{10} h_{\mathrm{bs}} + \left(44.9 - 6.55 \log_{10} h_{\mathrm{bs}}\right) \log_{10} R - a\left(h_{\mathrm{ms}}\right) \qquad (11.12)$$

where f_c is the carrier frequency (MHz), h_{bs} is the base station antenna height (m), R is the cell radius (km), h_{ms} is the mobile station antenna height (m), and $a(h_{\mathrm{ms}})$ is the correction factor for the effective mobile station antenna height. From (11.12), the cell radius is calculated as follows:

$$R = 10^{\left(\mathrm{MAPL} - 69.55 - 26.16 \log_{10} f_c + 13.82 \log_{10} h_{\mathrm{bs}} + a\left(h_{\mathrm{ms}}\right)\right) / \left(44.9 - 6.55 \log_{10} h_{\mathrm{bs}}\right)}. \qquad (11.13)$$

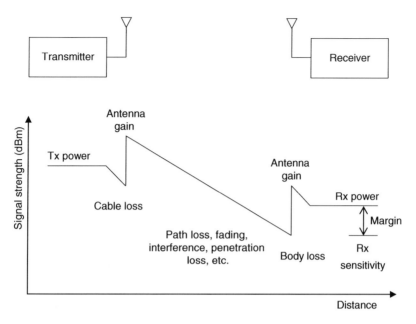

Figure 11.3   General link budget in wireless communication system

For a small size cell,

$$a\left(h_{\mathrm{ms}}\right)=0.8+\left(1.1\log_{10}f_{\mathrm{c}}-0.7\right)h_{\mathrm{ms}}-1.56\log_{10}f_{\mathrm{c}} \tag{11.14}$$

and for a large cell,

$$a\left(h_{\mathrm{ms}}\right)=\begin{cases}8.29\left(\log_{10}1.54h_{\mathrm{ms}}\right)^{2}-1.1, & \text{when } 150\le f_{\mathrm{c}}\le 200 \\ 3.2\left(\log_{10}11.75h_{\mathrm{ms}}\right)^{2}-4.97, & \text{when } 200< f_{\mathrm{c}}\le 1500.\end{cases} \tag{11.15}$$

This cell radius means the signal beyond this radius is not reliable. Each cell has different correction factor and a propagation model is adopted according to the type of surroundings (urban, suburban, and rural) and cell deployments (macrocell, microcell, and femtocell). Thus, the MAPL and cell radius should be separately calculated for each area. Once the cell radius is decided, the cell area A_{cell} is calculated as follows:

$$A_{\mathrm{cell}}=\alpha R^{2} \tag{11.16}$$

where α is 2.6 (omnidirectional site), 1.3 (2 sector site), and 1.95 (3 sector site) in the hexagonal cell model [8]. Figure 11.4 illustrates different types of sites.

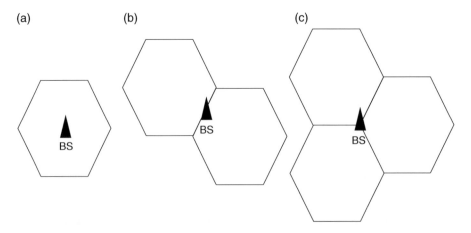

Figure 11.4 Omnidirectional site (a), 2 sector site (b), and 3 sector site (c)

Example 11.1   Link budget calculation
Consider the following system parameters and calculate the MAPL, cell radius and cell area.

System parameters	Values
Type of surrounding	Medium-sized urban area
Site type	Omnidirectional site
Base station height	30 m
Mobile station height	1.8 m
Bandwidth	20 MHz
Carrier frequency	1.5 GHz
Noise figure	7 dB
Throughput	3 Mbps
Receiver antenna gain	0 dB
Base station maximum transmit power	46 dBm
Transmitter antenna gain	15 dBi
Transmitter loss including cable loss	5 dB
Thermal noise	−104.5 dBm
Implementation margin	1.1 dB
Fading margin	15 dB
Interference margin	3 dB
Receiver loss including body loss	3 dB

Solution
From (11.1), MAPL is calculated as follows:

$$\text{MAPL} = \text{EIRP} - P_{\text{Rx}} - M_{\text{F}} - M_{\text{ITF}}.$$

The first term, EIRP of (11.1), is found from (11.2) as follows:

$$\begin{aligned}\text{EIRP(dBm)} &= P_{\text{Tx}}(\text{dBm}) + G_{\text{Tx}}(\text{dBi}) - L_{\text{Tx}}(\text{dB}) \\ &= 46(\text{dBm}) + 15(\text{dBi}) - 5(\text{dB}) = 56 \ \text{dBm}.\end{aligned}$$

The second term, the received power of (11.1), is calculated from (11.4) as follows:

$$P_{Rx}(dBm) = S_{Rx}(dBm) - G_{Rx}(dBi) + L_{Rx}(dB).$$

From (11.5), we calculate the receiver sensitivity as follows:

$$S_{Rx}(dBm) = N_T(dBm) + SINR(dB) + F_{Rx}(dB) + M_i(dB)$$

The second term, SINR of (11.5), can be estimated by using Shannon's channel capacity formula as follows:

$$SINR(dB) = 10\log(2^{throughput/b\,and\,width} - 1)$$
$$= 10\log_{10}(2^{3\ Mbps/20\ MHz} - 1) = -9.6\ dB.$$

We calculate the receiver sensitivity as follows:

$$S_{Rx}(dBm) = -104.5\ dBm - 9.6\ dB + 7\ dB + 1.1\ dB = -106\ dBm.$$

Thus, the receiver power is

$$P_{Rx}(dBm) = -106(dBm) - 0(dBi) + 3(dB) = -103\ dBm.$$

Finally, we calculate the MAPL as follows:

$$MAPL(dB) = 56(dBm) + 103(dBm) - 15(dB) - 3(dB) = 141\ dB.$$

From (11.13) and (11.14), the cell radius is calculated as follows:

$$a(1.8) = 0.8 + (1.1\log_{10} 1.5 \times 10^9 - 0.7)1.8 - 1.56\log_{10} 1.5 \times 10^9 = 3.4$$

$$R = 10^{(141-69.55-26.16\log_{10} 1.5\times10^9 + 13.82\log_{10} 30 + 3.4)/(44.9-6.55\log_{10} 30)} = 2.2\ km.$$

From (11.6), the cell area is found for the omnidirectional site as follows:

$$A_{cell} = \alpha R^2 = 2.6(2.2)^2 = 12.6\ km^2.$$

Figure 11.5 illustrates link budget analysis of Example 11.1   ∎

Based on the link budget analysis, we estimated the coverage which guarantees the signal reliability. Now, we take a look into capacity planning. It provides us with demanded resources and available traffics and estimates subscribers supported per a site. In order to calculate the capacity, we need to define several terms such as busy hour, overbooking factor, and interference. The busy hour is the traffic engineering measurement representing the greatest traffic and call attempts in a day. As a rule of thumb, the busy hour average loading is 50% and it carries 15% daily traffic. The overbooking factor is based on the fact that only small number of subscribers uses a radio resource simultaneously.

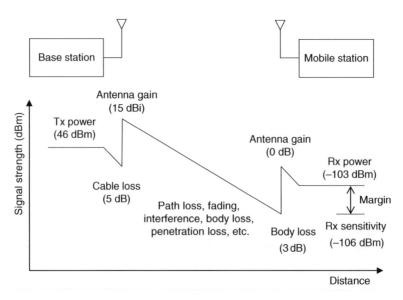

Figure 11.5   Link budget of Example 11.1

Table 11.5   Subscribers per site calculation based on data rate

System parameters	Variables
Bandwidth	a (Hz)
Spectral efficiency	b (bps/Hz)
Cell average capacity	$c=ab$ (bps)
Busy hour average loading	d (%)
Required user data rate	e (bps/subscriber)
Overbooking factor	f
Average busy hour data rate per subscriber	$g=e/f$ (bps)
The number of sectors per site	h
Subscribers per site	cdh/g (subscriber/site)

For example, if a user data rate is 1 Mbps and an average data rate per a subscriber during a busy hour is 20 kbps, the overbooking factor is 50 (=1 Mbps/20 kbps). Once system requirements are given, we can calculate the number of subscribers per site. In order to meet the target user data rate, the number of subscribers per site is calculated as shown in Table 11.5.

In order to meet the traffic requirements and find suitable network configurations, the subscribers per site should be calculated in terms of traffic demands. The number of subscribers per site is calculated as shown in Table 11.6.

Table 11.6 Subscribers per site calculation based on traffic

System parameters	Variables
Bandwidth	a (Hz)
Spectral efficiency	b (Mbps/Hz)
Cell average capacity	$c = ab$ (Mbps)
Required traffic	$d = c \times (3600/8192)$ (GB/h)
Busy hour average loading	e (%)
Busy hour carries of daily traffic	f (%)
Convert to month	30
The number of sectors per site	g
The required user traffic	h
Subscribers per site	$30deg/fh$ (subscriber/site/month)

Example 11.2   Capacity planning

Consider the following system parameters and calculate the subscribers per site when the required user data rate is 1 Mbps/subscriber and the required user traffic is 4 GB/month.

System parameters	Values
Bandwidth	20 MHz
Spectral efficiency	1.5 bps/Hz
Cell average capacity	30 Mbps
Busy hour average loading	50%
Overbooking factor	20
Busy hour carries of daily traffic	15%
The number of sectors per site	3

Solution

From Table 11.5, we calculate the subscriber per site when the required user data rate is 1 Mbps/subscriber. The calculation is summarized in Table 11.7.

From Table 11.6, we calculate the subscriber per site when the required user traffic is 4 GB/month. The calculation is summarized in Table 11.8.   ∎

11.2   Traffic Engineering

A customer visiting a bank wants to avoid a long queue. A car on a road does not want to be stuck in a traffic jam. Likewise, a mobile user does not want to wait a phone connection for a long time and expects a high quality service all the time. Thus, a wireless communication and network designer should manage a radio resource and provide a mobile user with adequate services. Traffic engineering deals with this issue. It is statistical techniques to predict the behavior of users and share network resources.

Table 11.7 Subscribers per site calculation based on user data rate

System parameters	Variable
Bandwidth	20 MHz
Spectral efficiency	1.5 bps/Hz
Cell average capacity	30 Mbps = 20 MHz × 1.5 bps/Hz
Busy hour average loading	50%
Required user data rate	1 Mbps/subscriber
Overbooking factor	20
Average busy hour data rate per subscriber	50 kbps = 1 Mbps/subscriber/20
The number of sectors per site	3
Subscribers per site	900 subscriber/site = 30 Mbps × 50% × 3/50 kbps

Table 11.8 Subscribers per site calculation based on user traffic

System parameters	Variable
Bandwidth	20 Hz
Spectral efficiency	1.5 Mbps/Hz
Cell average capacity	30 Mbps
Required traffic	13.18 GB/h
Busy hour average loading	50%
Busy hour carries of daily traffic	15%
Convert to month	30
The number of sectors per site	3
Subscribers per site	989 subscriber/site/month = (30 × 13.18 GB/h × 0.5 × 3)/(0.15 × 4)

Summary 11.1   Radio planning

1. The goal of radio planning is to find sufficient coverage, capacity, and service quality in a cell.
2. The radio planning process is composed of three steps: dimensioning, detailed planning, and optimization.
3. After collecting the required system coverage, capacity, and service quality, and on considering population, geographical area (dense urban area, urban area and rural area), frequency, and propagation model, the dimensioning step provides the initial estimation about the number and placement of base stations to meet the network requirements.
4. Based on the results of the dimensioning step, the detailed planning step obtains further detailed information such as site selection and acquisition, coverage calculation, capacity planning, and base station configuration and then decides a base station location and configuration using software tools.
5. In the optimization step, we use measurements from the networks and adjust them to achieve a better coverage and capacity. Basically, radio and network planning is very complex and there is no optimal solution. The only way we can do is to find a better parameter iteratively.

A Danish mathematician, A. K. Erlang, developed the fundamentals of trunking theory [9]. Many traffic models are based on his theory. Before discussing Erlang theory, we need to define several terms. Trunking is a concept in which many users share a limited radio resource. It allows n users to share the relatively small number of channels in a cell. The traffic unit, Erlang, means channel occupation ratio. For example, 0.5 Erlang means 30 min call for 1 h. The traffic intensity is a measure of the average channel occupancy during a specific time. It is measured in Erlang and denoted by A. The traffic intensity for one user is defined as follows:

$$A_u = \lambda H \tag{11.17}$$

where λ and H are the average number of channel/call request per unit time and the average occupancy duration of a channel/call, respectively. The total traffic intensity A is

$$A = nA_u. \tag{11.18}$$

The Grade of Service (GOS) is a measure of congestion, which is specified as the probability that a call is blocked or that a call is delayed. A blocked/lost call means inability to allocate a channel/call due to congestion. Erlang B is a measure of the GOS and simply defines the call blocking probability. It is formulated as follows:

$$P_b(N,A) = \frac{\dfrac{A^N}{N!}}{\displaystyle\sum_{i=0}^{N} \dfrac{A^i}{i!}} \tag{11.19}$$

where N is the number of resources (or the trunked channels). The Erlang B model has the following assumptions:

- There is no set-up time when a service is requested.
- The requesting users are blocked when there is no available channel and try again later.
- It offers no queuing for call requests.
- The arrival of calls is determined by the Poisson process.
- We have an infinite number of users.
- All users may request a channel resource at any time.
- The holding time is exponentially distributed.
- A finite number of channels are available in the trunking pool.

Example 11.3   Erlang B model
Consider that the total traffic intensity A is 2 Erlangs and compare the blocking probabilities when the number of resources N is 2 and 6.

Solution
From (11.19), the blocking probability for $N=2$ is

$$P_b(2,2) = \frac{\dfrac{2^2}{2!}}{\displaystyle\sum_{i=0}^{2} \dfrac{2^i}{i!}} = \frac{2}{1+2+\dfrac{2^2}{2!}} = 0.4 = 40\%$$

and the blocking probability for $N=6$ is

$$P_b(6,2) = \frac{\dfrac{2^6}{6!}}{\displaystyle\sum_{i=0}^{6}\dfrac{2^i}{i!}} = \frac{\dfrac{64}{720}}{1+2+\dfrac{2^2}{2!}+\dfrac{2^3}{3!}+\dfrac{2^4}{4!}+\dfrac{2^5}{5!}+\dfrac{2^6}{6!}} = 0.012 = 1.2\%.$$

As easily expected, we can get a lower blocking probability when we have many resources.
∎

Example 11.4 Supported users calculation using Erlang B model

Consider that the traffic intensity for one user A_u is 0.1 Erlangs and the number of resources are 5 and 15. How many users can be supported for 1% blocking probability?

Solution

Firstly, we calculate the supported user for $N=5$. From (11.19), we can formulate the equation for n supported users as follows:

$$P_b(5, n\times 0.1) = \frac{\dfrac{(n\times 0.1)^5}{5!}}{\displaystyle\sum_{i=0}^{5}\dfrac{(n\times 0.1)^i}{i!}} = 1\%,$$

$$\frac{\dfrac{(n\times 0.1)^5}{5!}}{1+(n\times 0.1)+\dfrac{(n\times 0.1)^2}{2!}+\dfrac{(n\times 0.1)^3}{3!}+\dfrac{(n\times 0.1)^4}{4!}+\dfrac{(n\times 0.1)^5}{5!}} = 0.01$$

This equation is not easy to solve. Thus, we calculate (11.9) for each value in advance and make Erlang B table as shown in Table 11.9. From Table 11.9 and (11.18), when $P_b(5, A) = 1\%$, $A \approx 1.36$ Erlangs and $A = nA_u$. Thus, we have

$$1.36 = n\times 0.1,$$

$$n \approx 14 \text{ users.}$$

Secondly, we calculate the supported user for $N=15$. From Table 11.9 and (11.18), we have

$$\text{when } P_b(15, A) = 1\%, A \approx 8.11 \text{ Erlangs,}$$

$$8.11 = n\times 0.1,$$

$$n \approx 81 \text{ users.}$$

As easily expected, we can obtain more supported users when we have many resources. ∎

Table 11.9 Erlang B Table

No. of resources (N)	Traffic intensity in Erlangs (A) for Pb(N, A)																
	0.1%	0.2%	0.5%	1%	1.2%	1.3%	1.5%	2%	3%	5%	7%	10%	15%	20%	30%	40%	50%
1	0.001	0.002	0.005	0.010	0.012	0.013	0.02	0.020	0.031	0.053	0.075	0.111	0.176	0.250	0.429	0.667	1.00
2	0.046	0.065	0.105	0.153	0.168	0.176	0.19	0.223	0.282	0.381	0.470	0.595	0.796	1.00	1.45	2.00	2.73
3	0.194	0.249	0.349	0.455	0.489	0.505	0.53	0.602	0.715	0.899	1.06	1.27	1.60	1.93	2.63	3.48	4.59
4	0.439	0.535	0.701	0.869	0.922	0.946	0.99	1.09	1.26	1.52	1.75	2.05	2.50	2.95	3.89	5.02	6.50
5	0.762	0.900	1.13	1.36	1.43	1.46	1.52	1.66	1.88	2.22	2.50	2.88	3.45	4.01	5.19	6.60	8.44
6	1.15	1.33	1.62	1.91	2.00	2.04	2.11	2.28	2.54	2.96	3.30	3.76	4.44	5.11	6.51	8.19	10.4
7	1.58	1.80	2.16	2.50	2.60	2.65	2.73	2.94	3.25	3.74	4.14	4.67	5.46	6.23	7.86	9.80	12.4
8	2.05	2.31	2.73	3.13	3.25	3.30	3.40	3.63	3.99	4.54	5.00	5.60	6.50	7.37	9.21	11.4	14.3
9	2.56	2.85	3.33	3.78	3.92	3.98	4.08	4.34	4.75	5.37	5.88	6.55	7.55	8.52	10.6	13.0	16.3
10	3.09	3.43	3.96	4.46	4.61	4.68	4.80	5.08	5.53	6.22	6.78	7.51	8.62	9.68	12.0	14.7	18.3
11	3.65	4.02	4.61	5.16	5.32	5.40	5.53	5.84	6.33	7.08	7.69	8.49	9.69	10.9	13.3	16.3	20.3
12	4.23	4.64	5.28	5.88	6.05	6.14	6.27	6.61	7.14	7.95	8.61	9.47	10.8	12.0	14.7	18.0	22.2
13	4.83	5.27	5.96	6.61	6.80	6.89	7.03	7.40	7.97	8.83	9.54	10.5	11.9	13.2	16.1	19.6	24.2
14	5.45	5.92	6.66	7.35	7.56	7.65	7.81	8.20	8.80	9.73	10.5	11.5	13.0	14.4	17.5	21.2	26.2
15	6.08	6.58	7.38	8.11	8.33	8.43	8.59	9.01	9.65	10.6	11.4	12.5	14.1	15.6	18.9	22.9	28.2
16	6.72	7.26	8.10	8.88	9.11	9.21	9.39	9.83	10.5	11.5	12.4	13.5	15.2	16.8	20.3	24.5	30.2
17	7.38	7.95	8.83	9.65	9.89	10.0	10.19	10.7	11.4	12.5	13.4	14.5	16.3	18.0	21.7	26.2	32.2
18	8.05	8.64	9.58	10.4	10.7	10.8	11.00	11.5	12.2	13.4	14.3	15.5	17.4	19.2	23.1	27.8	34.2
19	8.72	9.35	10.3	11.2	11.5	11.6	11.82	12.3	13.1	14.3	15.3	16.6	18.5	20.4	24.5	29.5	36.2
20	9.41	10.1	11.1	12.0	12.3	12.4	12.65	13.2	14.0	15.2	16.3	17.6	19.6	21.6	25.9	31.2	38.2
21	10.1	10.8	11.9	12.8	13.1	13.3	13.48	14.0	14.9	16.2	17.3	18.7	20.8	22.8	27.3	32.8	40.2
22	10.8	11.5	12.6	13.7	14.0	14.1	14.32	14.9	15.8	17.1	18.2	19.7	21.9	24.1	28.7	34.5	42.1
23	11.5	12.3	13.4	14.5	14.8	14.9	15.16	15.8	16.7	18.1	19.2	20.7	23.0	25.3	30.1	36.1	44.1
24	12.2	13.0	14.2	15.3	15.6	15.8	16.01	16.6	17.6	19.0	20.2	21.8	24.2	26.5	31.6	37.8	46.1
25	13.0	13.8	15.0	16.1	16.5	16.6	16.87	17.5	18.5	20.0	21.2	22.8	25.3	27.7	33.0	39.4	48.1
26	13.7	14.5	15.8	17.0	17.3	17.5	17.72	18.4	19.4	20.9	22.2	23.9	26.4	28.9	34.4	41.1	50.1
27	14.4	15.3	16.6	17.8	18.2	18.3	18.59	19.3	20.3	21.9	23.2	24.9	27.6	30.2	35.8	42.8	52.1
28	15.2	16.1	17.4	18.6	19.0	19.2	19.45	20.2	21.2	22.9	24.2	26.0	28.7	31.4	37.2	44.4	54.1

(Continued)

Table 11.9 *(Continued)*

No. of resources (N)	Traffic intensity in Erlangs (A) for Pb(N, A)																
	0.1%	0.2%	0.5%	1%	1.2%	1.3%	1.5%	2%	3%	5%	7%	10%	15%	20%	30%	40%	50%
29	15.9	16.8	18.2	19.5	19.9	20.0	20.32	21.0	22.1	23.8	25.2	27.1	29.9	32.6	38.6	46.1	56.1
30	16.7	17.6	19.0	20.3	20.7	20.9	21.19	21.9	23.1	24.8	26.2	28.1	31.0	33.8	40.0	47.7	58.1
31	17.4	18.4	19.9	21.2	21.6	21.8	22.07	22.8	24.0	25.8	27.2	29.2	32.1	35.1	41.5	49.4	60.1
32	18.2	19.2	20.7	22.0	22.5	22.6	22.95	23.7	24.9	26.7	28.2	30.2	33.3	36.3	42.9	51.1	62.1
33	19.0	20.0	21.5	22.9	23.3	23.5	23.83	24.6	25.8	27.7	29.3	31.3	34.4	37.5	44.3	52.7	64.1
34	19.7	20.8	22.3	23.8	24.2	24.4	24.72	25.5	26.8	28.7	30.3	32.4	35.6	38.8	45.7	54.4	66.1
35	20.5	21.6	23.2	24.6	25.1	25.3	25.60	26.4	27.7	29.7	31.3	33.4	36.7	40.0	47.1	56.0	68.1
36	21.3	22.4	24.0	25.5	26.0	26.2	26.49	27.3	28.6	30.7	32.3	34.5	37.9	41.2	48.6	57.7	70.1
37	22.1	23.2	24.8	26.4	26.8	27.0	27.39	28.3	29.6	31.6	33.3	35.6	39.0	42.4	50.0	59.4	72.1
38	22.9	24.0	25.7	27.3	27.7	27.9	28.28	29.2	30.5	32.6	34.4	36.6	40.2	43.7	51.4	61.0	74.1
39	23.7	24.8	26.5	28.1	28.6	28.8	29.18	30.1	31.5	33.6	35.4	37.7	41.3	44.9	52.8	62.7	76.1
40	24.4	25.6	27.4	29.0	29.5	29.7	30.08	31.0	32.4	34.6	36.4	38.8	42.5	46.1	54.2	64.4	78.1
41	25.2	26.4	28.2	29.9	30.4	30.6	30.98	31.9	33.4	35.6	37.4	39.9	43.6	47.4	55.7	66.0	80.1
42	26.0	27.2	29.1	30.8	31.3	31.5	31.88	32.8	34.3	36.6	38.4	40.9	44.8	48.6	57.1	67.7	82.1
43	26.8	28.1	29.9	31.7	32.2	32.4	32.79	33.8	35.3	37.6	39.5	42.0	45.9	49.9	58.5	69.3	84.1
44	27.6	28.9	30.8	32.5	33.1	33.3	33.69	34.7	36.2	38.6	40.5	43.1	47.1	51.1	59.9	71.0	86.1
45	28.4	29.7	31.7	33.4	34.0	34.2	34.60	35.6	37.2	39.6	41.5	44.2	48.2	52.3	61.3	72.7	88.1
46	29.3	30.5	32.5	34.3	34.9	35.1	35.51	36.5	38.1	40.5	42.6	45.2	49.4	53.6	62.8	74.3	90.1
47	30.1	31.4	33.4	35.2	35.8	36.0	36.42	37.5	39.1	41.5	43.6	46.3	50.6	54.8	64.2	76.0	92.1
48	30.9	32.2	34.2	36.1	36.7	36.9	37.34	38.4	40.0	42.5	44.6	47.4	51.7	56.0	65.6	77.7	94.1
49	31.7	33.0	35.1	37.0	37.6	37.8	38.25	39.3	41.0	43.5	45.7	48.5	52.9	57.3	67.0	79.3	96.1
50	32.5	33.9	36.0	37.9	38.5	38.7	39.17	40.3	41.9	44.5	46.7	49.6	54.0	58.5	68.5	81.0	98.1

Example 11.5   Average phone usage calculation using Erlang B model

Consider that the number of resource is 8, the total number of users is 100, and the blocking probability is 0.1%. How much average time does each subscriber use a phone during a peak hour?

Solution

From Table 11.9, we find the traffic intensity as follows:

$$A \approx 2.05\,\text{Erlangs}.$$

The traffic intensity per user is

$$2.05 = 100 A_u,$$

$$A_u = 0.0205\,\text{Erlangs}.$$

This traffic intensity per user means that each subscriber uses a phone call (or occupies the channel) for 0.0205 Erlangs. We express the average phone usage of each subscriber during a peak hour as follows:

$$\text{Average phone usage of each subscriber} = 0.0204 \times 3600 (\text{s/h})$$
$$= 73.44\ (\text{s/h})$$

Thus, we can say that each subscriber uses a phone call for 73.44 s during a peak hour.   ∎

Example 11.6   Radio planning

A city needs a cellular network. A wireless network designer has the following requirements:

System parameters	Values
System bandwidth	10 MHz
Channel bandwidth	30 kHz
Site type	Omnidirectional site
City area	500 km²
Cell radius	2 km
Cluster size	7
Average number of channel per unit time	0.5 calls/h
Average occupancy duration of a channel	1 min
Blocking probability	0.5%
Population	1 million

When the wireless network is based on Erlang B model, calculate the maximum carried traffic intensity over the network and find how many people can subscribe the network.

Solution

From (11.16), we calculate the cell area A_{cell} of the omnidirectional site as follows:

$$A_{cell} = \alpha R^2 = 2.6(2)^2 = 10.4 \ \text{km}^2.$$

The required number of cells for the city is:

$$\text{The required number of cells} = \frac{500}{10.4} \approx 48 \ \text{cells}.$$

The total number of channels in the network is calculated as follows:

$$\text{The number of channels per cell} = \frac{\text{system bandwidth}}{\text{channel bandwidth} \times \text{cluster size}}$$

$$= \frac{10 \ \text{MHz}}{30 \ \text{kHz} \times 7} \approx 48 \ \text{channels/cell}$$

$$\text{The total number of channels in the network}$$
$$= (48 \ \text{channels/cell}) \times (48 \ \text{cells}) = 2304 \ \text{channels}.$$

From Table 11.9,

$$\text{when } P_b(48, A) = 0.5\%, \quad A \approx 34.2 \ \text{Erlangs}.$$

Thus, the maximum carried traffic intensity over the network is

$$34.2 \ \text{Erlangs} \times 48 \ \text{cells} \approx 1642 \ \text{Erlangs}.$$

From (11.17), the traffic intensity for one subscriber is

$$A_u = \lambda H = \frac{0.5}{60} = 0.0084 \ \text{Erlangs}$$

and the subscriber for one cell is

$$\frac{34.2}{0.0084} \approx 4071 \ \text{subscribers/cell}.$$

Thus, the total subscriber in the network is

$$(4071 \ \text{subscribers/cell}) \times (48 \ \text{cells}) = 195 \ 408 \ \text{subscribers}.$$

From above result, we can verify that the city subscribes 195 408 people.   ∎

The Erlang B model assumes the blocked requests are lost. However, Erlang C model describes the probability that a service requester will have to wait for a resource. If all servers are busy, the request is queued. Like the Erlang B, the Erlang C model is formulated as follows:

$$P_c(N,A) = \frac{\dfrac{A^N N}{N!(N-A)}}{\displaystyle\sum_{i=0}^{N} \dfrac{A^i}{i!} + \dfrac{A^N N}{N!(N-A)}}.$$

(11.20)

The probability $P_c(N,A)$ means that a service requester has to wait for service. The average delay is given by

$$D = P_c(N,A)\frac{H}{N-A}$$

(11.21)

where H is an average call holding time. The average number of requests in the queue is given by

$$N_q = P_c(N,A)\frac{A}{N-A}.$$

(11.22)

The probability of the delay exceeding time t is given by

$$P_c(\text{delay} > t) = P_c(N,A)e^{-(N-A)t/H}.$$

(11.23)

The Erlang B model has the following assumptions:

- An unlimited number of requests can be held in the queue.
- A service requester is willing to wait in the queue
- The arrival of calls can be modeled by the Poisson process.
- Service requesters are independent of each other.
- Service time is exponentially distributed.

Example 11.7   Erlang C model

A company has five private phone lines between a head office and a research center. The number of employees is 200 and each employee makes 1 min call/h on average. Find the probability that a caller has to wait for service, the average delay, the average number of requests in the queue, and the probability of the delay exceeding 2 min.

Solution

The total traffic intensity A is calculated as follows:

$$A = 200 \times \frac{1}{60} = 3.34 \text{ Erlangs.}$$

From (11.20), the probability that a caller has to wait for service is

$$P_c(5,3.34) = \frac{\dfrac{A^N N}{N!(N-A)}}{\displaystyle\sum_{i=0}^{N}\dfrac{A^i}{i!} + \dfrac{A^N N}{N!(N-A)}}$$

$$= \frac{\dfrac{(3.34)^5(5)}{5!(5-3.34)}}{\displaystyle\sum_{i=0}^{5}\dfrac{(3.34)^i}{i!} + \dfrac{(3.34)^5(5)}{5!(5-3.34)}} = 0.29.$$

From (11.21), the average delay is

$$D = P_c(N,A)\frac{H}{N-A} = 0.29\frac{1}{5-3.34} = 0.17 \text{ min}.$$

From (11.22), the average number of requests in the queue is

$$N_q = P_c(N,A)\frac{A}{N-A} = 0.29\frac{3.34}{5-3.34} = 0.57.$$

From (11.23), the probability of the delay exceeding 2 min is

$$P_c(\text{delay} > 2 \text{ min}) = P_c(N,A)e^{-(N-A)t/H} = 0.29e^{-(5-3.34)2/1} = 0.01. \qquad\blacksquare$$

The Erlang models are designed for the telephony industry but some assumptions do not fit into the real world. For example, the Erlang model assumes that a subscriber waits until speaking with another subscriber. However, some subscriber hangs up immediately if a connection failure occurs. Some subscriber waits for only few seconds. In addition, the Erlang model assumes that we have unlimited queuing capacity. However, it is limited.

Summary 11.2   Traffic engineering

1. The traffic engineering deals with statistical techniques to predict the behavior of users and share network resources.
2. The Erlang B model is formulated as follows:

$$P_b(N,A) = \frac{\dfrac{A^N}{N!}}{\displaystyle\sum_{i=0}^{N}\dfrac{A^i}{i!}}$$

where N is the number of resources (or the trunked channels) and the total traffic intensity A is

$$A = nA_u = n\lambda H$$

where λ and H are the average number of channel/call request per unit time and the average occupancy duration of a channel/call, respectively.

3. The Erlang C model is formulated as follows:

$$P_c(N,A) = \frac{\dfrac{A^N N}{N!(N-A)}}{\displaystyle\sum_{i=0}^{N} \dfrac{A^i}{i!} + \dfrac{A^N N}{N!(N-A)}}.$$

11.3 Problems

11.1. Consider the following system parameters and calculate the MAPL, cell radius, and cell area.

System parameters	Values
Type of surrounding	Rural area
Site type	Omnidirectional site
Base station height	30 m
Mobile station height	1.8 m
Bandwidth	10 MHz
Carrier frequency	3 GHz
Noise figure	7 dB
Throughput	5 Mbps
Receiver antenna gain	0 dB
Base station maximum transmit power	46 dBm
Transmitter antenna gain	20 dBi
Transmitter loss including cable loss	5 dB
Thermal noise	−104.5 dBm
Implementation margin	1.1 dB
Fading margin	10 dB
Interference margin	3 dB
Receiver loss including body loss	5 dB

11.2. Consider the following system parameters and calculate the subscribers per site when the required user data rate is 1 Mbps/subscriber and the required user traffic is 5 GB/month.

System parameters	Values
Bandwidth	30 MHz
Spectral efficiency	3 bps/Hz
Cell average capacity	20 Mbps
Busy hour average loading	50%
Overbooking factor	25
Busy hour carries of daily traffic	20%
The number of sectors per site	3

11.3. Consider that the total traffic intensity A is 4 Erlangs and compare the blocking probabilities when the number of resources N is 2 and 20.

11.4. Find the number of channels required to support 10 Erlangs of traffic at GOS of 1%.

11.5. Consider that the traffic intensity for one user A_u is 0.5 Erlangs and the number of resources are 2 and 20. How many users can be supported for 1% blocking probability?

11.6. Consider that the number of resource is 10, the total number of users is 500, and the blocking probability is 0.1%. How much average time does each subscriber use a phone during a peak hour?

11.7. A small town needs a cellular network. A wireless network designer has the following requirements:

System parameters	Values
System bandwidth	1 MHz
Channel bandwidth	20 kHz
Site type	Omnidirectional site
City area	50 km²
Cell radius	0.2 km
Cluster size	7
Average number of channel per unit time	0.5 calls/h
Average occupancy duration of a channel	1 min
Blocking probability	0.1%
Population	10000

When the network is based on the Erlang B model, calculate the maximum carried traffic intensity over the network and find how many people can subscribe the network.

11.8. A company operates franchises in London and Paris. It has five private phone lines between them. The number of employees is 100 and each employee makes 2 min call/h on average. Find the probability that a caller has to wait for service, the average delay, the average number of requests in the queue, and the probability of the delay exceeding 5 min.

11.9. The Reference Signal Receive Power (RSRP) is used to measure the coverage of the cell in LTE standard. The mobile station sends RSRP measurements reports to the base station. Describe how to calculate the RSRP and set the RSRP threshold.

11.10. Compare LTE with LTE-Advanced in terms of coverage, latency, and capacity.

11.11. A drive test is necessary to optimize configurations of a cell. It provides us with important measurements. Survey the drive test tools and compare them.

References

[1] The 3rd Generation Partnership Project; Technical Specification Group GSM/EDGE Radio Access Network; Radio Transmission and Reception (Release 1999) 3GPP TS 05.05 V8.11.0. http://www.3gpp.org/ftp/Specs/html-info/0505.htm (accessed April 8, 2015).
[2] J. Lempiäinen and M. Manninen, *UMTS Radio Network Planning, Optimization and QoS Management*, Kluwer Academic Publishers, Boston, MA, 2003.
[3] The 3rd Generation Partnership Project; Technical Specification Group Radio Access Network; Evolved Universal Terrestrial Radio Access (E-UTRA); User Equipment (UE) radio transmission and reception (Release 10), 3GPP TR 36.807 V10.0.0.
[4] 3rd Generation Partnership Project; Technical Specification Group Radio Access Network; Evolved Universal Terrestrial Radio Access (E-UTRA); Base Station (BS) radio transmission and reception (Release 12), 3GPP TS 36.104 V12.3.0.
[5] C. Chevallier, C. Brunner, A. Garavaglia, K. Murray, and K. Baker, *WCDMA (UMTS) Deployment Handbook: Planning and Optimization Aspects*. John Wiley & Sons, Ltd, Chichester, 2006.
[6] H. Holma and A. Toskala, *WCDMA for UMTS: HSPA Evolution and LTE*, John Wiley & Sons, Ltd, Chichester, 2010.
[7] M. Hata, "Empirical Formula for Propagation Loss in Land Mobile Radio Services," *IEEE Transactions on Vehicular Technology*, vol. **29**, no. 3, pp. 317–325, 1980.
[8] J. Laiho, A. Wacker, and T. Novosad, *Radio Network Planning and Optimisation for UMTS*, John Wiley & Sons, Ltd, Chichester, 2005.
[9] T. S. Rappaport, *Wireless Communications—Principles and Practice*, Prentice Hall, Upper Saddle River, NJ, 2nd Edition, 2002.

12

Wireless Communications Systems Design and Considerations

A wireless communication system designer requires a wide range of requirements and conditions. One decision in one design step is closely related to another decision in the next design step. Thus, it is very difficult to optimize many metrics such as complexity (or area in a chip), throughput, latency, power consumption, and flexibility because many of them are trade-off relationship. A wireless communication system designer often makes a decision subjectively and empirically. In this chapter, we take a look at considerations when designing wireless communications systems. We mainly deal with wireless communication system design flows and considerations, a high level view of wireless communication systems, several implementation techniques, and hardware/software codesign.

12.1 Wireless Communications Systems Design Flow

A brief design flow of wireless communications systems is as follows: Firstly, the international regulator such as the International Telecommunication Union (ITU) defines the spectrum allocation for different services such as satellite services, land mobile services, and radar services. The regional regulatory bodies such as European Conference of Postal and Telecommunications Administrations (CEPT) in Europe and Federal Communications Commission (FCC) in the United States play quite important role in the decision making process in the ITU. They build a regional consensus even if there are national representatives in ITU meetings. After spectrum allocation for a certain service, the ITU may label certain allocations for a certain technology and determine the requirements for the technology. For the requirements, some standardization bodies such as the 3rd Generation Partnership Project (3GPP) and the Institute of Electrical and Electronics Engineers (IEEE) make their contributions to the ITU, where they guarantee that their technology will meet the requirements. The actual technology is specified in the internal standardization such as 3GPP and IEEE.

Wireless Communications Systems Design, First Edition. Haesik Kim.
© 2015 John Wiley & Sons, Ltd. Published 2015 by John Wiley & Sons, Ltd.

Some standards are established at the regional level. The regulators are responsible for spectrum licensing and protection. Secondly, a wireless channel is defined. Its mathematical model is developed and the test scenario is planned for the evaluation of the proposed technical solutions or algorithms. For example, short-range wireless channel models are different from cellular channel models. In cellular systems, we basically use empirical channel models such as ITU channel models. Their test scenarios are different as well. Thirdly, each block of wireless communication systems is selected to satisfy the requirements of wireless communication systems. This step is discussed and defined in wireless communication standards such as 3GPP and IEEE. The next steps are wireless communication system implementation. They are carried out by enterprises. Fourthly, receiver algorithms under the given standard are developed (or selected) and their performances are confirmed by a floating point design. Basically, standard bodies define interfaces, signaling protocols, and data structures. They give us the requirements of the transmitter and receiver. In addition, we can roughly obtain key algorithms of the transmitter and receiver from the given standard. However, many freedoms for actual design and implementation are left. Fifthly, a fixed point design is performed. In this step, the wireless communication system architecture is defined. The performance (complexity, latency, etc.) of the wireless communication system can be roughly calculated. Lastly, conventional chip design processes as the remaining steps (RTL design and back-end design) are performed.

We dealt with key wireless communications algorithms in Part II and each algorithm can be implemented in widely different ways. We have many options to assemble these algorithms in various architectures from a general purpose microprocessor to a full-custom chip. Depending on system requirements such as latency, power consumption, and cost, we decide the wireless communication system structure including flexibility, parallelism, pipelining, and time sharing and choose the system parameters including supply voltage, clock rate, and word lengths. The comprehensive design requires a wide range of data and one decision in one design step is closely related to another decision in the next design step. We need to optimize many metrics such as complexity (or area in a chip), throughput, latency, power consumption, and flexibility. Many of them are trade-off relationship. It is very complex and difficult to find an optimal solution. Thus, mobile phone vendors or network equipment vendors optimize their products based on their own design know-how.

The hardware system design can be classified into *front-end design* and *back-end design*. The front-end design deals with all logical design and verification. Each algorithm is verified by a floating point simulation. In this step, the full-chain wireless communication system models (including error correction coding, modulation, FFT/IFFT, etc.) provide us with their system performances such as BER or capacity. As the next step, a fixed-point simulation is carried out in consideration of a bit level. We can obtain a further refined result of their system performances. These simulations are developed by software tools such as Matlab or C++. Sometime System-C or System Verilog can be used for a cycle-accurate model. After verifying system performances, a Register-Transfer Level (RTL) design is performed using hardware description languages such as Verilog or VHSIC Hardware Description Language (VHDL). In this step, we implement each algorithm as hardware registers and logic operations and then integrate all blocks. If the function and logic works correctly, we move to the back-end design which deals with system development and fabrication. An RTL from the front-end design is converted to a gate level by synthesis tools such as Cadence, Synopsys, and Mentor Graphics. A netlist as a result of synthesis contains the used cells and area and their

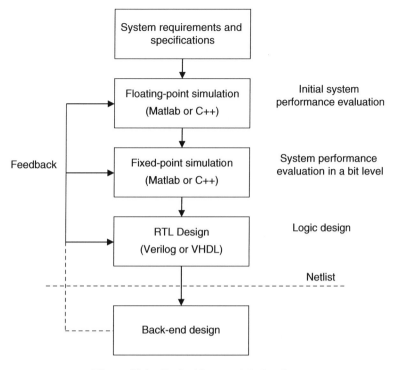

Figure 12.1 Typical front-end design flow

interconnections. It is used to create a physical layout by placement and routing tools. The back-end design process is too costly and time consuming. Many important parts of the back-end design are automated. Thus, the front-end design is getting more important in the modern hardware design. Figure 12.1 illustrates a typical front-end design flow.

In floating point simulation, a module-based approach is used. The module means any arithmetic operation, memory, interconnection, or each algorithm block. It provides us with accurate estimation of hardware complexity and cost. In addition, it is very helpful for optimization of the complex wireless communication system. Each wireless communication block is designed and pairwise verified at the transmitter and the receiver. If each block works correctly, we integrate and evaluate system performances. In fixed-point simulation, we start from bit level (or bit resolution/input bit quantization) decision of Analogue-to-Digital Converter (ADC) and Digital-to-Analogue Converter (DAC). The bit level decision affects the next blocks. The ADC/DAC is the first block of a whole baseband receiver/transmitter chain. Thus, the ADC bit level becomes a basis of all bit levels for each block. As the results of fixed-point simulation, the input/output bit level of each block and rough architecture become important information at the RTL design. The RTL design including the datapath flow and the control flow follows four steps. The first step is to capture behavior using a high-level state machine. The second step is to construct a data path with parameterized modules on the high-level state machine. The third step is to connect the data path to the controller. The last step is to convert the high-level state machine to the finite state machine. In the RTL design, we can find the critical path to meet system requirements and predict latency, complexity, power consumption, etc.

The critical path enables us to identify system bottlenecks and share resources to improve system performances. In addition, a hardware designer decides the system architecture including parallelism, pipelining, and time sharing. In each step of the front-end design, it is very important to provide a hardware designer with a feedback. Thus, the architecture and parameter should be iteratively improved to meet the system requirements. If we obtain a netlist as the result of the RTL design, we can proceed to the back-end design step. Figure 12.2 illustrates a typical back-end design flow. The detailed back-end design flow depends on software tool, methodology, and the technology libraries by a fabrication house. Basically, the design period at a floating point simulation step is much shorter than at a lower level design step. We make further efforts on this step and analyze high-level decisions. If we consider a rapid prototyping, the optimization as well as the system design is carried out in this step. In this step, it is difficult to estimate accurate complexity and power consumption. There are many constraints to optimize performance, complexity, latency, power consumption, etc. Thus, the trade-off of high-level decision is inevitable. For example, the trade-off between flexibility and efficiency is one important decision point. A general purpose device such as a Central Processing Unit (CPU) is flexible but not efficient. On the other hand, a dedicated device such as a baseband

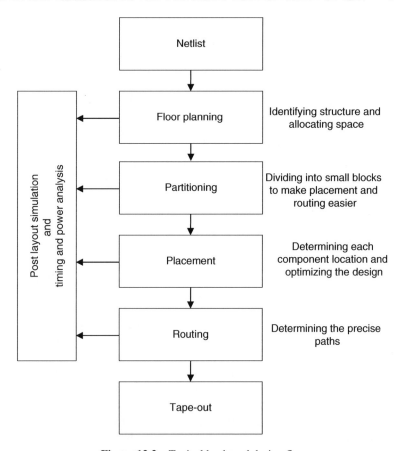

Figure 12.2 Typical back-end design flow

Summary 12.1   Wireless communications system design flow

1. Front-end design (logic design and verification)
   a. System requirements and specifications
   b. Floating point simulation (initial system performance evaluation)
   c. Fixed point simulation (system performance evaluation in a bit-level)
   d. RTL design (logic design)
2. Back-end design (development and fabrication)
   a. Netlist from RTL design
   b. Floor planning (identifying structure and allocation space)
   c. Partitioning (dividing into small blocks to make placement and routing easier)
   d. Placement (determining each component location and optimizing the design)
   e. Routing (determining the precise paths)

chipset is efficient but not flexible. The general purpose device is useful when we need to compute a general task. However, the baseband chipset has an inherent property of the wireless communication system and does not need the full flexibility. Most of wireless communication algorithms are datapath intensive design block. Thus, we exploit many signal processing optimization techniques such as parallel processing, pipelining, and windowing.

If a target application is decided, a high level-design starts from algorithm choice, architecture selection, and implementation parameter decision. For example, when designing the turbo decoder with a low complexity and latency, we may choose Max-log MAP decoders or soft output Viterbi decoders as a component decoder instead of log MAP decoders, select a sliding window decoder architecture, and adopt the 3bits soft input/output. As we discussed in Chapter6, there is a trade-off between complexity and performance. In terms of BER performance, the log MAP algorithm and 4bits soft input/output would be a better choice. Basically, there are many algorithm blocks and a designer should make a decision in consideration of system requirements and performance. Thus, a design methodology for high-level design is needed. There are some design guidelines and methodologies in public but many chip vendors have their own design guidelines based on their know-how.

12.2   Wireless Communications Systems Design Considerations

Now, we take a look into wireless communications system design considerations and several implementation techniques. The first design consideration is *bit-level* (bit resolution or word length) *decision*. The bit-level directly affects to system complexity and performance. For example, if we increase one bit from 3 bits soft input/output to 4 bits soft input/output in the turbo decoder, the memory size and each sub-block complexity such as branch metric calculation and path metric calculation must be increased. In addition, the bit-level is highly related to vulnerability and sensitivity regarding noise and nonlinearity. Thus, it results in performance degradation. The floating point representation uses an integer but the fixed point representation supports a fixed range. When dealing with the fixed point representation, we should manage a finite precision error caused by truncation and overflow. We start making the bit-level decision from ADC/DAC to the next blocks sequentially. The suitable decision of the bit-level

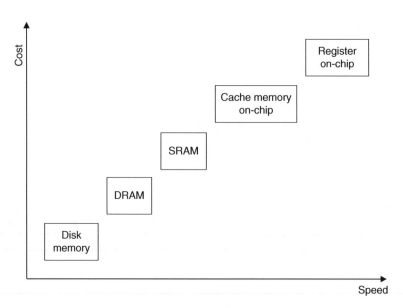

Figure 12.3   Memory comparisons

is based on a trade-off between performance and complexity. The final decision should be made through a full-chain simulation and feedback from other design steps.

The second design consideration is *the number of computation*. It is tightly coupled with power consumption and latency as well as complexity. This is the basic step to estimate the number of operations such as multiplication, addition, shift register, and memory. Based on this estimation, we can have the initial approximation about system complexity. In addition, the initial power consumption and latency can be estimated if we have a basic architecture and component. The architecture and component choice is highly related to the system complexity and performance. For example, when we need a memory for a baseband chip, a register is fast but expensive. On the other hands, Static Random Access Memory (SRAM) is less fast and cheaper than a register. Figure 12.3 illustrates simple comparison of memory types.

The third design consideration is *algorithm type and property*. There are many algorithm types such as recursive algorithms, backtracking algorithms, greedy algorithms, divide and conquer algorithms, dynamic programming algorithms, and brute force algorithms. Depending on algorithm types, we use different resources and it affects the process of building the corresponding architecture. For example, a recursive algorithm requires simple computation and high interconnection power. Thus, we focus on global interconnection, connectivity ratio, and memory management rather than computational logic. In Ref. [1], *property metrics* are investigated and classified as follows: size, topology, timing, concurrency, uniformity, temporal locality, spatial locality, and regularity.

1. Size is the amount of computation and includes the number of data transfers, operations, and variables.
2. Topology is the relative arrangement of operations.
3. Timing includes latency, critical path delay, and scheduling.
4. Concurrency is the ability of parallel processing.

5. Uniformity is how evenly resource accesses are distributed.
6. Temporal locality is the persistence of computation's variables.
7. Spatial locality is the communication pattern among blocks and how well the computation is partitionable into clusters of operations with interconnections.
8. Regularity is how often common patterns appear.

Those metrics are highly related to system performance and help us to optimize system design. Among them, the concurrency is very important and measures how many operations can be simultaneously executed. Depending on the level of parallelism, the effective sampling rate is decided. For example, we consider a discrete time Finite Impulse Response (FIR) filter as follows:

$$y[n] = ax[n] + bx[n-1] + cx[n-2] \qquad (12.1)$$

where $y[n]$ and $x[n]$ are the output signal and the input signal, respectively. The coefficients a, b, and c are the values of the impulse response. We can convert the sequential FIR filter into the three parallel FIR filter as follows:

$$y[3k] = ax[3k] + bx[3k-1] + cx[3k-2]$$
$$y[3k+1] = ax[3k+1] + bx[3k] + cx[3k-1] \qquad (12.2)$$
$$y[3k+2] = ax[3k+2] + bx[3k+1] + cx[3k].$$

The parallel systems receive three inputs ($x[3k]$, $x[3k+1]$, and $x[3k+2]$) at the kth clock cycle and produces three outputs ($y[3k]$, $y[3k+1]$, and $y[3k+2]$) as shown in Figure 12.4.

In the parallel system, the same operation is duplicated in order to achieve faster processing. Thus, we should increase the area (or complexity). Another important property metric is the timing metrics. Especially, this is critical metric when we develop a high-speed baseband chipset. There are many metrics about timing. The most important metric is the critical path which creates the longest delay. The critical path can be found in linear time with a modified shortest path algorithm [2]. For example, Figure 12.5 illustrates one example of data flow graph. In this figure, the critical path is six clock cycles. The inverse of the critical path is tightly coupled with the maximum throughput of the system.

Another important metric is the latency. The definition of the latency in Very-Large-Scale Integration (VLSI) design is a little bit different from its wireless communication systems.

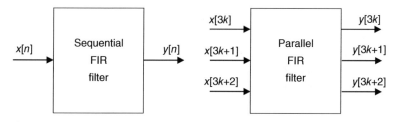

Figure 12.4   Sequential system and parallel system

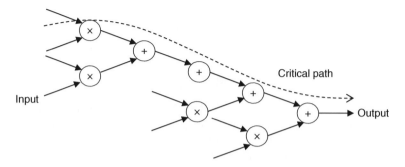

Figure 12.5   Example of critical path in data flow graph

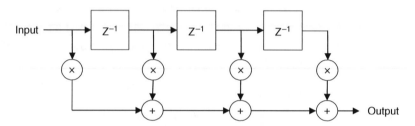

Figure 12.6   Example of regularity metric

In wireless communication systems, latency is defined as the time a signal takes to travel between a transmitter and a receiver. On the other hand, the latency in VLSI design means the time between input data and processed output data. It is often used to mean waiting time that increases actual response time beyond desired response time. The iteration of the system is key metric as well. It provides us with estimation of potential throughput improvement by pipelining and parallel processing. In Ref. [1], the iteration bound is defined as the maximum ratio (over all feedback cycles) of the total execution time, $T(c)$, divided by the number of sample delays in the cycle, $D(c)$, as follows:

$$\max_{c \in \text{cycles}} \frac{T(c)}{D(c)}. \tag{12.3}$$

If the system is non-recursive, the iteration bound is zero. The other metrics about timing are the longest path, maximum operator delay, and slacks for each operator type. The regularity is helpful for designing a structure. It is widely investigated in Refs[3–5]. Especially, the structured design is highly desirable when developing in a modular concept. In order to quantify the regularity, the percentage operation coverage by common patterns is defined in Ref. [1]. For example, it can be a set of patterns containing all chained pairs of associative operators. Figure 12.6 illustrates one example of FIR filter to quantify the regularity. In this figure, there are 10 operators (3 flip-flops, 4 multipliers, and 3 adders). We can observe 4 multiplication–addition patterns and they cover 7 out of 10 operations. Thus, the coverage of the multiplication–addition patterns is 70%.

Example 12.1 Property metrics

Consider the following discrete time FIR filter of order 4:

$$y[n] = ax[n] + bx[n-1] + cx[n-2] + dx[n-3] + ex[n-4]$$

where $x[n]$ and $y[n]$ are the input signal and the output signal, respectively. The coefficients a, b, c, d and e are the values of the impulse response. Find the critical path, the iteration bound, and the regularity.

Solution

We can illustrate the discrete time FIR filter as shown in Figure 12.7.

From Figure 12.7, the critical path is seven clock cycles. The FIR filter is non-recursive. Thus, the iteration bound is 0. In Figure 12.7, there are 13 operators (4 flip-flops, 5 multipliers, and 4 adders). We can observe 5 multiplication–addition patterns and they cover 9 out of 13 operations. Thus, the coverage of the multiplication–addition patterns is 70%. ■

The fourth design consideration is *architecture selection*. There are many types of design architectures such as Digital Signal Processor (DSP), full-custom hardware, Application-Specific Integrated Circuit (ASIC), and Field Programmable Gate Arrays (FPGA). The DSP can be simply defined as a specialized microprocessor for signal processing applications. When comparing a DSP with a general purpose microprocessor, the DSP supports specialized instructions and multiple operations per instruction. Thus, it is more suitable for signal processing which requires high memory bandwidth, high energy efficiency, and math centric complex computation. In addition, the DSP is flexible because it is software that is programmed to perform a task. On the other hand, a full-custom chip is not a programmable hardware but a dedicated hardware. It supports a high degree of optimization in both area and performance. However, it is too much time-consuming due to manual layout design. Thus, it is almost extinct in the modern hardware design. The ASIC is a customized chip to perform a particular operation. It is suitable for communication or multimedia applications. There are mainly two types of ASICs: standard cell-based ASICs and gate-array based ASICs. The standard cell-based ASICs are also called cell-based IC (CBIC) or semi-custom where standard cell means a predesigned logic cell such as

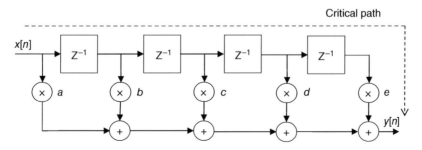

Figure 12.7 FIR filter

flip-flop, multiplexer, AND gate, and XOR gate. The standard cells are usually synthesized from an RTL description and inserted into a chip vendor own library. An ASIC designer uses the standard cells in a schematic and they are placed and routed by backend tools. Some standard cells such as memory and datapath are tiled to create macrocells. In addition, an ASIC vendor provides electrical characteristics such as propagation delay and capacitance to an ASIC designer. The ASIC is the most popular design architecture because it has much quicker design time than the full custom hardware design. However, one disadvantage is that the mask cost is increasingly expensive. The gate-array-based ASICs enable us to cut mask costs because the transistor level masks are fully defined and an ASIC designer can customize the metal connection for each design. Thus, it has a short fabrication time and low production cost. However, their disadvantages are (i) inefficient memory usage and (ii) slow and big logics due to fixed transistors and limited wiring space. The FPGAs can be classified into one of ASIC types. It is cheaper solution for small volumes. A FPGA designer programmes the basic logic cells and the interconnection to implement combinational logics and sequential logics. It contains the programmable logic cells and surrounds them by the programmable interconnection. The FPGA has two important advantages: flexibility and short design process. The flexibility allows us to reconfigure our design. We do not need layouts or masks. Thus, it provides us with simple design cycles and faster time to market. Most importantly, the development cost is low. On the other hand, the efficiency is low. Namely, unused logic cells and interconnections can exist. In addition, an internal memory is limited, power consumption is higher than the ASIC, analogue interface is challenging, and optimization is difficult. Thus, the FPGA is used as the prototype of the design before the ASIC implementation. Table 12.1 summarizes comparison of the design architectures.

Once a wireless communication system design works satisfactorily, it can be implemented on one of the above architectures. When we select one of the above architectures, the flexibility and efficiency is the key criterion. Basically, the flexibility and efficiency is trade-off relationship. Figure 12.8 represents their relationship in terms of the flexibility and efficiency.

The full-custom chip is implemented as the dedicated hardware. The efficiency of the full-custom chip design is high. The full-custom chip design can be significantly optimized. However, its flexibility is poor. On the other hand, a software design provides us with high flexibility but low efficiency. Hardware and software partition is very important matter in the wireless communication system design. Hardware and software codesign settles down in common design methodology. We will discuss this topic in the next section. When implementing wireless communication systems, a flexible architecture is one of the key issues because many wireless communication systems support multiple modes and variable parameters.

Table 12.1 Comparison of design architectures

	Full-custom	Standard cell ASIC	FPGA
Area	Compact	Compact to moderate	Large
Performance	High	High to moderate	Low
Cell size	Variable	Variable/fixed	Fixed
Interconnection	Variable	Variable	Programmable
Time to market	Long	Middle	Short
Development cost	High	Moderate	Low

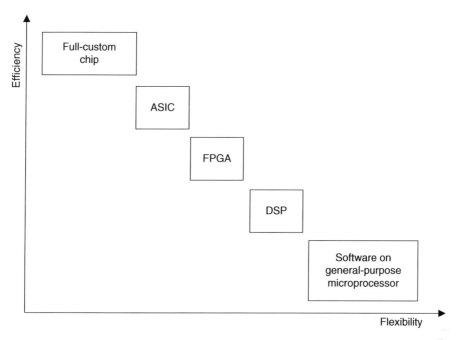

Figure 12.8   Flexibility vs. efficiency

For example, LTE downlink physical layer supports the multiple channel bandwidths (1.25, 2.5, 5, 10, 15, and 20 MHz) and the corresponding multiple FFT sizes (128, 256, 512, 1024, 1536, and 2048). Adaptive modulation coding is one essential technique of modern wireless broadband communications. Thus, the flexibility of the architecture is very important consideration and the receiver should be flexible enough to accommodate various transmission modes. Naturally, it is essential to implement efficiently. The area (or complexity) efficiency is directly related to cost. The energy efficiency is one key parameter to mobile devices. In terms of both flexibility and efficiency, the ASIC is good choice and widely used for the wireless communication system design. Basically, multiple functional blocks of a wireless communication system should be developed using an ASIC library. Several different teams are involved in each block design and they individually develop it using implementation techniques such as pipelining and parallel processing. In order to integrate each block, a top level system designer should figure out all inputs/outputs, timing, power consumption, and requirements of each block.

The fifth design consideration is *implementation-type selection*. The implementation type is selected according to the target application. If we develop a mobile device, we should focus on a low-power implementation in order to support a long battery life. If we consider a broadband communication system, we have to implement using a high-speed technique. If we develop Software Defined Radio (SDR), the flexibility is the most important design consideration. In this section, we investigate a low-power implementation and its techniques. A low-power implementation requires a holistic approach including collaboration with architecture level design, algorithm level design, physical design, power verification, and so on. First of all, we investigate the physical or circuit level design. In digital

Complementary Metal Oxide Semiconductor (CMOS) design, the power consumption can be represented as follows:

$$P_{total} = P_{active} + P_{leakage} \qquad (12.4)$$

$$= P_{switching} + P_{internal} + P_{leakage} \qquad (12.5)$$

$$= \alpha CV^2 f + I_{scc}V + I_{leakage}V. \qquad (12.6)$$

In (12.5) and (12.6), the first term is the switching power consumption, where α is the activity factor (which means the probability a transition occurs) and f is the system clock frequency. If the signal rises and falls every cycle, $\alpha = 1$. If it switches once per cycle, $\alpha = 0.5$. In static gates, the activity factor typically is 0.1 [6]. C is the capacitive load. V is the voltage swing which is typically same as supply voltage. The second term is the internal power consumption. I_{scc} is the short circuit current. When transistors switch, the short circuit current arises. The third term is the leakage power consumption. $I_{leakage}$ is the leakage current. It arises from subthreshold and gate. The typical value of the subthreshold leakage is 20 nA/μm for low V and 0.02 nA/μm for high V [6]. The typical value of the gate leakage is 3 nA/μm for thin oxide and 0.002 nA/μm for thick oxide [6]. In (12.6), the switching part represents the dynamic dissipation associated with charging and discharging of load capacitances. This is the dominant term of the power dissipation. Thus, the low-power implementation in circuit level is based on minimizing the capacitive load C and the voltage swing (supply voltage) V. In order to minimize them, we can use small transistors and short wires and adopt the lowest supply voltage. In addition, we can reduce the dynamic power using the lowest clock frequency. However, the lowest frequency increases a delay so that it is not suitable way to meet a high speed processing. Another approach is to reduce the static power by minimizing the leakage in a fabrication stage. The leakage power consumption is getting more dominant term of the total power dissipation. The power gating technique [7] is used to reduce the leakage power by shutting off the current to blocks which are not in use. The key design goal of this technique is to switch between a low power mode and an active mode at the appropriate time and in the appropriate way while minimizing the performance degradation. It is widely used in many mobile devices. Another important low-power technique is *clock gating* [7]. It is extensively used to reduce the dynamic power dissipation by reducing unnecessary clock activities. In synchronous digital circuits, the clock takes significant part (up to 40%) of power dissipation [8]. The supply voltage can be reduced by Adaptive Voltage Scaling (AVS) technique. The multiple critical paths are implemented on hardware (integrated circuit). We compare the delay on these critical paths with the required delay. If one delay is smaller than the required delay, the AVS reduces the supply voltage gradually with a given step until the value between the required delay and the delay of the critical path is in the AVS error margin. We can adjust both the clock frequency and the supply voltage by Dynamic Voltage and Frequency Scaling (DVFS) technique. In the first term of (12.6), the required supply voltage is determined by the frequency at which the circuit is clocked. Thus, a look-up table about the relationship between the supply voltage and the frequency is created by silicon measurement. We can define suitable pairs of the frequency and the supply voltage. A large number of voltage levels are dynamically switched according to the workloads. This technique is also commonly used to reduce power dissipation on a wide range of mobile devices.

Secondly, we take a look at a low-power approach in terms of algorithm design. There are several ways to reduce the power consumption at the algorithm development stage: algorithm modifications, structure transformations of an algorithm, and appropriate arithmetic operation selection. In Part II, we discussed algorithm modifications. For example, the log-MAP algorithm is very complex so that it is very difficult to implement. Especially, the log term and the exponential term are very difficult to compute. Thus, we changed the log-MAP algorithm to the max-log-MAP algorithm. The simplified max-log-MAP algorithm has only compare operation even if slight performance degradation occurs. The structure transformations do not change the functional behavior of algorithm blocks and optimize computational modules/operations and their interconnection. In Ref. [9], several transformation techniques are introduced to reduce power consumption. They basically pay attention to the first term (switching part) of (12.6). They approach to reduce the power consumption for a fixed throughput by reducing the supply voltage through speed-up transformation and by minimizing the effective capacitance being switched through several transformations. The speed-up transformations reduce the required control steps so that it increases the degree of concurrency. It includes *retiming*, *pipelining*, *unfolding*, and *algebraic transformation*. The transformation to minimize the effective capacitance reduces the total required operations by optimizing resource usages. For example, when one feedback loop is the bottleneck of the system, we can use the retiming technique so that the feedback loop shortens the critical path and increases the degree of concurrency. This transformation results in the reduced voltage through pipelining. In Ref. [10], algorithm transformation techniques for a low-power wireless communication system design are described. We take a look at retiming, pipelining, folding, and so on. In order to apply the transformation techniques, we need to employ Data Flow Graph (DFG) representation. It is a directed graph to show the data dependencies between functions. The DFG is defined as follows:

$$G = (V, E, q, d) \tag{12.7}$$

where V is the set of graph nodes representing computation (function or subtask), E is the set of directed edges (or arcs) representing data path (connection between nodes), q is the set of node delays representing execution time of each node, and d is the set of edge weights representing associated delay. Figure 12.9 illustrates an example of one block diagram and corresponding DFG representation of FIR filter.

In Figure 12.9b, the node A and B represent multiplication and addition, respectively. Each set and element is defined as follows:

$$V = \{v_1, v_2, v_3\} \tag{12.8}$$

$$E = \{e_0, e_1, e_2, e_3, e_4\} \tag{12.9}$$

$$q(v_1) = q(v_2) = 20 \quad \text{and} \quad q(v_3) = 10. \tag{12.10}$$

$$d(e_0) = d(e_2) = d(e_3) = 0 \text{ and } d(e_1) = d(e_4) = 1 \tag{12.11}$$

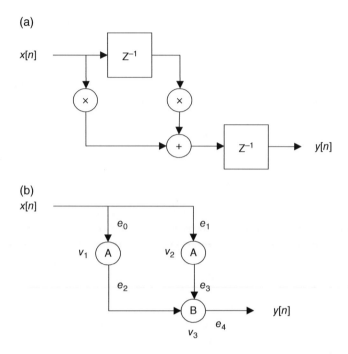

Figure 12.9   Block diagram (a) and DFG representation (b) of FIR filter

The node delay represents the required time to produce one sample output of the block. Now, we define several properties of the DFG as follows:

1. The iteration represents an execution of each DFG node.
2. Iteration period is the required time to perform the iteration. Intra-iteration and inter-iteration represent a direct edge without any delay and a direct edge with at least one delay, respectively. Figure 12.10 illustrates the iteration period.
3. The critical path of the DFG is the path with the longest computation time among all paths containing no delay.
4. The loop (cycle) means the directed path that begins and ends at the same node.
5. The loop bound means the minimum time to execute one loop in the DFG and it is defined as follows:

$$\text{Loop bound} = \frac{q(v)}{d(e)} \tag{12.12}$$

   where $q(v)$ and $d(e)$ represent the loop computation time and the number of delays in the loop, respectively.
6. The critical loop represents the loop in which has the maximum loop bound.
7. The iteration period bound $\left(T_\infty\right)$ is defined as follows:

$$T_\infty = \max_{\forall L} \frac{\sum_{v \in L} q(v)}{\sum_{e \in L} d(e)} \tag{12.13}$$

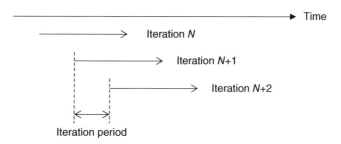

Figure 12.10 Iteration period

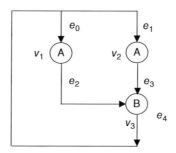

Figure 12.11 DFG of Example 12.2

where L is a loop in the DFG. It is not possible to achieve an iteration period less than the iteration period bound even if we have infinite hardware resource.

Example 12.2 Data flow graph

Consider the following DFG:

$$q(v_1) = q(v_2) = 20 \text{ and } q(v_3) = 10$$

$$d(e_0) = d(e_2) = d(e_3) = 0 \text{ and } d(e_1) = d(e_4) = 1.$$

Find the iteration period bound.

Solution

In Figure 12.11, we can find two loops as shown in Figure 12.12.

We calculate the loop bound for loop 1 and loop 2 of the DFG as follows:

$$T_{\text{loop1}} = \frac{20+10}{1} \text{ and } T_{\text{loop2}} = \frac{20+10}{1+1}.$$

From (12.13), we calculate the iteration period bound as follows:

$$T_\infty = \max_{\forall L} \left\{ T_{\text{loop1}}, T_{\text{loop2}} \right\} = 30 \text{ time units.}$$

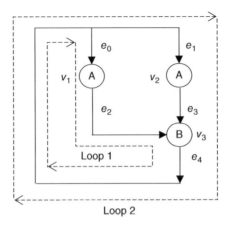

Figure 12.12   Loops of DFG example

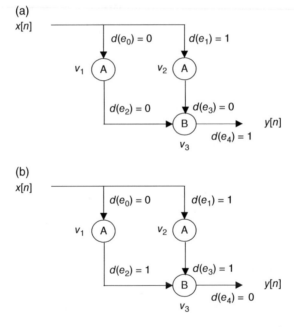

Figure 12.13   Before retiming (a) and after retiming (b)

The loop 1 is the critical loop.   ∎

The retiming technique is a transformation technique moving delay from one edge to another edge in a DFG. It is used to reduce the total number of registers and improve scheduling of the DFG. Thus, we can reduce the power consumption and the critical path delay of the system. Figure 12.13 illustrates an example of retiming.

In Figure 12.13b, one delay from the output of the node B is transferred to its inputs. In Figure 12.13a, the iteration period is 30 time units ($=T_m+T_a$) when the multiplication computation (T_m) and the addition computation (T_a) are 20 time units and 10 time units, respectively. After retiming as in Figure 12.13b, the iteration period is 20 time units ($=\max(T_m,T_a)$).

The pipelining technique, as the name suggests, comes from the idea that a pipe can send water without waiting. It allows us to process new data before a prior data processing finishes. It reduces the critical path and increases sample speed by adding delays and retiming. Basically, parallel processing duplicates the same operation to achieve faster processing. However, the pipelining technique inserts latches or registers along the data path so that it can shorten the critical path. Figure 12.14 illustrates comparison of one logic system and its pipelined version. The pipelined system inserts one delay in the system and can achieve a higher throughput by reducing the critical path. However, the system has a longer latency due to the additional delay.

Consider simple 2 adders with and without one delay as shown in Figure 12.15. Figure 12.15a does not show any delay (pipeline depth is 0) and the latency is zero. On the other hand, Figure 12.15b has one pipeline delay (pipeline depth is 1) and the latency is one. Their timing diagrams are shown in Figure 12.16.

If the pipeline depth increases, the system throughput increases and the clock period decreases. Now, we need to consider where to put the delay. Basically, we place the pipeline delay across feed-forward cutsets. Cutset is a set of edges in the DFG such that if the edges are removed from the original DFG, the remaining DFG becomes two disjoint graphs. The feed-forward is the cutset when the data move in the forward direction on all the edge of the cutset. Figure 12.17 illustrates three feed-forward cutsets and delay placements in the feed-forward cutsets. In Figure 12.17a, the dash lines represent feed-forward cutsets but the dash-dot line is not feed forward cutset. When comparing the original FIR filter (Figure 12.17a) and the

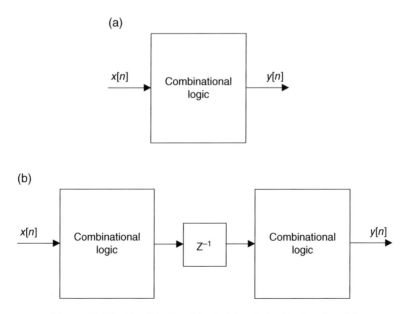

Figure 12.14   Combinational logic (a) and pipelined system (b)

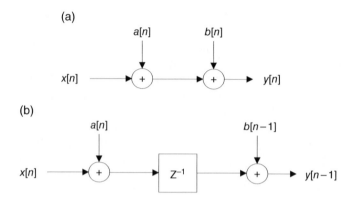

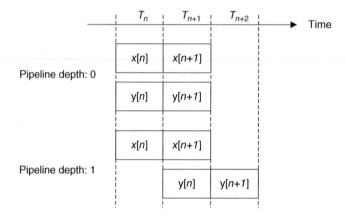

Figure 12.15   Simple 2 adders system without one delay (a) and with one delay (b)

Figure 12.16   Timing diagram of 2 adders system with/without one delay

pipelined FIR filter (Figure 12.17b), the critical path has been changed from 40 time units $(=T_m + 2T_a)$ to 30 time units $(=T_m + T_a)$.

The pipelining technique is very useful in a low power design. From(12.6), the switching (dynamic) power dissipation of a serial architecture is defined as follows:

$$P_{sr} = C_{tot} V^2 f \tag{12.14}$$

where C_{tot} is the total capacitance of the serial architecture. The propagation delay T_{pd} is defined as follows:

$$T_{pd} = \frac{C_c V}{k(V - V_t)^2} \tag{12.15}$$

where C_c is a capacitance to be charged and discharged in the critical path, k is a transistor constant, and V_t is a threshold voltage. When we consider an M pipeline depth, the pipelining technique reduces the critical path to $1/M$. Thus, the capacitance C_c is reduced to C_c/M

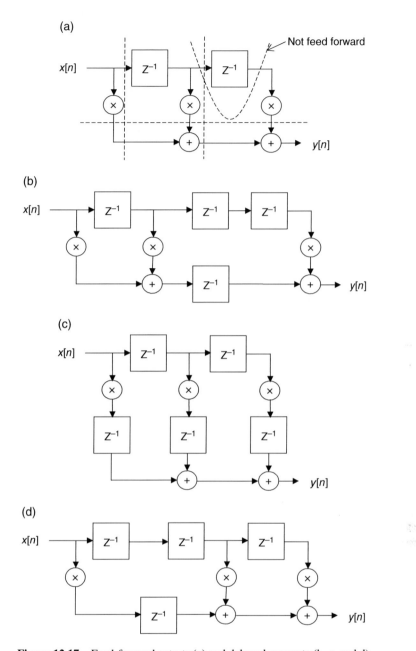

Figure 12.17   Feed-forward cutsets (a) and delay placements (b, c, and d)

which means only C_c/M is now charged and discharged. If the frequency f is maintained, the supply voltage can be reduced to βV where $0 < \beta < 1$. The power dissipation of the pipelined system is

$$P_{\text{pip}} = C_{\text{tot}} \beta^2 V^2 f = \beta^2 P_{\text{sr}} \qquad (12.16)$$

where β is determined by the propagation delay. The propagation delay of the pipelined system is

$$T_{\text{pip}} = \frac{(C_c/M)V}{k(\beta V - V_t)^2}. \qquad (12.17)$$

The same clock speed is maintained for both serial architecture and pipelined architecture. Thus, we can find β using the following equation:

$$M(\beta V - V_t)^2 = \beta(V - V_t)^2. \qquad (12.18)$$

From (12.16), we can observe that the power dissipation of the pipelined system is lower than the corresponding serial system.

The folding technique is very useful technique to improve area efficiency by time-multiplexing many operations to single functional units. The unfolding technique is the reverse of the folding technique. The folding technique inserts delays (registers) and reuse logics so that it saves areas. Figure 12.18 illustrates an example of two folding transformation. As we can see in the figure, we can reduce the number of operation.

When we have N same patterns in the system, we can apply the folding transformation and reduce the hardware complexity by a factor of N. Instead, the computation time is increased by a factor of N.

In addition, algebraic transformations such distributivity, associativity, and commutativity are very useful for improving the efficiency. For example, one computation $(X_1 \times X_2) + (X_1 \times X_3)$ can be reorganized as $X_1 \times (X_2 + X_3)$ using distributivity. We can reduce from two multipliers and one adder to one multiplier and one adder. Figure 12.19 illustrates algebraic transformations.

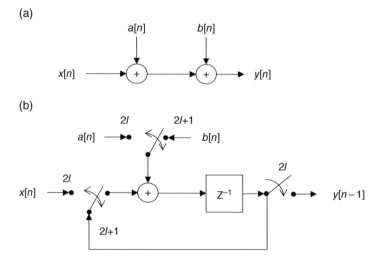

Figure 12.18 Simple two adders system (a) and two folding transformation (b)

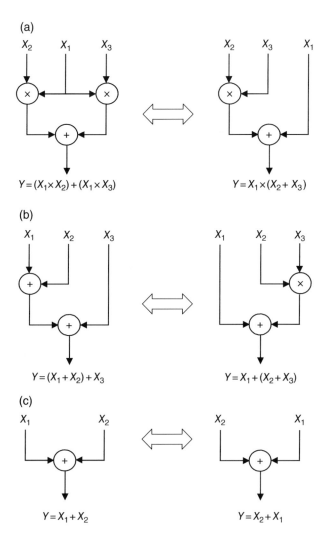

Figure 12.19 Algebraic transformations: distributivity (a), associativity (b), and commutativity (c)

Summary 12.2   Wireless communications system design considerations

1. The first design consideration is bit-level (bit resolution or word length) decision. The bit-level directly affects the system complexity and performance. We start making the bit-level decision from ADC/DAC to the next blocks sequentially.

2. The second design consideration is the number of computation. It is tightly coupled with power consumption and latency as well as complexity. It is the basic step to estimate the number of operations such as multiplication, addition, shift register, and memory.

3. The third design consideration is algorithm type and property. There are many algorithm types such as recursive algorithms, backtracking algorithms, greedy algorithms, divide and conquer algorithms, dynamic programming algorithms, and brute force algorithms. Depending on algorithm types, we use different resources and it affects the process of building the corresponding architecture.
4. The fourth design consideration is architecture selection. There are many type of design architectures such as digital signal processor (DSP), full-custom hardware, application-specific integrated circuit (ASIC), and field programmable gate arrays (FPGA).
5. The fifth design consideration is implementation-type selection. The implementation type is selected according to the target applications such as a mobile device and a high speed device. The transformation (retiming, pipelining, parallelism, folding, etc.) is very useful technique.

12.3   Hardware and Software Codesign

We mainly dealt with a hardware design in Sections 12.1 and 12.2. However, the role of software is getting more important in the modern wireless communication system. Before 1990s, hardware and software (HW/SW) are developed independently and then integrated. The role of software in wireless communication devices was not significant and the interaction between hardware and software was very limited. However, as the performance of DSPs and microprocessors is improved, many tasks are implemented in software. HW/SW codesign emerged as a new system design discipline. The HW/SW codesign is the most efficient implementation. It improves an overall system performance and provides us with lower costs and smaller development cycle. Figure 12.20 illustrates classical system design flow and timeline. After receiving system specifications, we decide a HW architecture and partition HW tasks and SW tasks. Namely, we develop hardware first and then design software after fixing the hardware architecture. When both HW and SW are developed, we integrate them and verify the whole system.

This design flow is inefficient and requires a long design period. Basically, the development time depends on a target application, resource, and fabrication. However, the architecture definition and HW/SW partition typically requires 6–12 months. The HW design, SW design, and system integration typically require 24–48 months. The field test typically requires 6–12 months. One important reason why it takes so long is that the SW design begins when the HW architecture is fixed. Recently, the time-to-market is very important because mobile devices are outmoded very quickly. In addition, the classical system design flow does not provide enough interaction between HW and SW so that the system optimization is limited. On the other hand, the HW/SW codesign allows us to develop HW and SW concurrently and save the development time. In addition, the HW/SW codesign is based on iterative process. Thus, we can optimize a system more efficiently. Figure 12.21 illustrates HW/SW codesign flow and timeline.

The main advantages of SW design are (i) short time-to-market and (ii) small design and verification costs. The main advantages of HW design are (i) good performance and (ii) high energy efficiency. The HW/SW codesign provides us with good balance between HW design and SW design and allows us to have their advantages.

As we can see in Figure 12.21, the first step of the HW/SW codesign is *system specification and architecture definition*. The specification is to define the functions of a system and

Time

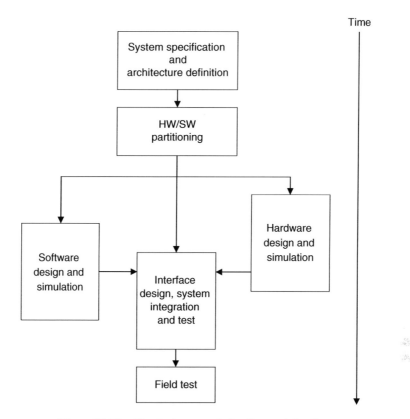

Figure 12.20   Classical system design flow and timeline

clearly describe the behavior including concurrency, hierarchy, state transition, timing, etc. The architecture definition includes system modeling to refine the specification and produce a HW and SW model. In this step, it is important to have a unified HW/SW representation. The unified representation describes tasks of the system which can be implemented in HW or SW. It helps a HW designer and a SW designer to communicate with each other. In addition, it defines an architectural model as shown in Figure 12.22.

Many unified models have been developed and used to represent heterogeneous systems in the literatures [11–13]. Thus, the system can be modeled as data flow graphs, communicating processes, finite state machines, discrete event system, and Petri Nets. As we discussed in Section 12.2, the data flow graphs are very popular in modeling data driven systems because it easily models data and control flow and concurrent operations. The nodes of the data flow graphs correspond to operations in HW or SW. In Ref. [14], a system is described as a set of interactive processes executing each other. The process corresponds to either HW computations or SW computations in the system. This modeling method is well matched with high level simulation and synthesis. The Finite State Machine (FSM) is one of the most well-known models to describe control flow. We define a set of states, inputs, outputs, and functions in the model. It changes from one state to another state when a transition happens. Although the FSM provides us with good mathematical foundation, it is not suitable for a complex system because the system complexity increases exponentially as the number of

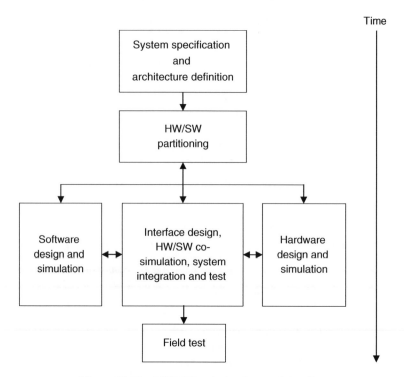

Figure 12.21 HW/SW codesign flow and timeline

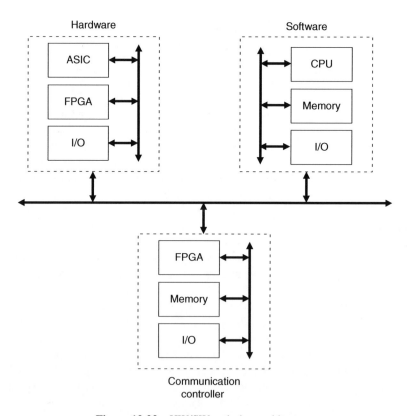

Figure 12.22 HW/SW codesign architecture

states increases. In addition, it is difficult to represent concurrency of states. Thus, the FSM is not appropriate to model modern communication systems. A discrete event system can be defined as a discrete-state event-driven system. Its state evolution entirely depends on the occurrence of asynchronous discrete events over time [15]. An event occurs instantaneously and causes a transition from one state to another state. The state transitions in the time driven systems are synchronized by the clock. However, the state transition in the discrete event system happens as a result of asynchronous and concurrent event process. Thus, the events are labeled as a time and analyzed in chronological order. The discrete event system is useful for the unified HW/SW representation. However, the computational complexity is so high because it requires global time sorting for all events. A Petri net is a directed bipartite graph consisting of places, tokens, transitions, and arcs. The place describes conditions and holds tokens. The token represents data flow in the system. The transition represents an event and a transition firing indicates that an event has occurred. The arc moves between a place and a transition.

The second step is *HW/SW partition*. It is the key part of HW/SW codesign process because the partition affects the overall system performance, development time, and cost. The main purpose of HW/SW partition is to design a system to meet the required performance under the given constraints (cost, power, complexity, etc.). In this step, we have to consider many design parameters. Basically, we want to automate HW/SW partition but it is still difficult to have optimal HW/SW partition. Thus, we approach either software-oriented partition or hardware-oriented partition. The software-oriented partition is to assign all functions in software and then move time-critical parts into hardware. Thus, we can develop a flexible system at low cost. On the other hand, the hardware-oriented partition is to assign all functions in hardware and then move sequential, not-time-critical and flexible parts into software. Thus, we can build a high performance and high speed system. Both approaches are based on functional partitioning. It defines all functions of the system and then implements the system. Thus, it provides us with a better trade-off between complexity and performance. In Ref. [16], the problem formulation of HW/SW partitioning is described as follows:

Given a set $O = \{o_1, o_2, ..., o_n\}$ of functional objects, a partition $P = \{p_1, p_2, ..., p_m\}$ of system components is sought that satisfies:

- $p_1 \bigcup p_2 \bigcup \cdots \bigcup p_m = 0,$
- $p_i \bigcap p_j = \emptyset, \ \forall i, j, i \neq j,$ and
- the cost function $f(P)$ is minimal.

A partitioning algorithm allocates all functional objects o_i to a system component p_j while the cost function is minimized. Cost function is defined as some parameter that a wireless communication system designer wants to minimize. For example, complexity, latency, power, and so on. This problem is known as a multivariate optimization problem and a Non-deterministic Polynomial-time hard (NP-hard) problem. In order to solve this problem, there are two main approaches: constructive algorithms and iterative algorithms. The constructive algorithms compute the closeness between p_i and p_j and then group objects, where the closeness metric often uses the communication cost between two objects. One of constructive algorithms is the hierarchical clustering algorithm. This algorithm groups the closest objects and then computes the closeness until a certain condition is reached. For example, four objects and their closeness are expressed as nodes and edges as shown in Figure 12.23. The termination condition is that we have one node.

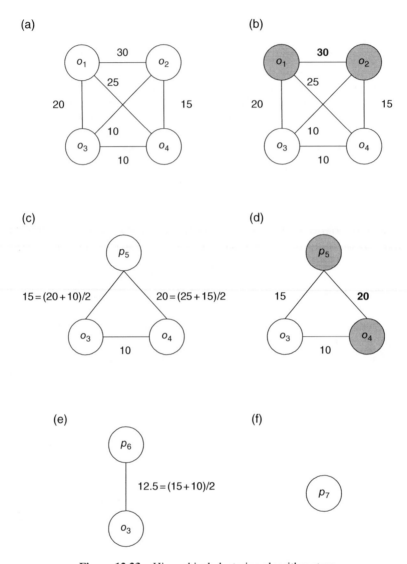

Figure 12.23   Hierarchical clustering algorithm steps

In Figure 12.23a, there are four nodes: $p_1 = \{o_1\}$, $p_2 = \{o_2\}$, $p_3 = \{o_3\}$, and $p_4 = \{o_4\}$. The highest closeness is 30 between p_1 and p_2 as shown in Figure 12.23b. Thus, we group them and have a new node $p_5 = \{o_1,o_2\}$. A new closeness is calculated as the average of edges between two nodes as shown in Figure 12.23c. Likewise, the highest closeness is 20 between p_5 and o_4 as shown in Figure 12.23d. We group them and have a new node $p_6 = \{p_5,o_4\}$. Next, we group p_6 and o_3 and have one node p_7 as shown in Figure 12.23e and f. Finally, we satisfy the termination condition and finish the process. The cluster tree of the hierarchical clustering algorithm steps is shown in Figure 12.24.

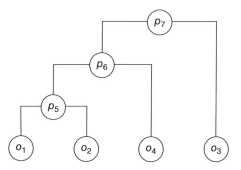

Figure 12.24   Cluster tree of hierarchical clustering algorithm

The iterative algorithms start from an initial partition (Sometimes a randomly generated partition is used as an initial partition.) and then improve it iteratively. The computation time of the iterative algorithm is very long because we have to evaluate a large number of partitions. However, it can find the global minimum cost. In Ref. [17], a simulated annealing algorithm is introduced and it is inspired from annealing in metals. The annealing process has two steps: (i) melting down the metal at a high temperature and (ii) cooling it down progressively. During the annealing process, the elements (molecules or atoms) of the material move slowly to attach each other in a pattern. The material has a stable structure representing a global minimum internal energy. Similarly, the cost function of the simulated annealing algorithm is regarded as the internal energy of the material. Just as the elements of the material can move at a high temperature, objects and components of the system are placed at a high tempera-ture where the temperature means an external parameter affecting the system modification. (The stimulated annealing algorithm still uses same term "temperature.") We try to minimize the cost function while the temperature decreases progressively. The temperature decrease rate significantly affects the system results such as complexity, etc. The steps of the simulated annealing algorithms are as follows: (i) the initialization step is to start with a random initial partition at a very high temperature; (ii) the move step is to perturb the partition while the tem-perature decreases slowly; (iii) the cost function calculation step is to calculate the cost function for the complete partition. We define the probability the move is possible from one state (cost function value f_1) to another state (cost function value f_2) at the temperature T as follows:

$$P(f_1, f_2, T) = \begin{cases} 1 & \text{when } f_2 < f_1 \\ e^{\frac{f_1 - f_2}{T}} & \text{otherwise} \end{cases}. \tag{12.19}$$

This equation means that the downhill move of the cost function is always allowed and the hill climbing move is accepted with the probability $e^{f_1 - f_2 / T}$; (iv) the choice step is to accept or reject any move; (v) the update and repeat step is to update the temperature and then repeat from step 2 until the termination condition is reached. Figure 12.25 illustrates one example of the simulated annealing algorithm.

As shown in Figure 12.25a, we start with a random initial partition at a very high initial temperature (T_i). The current position (the black dot) has a high cost function value and the

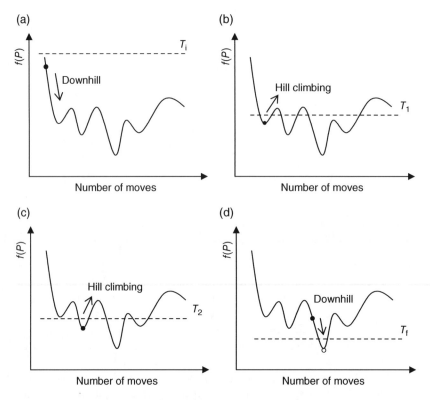

Figure 12.25   Simulated annealing algorithm steps

downhill move is allowed. In the next iteration as shown in Figure 12.25b and c, we can move in its close neighbourhood and the hill climbing move is allowed while the temperature decreases to T_1 and T_2. In the final temperature (T_f) as shown in Figure 12.25d, the only downhill move is allowed and we have the final cost function value (the circle). Practically, we use both the iterative algorithms and the constructive algorithms. The initial partition is performed by the constructive algorithms and then we modify the complete partition iteratively. When the HW/SW partition is finished, we evaluate the result using performance metrics (clock cycle, control steps, communication rates, execution time, and power consumption) and cost metrics (hardware manufacturing cost, program size, and memory size).

The next step is *HW/SW interface design and cosynthesis*. A wireless communication chipset is based on a multiprocessor system-on-chip (MPSoC). It integrates microprocessors (CPU), hardware intellectual property blocks (RF device, baseband device, etc.), memories, and communication interconnects. The major tasks of HW/SW interface include data/control access refinements, bus selection to reduce interfacing cost, scheduling software behaviour, etc. Thus, HW/SW interfacing becomes a major challenge. Cosynthesis is HW/SW/Interface implementation step under the constraints and optimization criteria. In HW synthesis, we design a specific hardware from specification described by hardware description language (VHDL or Verilog). In SW synthesis, we develop and optimize software components for a target processor in order to satisfy specifications. In interface

Summary 12.3   Hardware and software co-design

1. HW/SW codesign allows us to develop HW and SW concurrently and save the development time.
2. The first step of HW/SW codesign is system specification and architecture definition. The specification is to define the functions of a system and clearly describe the behavior including concurrency, hierarchy, state transition, timing, and so on. The unified representation (data flow graphs, communicating processes, finite state machines, discrete event system, and Petri Nets) describes tasks of the system which can be implemented in HW or SW.
3. The second step is HW/SW partition. It is the key part of HW/SW codesign process because the partition affects the overall system performance, development time, and cost. The main purpose of HW/SW partition is to design a system to meet the required performance under the given constraints (cost, power, complexity, etc.). The HW/SW partition problem is known as a multivariate optimization problem and a Non-deterministic Polynomial-time hard (NP-hard) problem. In order to solve this problem, there are two main approaches: constructive algorithms and iterative algorithms.
4. The next steps are HW/SW interface design, cosynthesis, and co-verification. Many EDA tools are used.

synthesis, we define HW/SW interface and make each module communicate. After that, we verify both HW and SW in simulation or emulation. Many Electronic Design Automation (EDA) tools are used in this step.

12.4   Problems

12.1. Survey the 5G standard activities of ITU and 3GPPP.

12.2. Describe the requirements of 5G standard by ITU.

12.3. In Ref. [18], the research challenges of HW/SW codesign are introduced: the wall of complexity, the wall of heterogeneity, the wall of dependability, the need for self-adaptivity, and the need for cross-layer co-verification. Select one topic and investigate the state of the art.

12.4. When implementing software part, errors in field test stage are more costly to fix than those in simulation stage. Sometimes, inadequate hardware allocation results in software cost increase. Survey and discuss the relationship between software cost and inadequate hardware allocation.

12.5. Compare the processor architectures: single instruction multiple data (SIMD), multiple instruction multiple data (MIMD), and very long instruction words (VLIW).

12.6. Retiming is generalization of pipelining. Describe their similarity.

12.7. Design the 5-tap FIR filter using parallel processing and pipelining.

12.8. HW implementation requires three-dimensional (speed, area, and Power) optimization. When achieving the required speed, both area and power should be trade-off. Discuss the design consideration when achieving the required speed.

12.9. Design the low complexity 8-point FFT using transformation technique.

References

[1] L. M. Guerra, *Behavioral Level Guidance Using Property-Based Design Characterization*, Ph.D. Thesis, University of California, Berkeley, CA, 1996.
[2] T. H. Cormen, C. E. Leiserson, and R. L. Rivest, *Introduction to Algorithms*, MIT Press, Cambridge, MA,1990.
[3] C. Mead and L. Conway, *Introduction to VLSI Systems*, Addison-Wesley, Reading, MA, 1980.
[4] S. Note, W. Geurts, F. Catthoor, and H. De Man, "Cathedral-III: Architecture Driven High-Level Synthesis for High Throughput DSP Applications," *Proceedings of ACM/IEEE 28th Design Automation Conference*, San Francisco, CA, pp. 597–602, June 17–21, 1991.
[5] D. Rao and F. Kurdahi, "Partitioning by Regularity Extraction," *Proceedings of ACM/IEEE 29th Design Automation Conference*, Anaheim, CA, pp. 235–238, June 8–12, 1992.
[6] N. Weste and D. Harris, *CMOS VLSI Design: A Circuits and Systems Perspective*, Addison-Wesley, Boston, MA, 4th edition, 2010.
[7] P. R. Panda, B. V. N. Silpa, A. Shrivastava, and K. Gummidipudi, *Power-Efficient System Design*, Springer, New York/London, 1st edition, 2010.
[8] D. Dobberpuhl and R. Witek, "A 200MHz 64b Dual-Issue CMOS Microprocessor," *Proceedings of IEEE International Solid State circuits Conference*, pp. 106–107, San Francisco, CA, USA, February 19–21, 1992.
[9] A. P. Chandrakasan, M. Potkonjak, R. Mehra, J. Rabaey, and R. W. Brodersen, "Optimizing Power Using Transformations," *IEEE Transactions on Computer Aided Design of Integrated Circuits and Systems*, vol. 14, no. 1, pp. 12–31, 1995.
[10] N. R. Shanbhag, "Algorithms Transformation Techniques for Low-Power Wireless VLSI Systems Design," *International Journal of Wireless Information Networks*, vol. 5, no. 2, p. 147, 1998.
[11] E. A. Lee and A. Sangiovanni-Vicentelli, "Comparing Models of Computation," *Proceedings of International Conference on Computer-Aided Design (ICCAD 1996)*, San Jose, CA, pp. 234–241, November 10–14, 1996.
[12] S. Edwards, L. Lavagno, E. A. Lee, and A. Sangiovanni-Vicentelli, "Design of Embedded Systems: Formal Models, Validation, and Synthesis," *Proceedings of the IEEE*, vol. 85, no. 3, pp. 366–390, 1997.
[13] J. Staunstrup and W. Wolf, *Hardware/Software Co-Design: Principles and Practice*, Springer, Boston, MA, 1997.
[14] C. A. R. Hoare, *Communicating Sequential Processes*, Prentice-Hall, Englewood Cliffs, NJ, 1985.
[15] C. G. Cassandras, *Discrete Event Systems: Modeling and Performance Analysis*, Aksen Associates, Homewood, IL, 1993.
[16] J. Teich, *Digitale Hardware/Software Systeme*, Springer Verlag, Berlin, 1997.
[17] S. Kirkpatrick, C. D. Gelatt, and M. P.Vecchi, "Optimization by Simulated Annealing," *Science*, vol. 220, no. 4598, pp. 671–680, 1983.
[18] J. Teich, "Hardware/Software Codesign: The Past, the Present, and Predicting the Future." *Proceedings of the IEEE*, vol. 100, pp. 1411–1430, 2012.

13

Wireless Communications Blocks Integration

We designed key wireless communication blocks and tried to optimize each parameter such as complexity, throughput, latency, and power consumption in Part II. Although each block is optimized, it is not easy to integrate them and find an optimal solution of wireless communication systems at high level. Thus, we discussed radio planning and link budget analysis in Chapter 11. We looked into the wireless communication system design flows and wireless communication system design considerations in Chapter 12. In this chapter, we will take a look at top level view of wireless communication systems, review 4G standards, and integrate each wireless communication block.

13.1 High Level View of Wireless Communications Systems

In order to exchange or broadcast data between communication systems, we need a standardized protocol. If not, it is difficult to understand the contents even if we receive a signal correctly. Thus, the International Standard Organisation (ISO) defined the Open Systems Interconnection (OSI) 7 layer model for network protocols. Although it is not widely used in industry and many standardization groups prefer a simpler model, it is good conceptual model characterizing the internal functions of wireless communication and network systems. The advantage of the OSI 7 layer model is that each layer is regarded as a black box and one layer transfers its input/output to another layer above/below the layer. Thus, we can develop each layer independently and update each layer without a whole system change. The OSI 7 layer model is organized in the specific order from the top layer to the bottom layer. The bottom layer is the *physical layer*. It defines how the data is transmitted and received over the communications media such as copper cable, fiber, and radio. The physical layer deals with physical properties of the media, mechanical properties of the data connection (pin layout, cable specification, etc.), signal presentation (waveform, encoding, etc.), and control signals.

Wireless Communications Systems Design, First Edition. Haesik Kim.
© 2015 John Wiley & Sons, Ltd. Published 2015 by John Wiley & Sons, Ltd.

The physical layer protocol includes Ethernet PHY, RS-232, IEEE802.11 PHY, 3GPP PHY, and so on. The layer 2 is the *data link layer*. It defines how to make a reliable link. The data link layer is composed of two sub-layers: Media Access Control (MAC) layer and Logical Link Control (LLC) layer. The MAC layer is to distinguish the physical devices using the MAC address and the LLC layer is to detect errors in received frames. The data link layer protocol includes IEEE802.11 MAC, 3GPP MAC, Point-to-Point Protocol (PPP), and so on. The layer 3 is the *network layer*. It is in charge of routing packets across the network. For example, one big video file is divided into many packets and then each packet is labeled by a network address. Each packet is transmitted from one node to another node in the same network. The network layer as one of the complicated layers divides packets and manages network addresses. The network layer protocol includes Internet Protocol (IP), Internet Control Message Protocol (ICMP), Internet Group Management Protocol (IGMP), and so on. The layer 4 is the *transport layer*. It is an intermediate layer between lower layers and higher layers. It hides the complexities of lower layers from the higher layers and provides a reliable link by data retransmission. The transport layer protocol includes Transmission Control Protocol (TCP), User Datagram Protocol (UDP), and so on. The layer 5 is the *session layer*. It manages the connection or session between nodes. The session layer protocol includes Secure Shell (SSH), Secure Sockets Layer (SSL), and so on. The layer 6 is the *presentation layer*. It allows the nodes/devices to exchange information in the same language. The presentation layer protocol includes Moving Picture Experts Group (MPEG), Joint Photographic Experts Group (JPEG), Server Message Block (SMB), and so on. The top layer is the *application layer*. It allows an end user to access the network resources and provides an interface between end users and networks. The application layer protocol includes Hypertext Transfer Protocol (HTTP), File Transfer Protocol (FTP), Domain Name System (DNS), and so on. Figure 13.1 illustrates OSI 7 layer.

In this book, the wireless communication system design mainly meant the physical layer system design. The upper layer is about network design issues. As we can see Figure 13.1, the physical layer is implemented mainly in hardware. The data link layer is implemented in both hardware (for a low MAC) and software (for a high MAC). The upper layers are implemented in software. The OSI 7 layer enables us to design wireless communication systems with interoperability. However, the disadvantage is the lack of collaboration between layers. The strict boundaries between layers prevent us to optimize a whole wireless communication system. We face the limitation of the system performance improvement. Thus, cross-layer design [1] was proposed in order to overcome the limitation. The key idea is to break the strict boundaries and permit collaboration and joint optimization between layers. The goal of the cross-layer design is to improve security, Quality of Service (QoS), and mobility [2, 3]. However, this approach is confronted with several challenges such as the lack of the unified cross-layer design, cross-layer signaling, and overhead for cross-layer signaling. Thus, the layered communication architectures are still widely used in 4G communication systems.

The physical layer of wireless communications is composed of Radio Frequency (RF) front-end and digital transmitter/receiver. The RF front-end is placed between an antenna and Digital to Analogue Converter (DAC)/Analogue to Digital Converter (ADC). It converts from a low/high frequency signal (baseband/passband signal) to a high/low frequency signal (passband/baseband signal). The RF front-end is composed of impedance matching circuit, Power Amplifier (PA), Low Noise Amplifier (LNA), Band Pass Filter (BPF), mixer, Local Oscillator (LO), and Phase Locked Loop (PLL). Figure 13.2 illustrates the typical RF front-end transmitter and receiver.

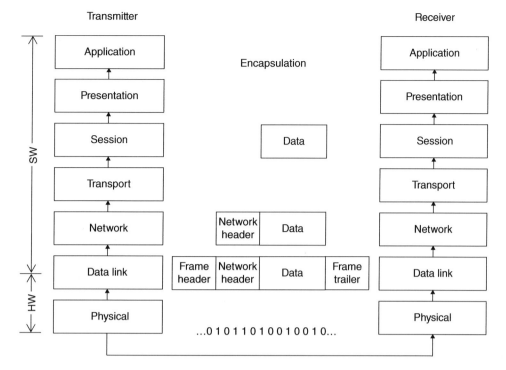

Figure 13.1 OSI 7 layer

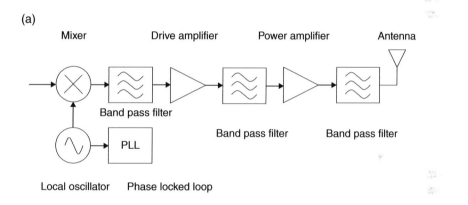

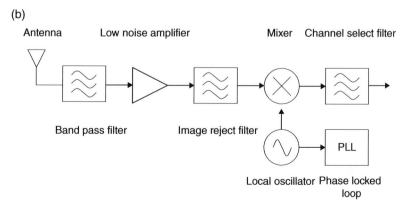

Figure 13.2 RF front-end transmitter (a) and receiver (b)

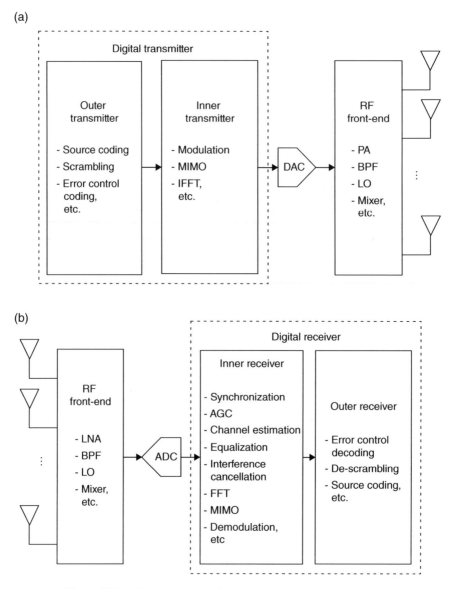

Figure 13.3   Abstracted transmitter (a) and receiver (b) architecture

The incoming signal of the RF front-end transmitter is converted to high frequency signal in a mixer and local oscillator and amplifies the signal in the power amplifier. Then, the antenna radiates the electromagnetic waves in the air. In the RF front-end receiver, the antenna receives the electromagnetic waves and the band pass filter selects the desired signal. The received signal basically includes a noise. Thus, LNA amplifies the received signal while it reduces the effect of the noise. The received signal is converted to a low-frequency signal in a mixer and local oscillator. After that, the analogue signal is converted to digital signal in ADC. Now, the received signal is processed in digital domain.

Summary 13.1   OSI 7 layer

1. The advantage of the OSI 7 layer model is that each layer is regarded as a black box and one layer transfers the input/output to another layer above/below the layer.
2. Layer 1: The physical layer deals with physical properties of the media, mechanical properties of the data connection, signal presentation, and control signals.
3. Layer 2: The data link layer defines how to make a reliable link and is composed of two sub-layers: Media Access Control (MAC) layer and Logical Link Control (LLC) layer.
4. Layer 3: The network layer is in charge of routing packets across the network.
5. Layer 4: The transport layer is an intermediate layer between lower layers and higher layers. It hides the complexities of lower layers from the higher layers and provides a reliable link by data retransmission.
6. Layer 5: The session layer manages the connection or session between nodes.
7. Layer 6: The presentation layer allows the nodes/devices to exchange information in the same language.
8. Layer 7: The application layer allows an end user to access the network resources and provides an interface between the end user and the networks.

The digital transmitter/receiver is divided into the inner part and the outer part. The outer transmitter/receiver includes source coding/decoding, scrambling/descrambling, formatting/deformatting, error correction coding/decoding, interleaving/deinterleaving, and so on. They deal with digital data representing digital signals. For example, the soft decision data of Turbo decoding comes from the inner receiver. The soft decision data express how likely the data is—0 or 1. After Turbo decoding, we have the hard decision data and cannot represent how likely it is. Thus, the outer receiver mainly deals with the digital data representing digital signals. On the other hand, the inner transmitter/receiver includes modulation/demodulation, IFFT/FFT, synchronization, channel estimation, equalization, MIMO, and so on. They deal with digital data representing analogue signals. Thus, the main purpose of the inner receiver compensates a noise and tries to recover the original transmitted signals. Then, the outer receiver corrects or detects an error. If the data integrity is confirmed, we send the data to an upper layer. If not, data retransmission is requested. Figure 13.3 illustrates the abstracted transmitter and receiver.

13.2   4G Physical Layer Systems

There have always been technology competitions to meet the requirements and secure a share of the new market. For example, Ultra-Wideband (UWB) development was divided into two technology camps: Direct Sequence (DS) UWB led by Motorola and Multi-Band OFDM led by Intel. Likewise, in order to meet the 4G system requirements (max. 1 Gbps for a low mobility communication and max. 100 Mbps for a high mobility communication) by International Telecommunication Union Radio communications sector (ITU-R), there are two technology camps: Long-Term Evolution (LTE) by the 3rd Generation Partnership Project (3GPP) and Worldwide Interoperability for Microwave Access (WiMAX) by the Institute of Electrical and Electronics Engineers (IEEE). Although both technologies do not meet these requirements, they are called 4G communication systems. Both physical layer technologies are quite similar. Their key physical layer technologies include MIMO, Orthogonal Frequency-Division Multiple Access (OFDMA), AMC, Turbo code, and so on. On the other hand, the differences are channel bandwidth, frame structure, uplink modulation, and so on. In this chapter, we will take a look at both LTE and WiMAX and compare their physical layer technologies.

13.2.1   LTE

The LTE was initiated in 3GPP Release 8 [4]. Some key technologies in the 3GPP Release 8 include Single Carrier Frequency Division Multiple Access (SC-FDMA) [5], Advanced MIMOs, and six channel bandwidths (1.4, 3, 5, 10, 15, and 20 MHz). The 3GPP continues the revision of the LTE standard from release 8 to release 12. In a cellular system, many users should share finite radio resources. Multiple access schemes are an essential part of the cellular system. They divide radio resources and allocate users to a predetermined time and frequency slot. The multiple access schemes allow us to manage multiple users without interferences and maximize the radio resource utilization. Roughly speaking, Frequency Division Multiple Access (FDMA) divides the total spectrum into narrow channels which are allocated to multiple users. The FDMA system has a simple hardware architecture and is used in analogue systems such as Advanced Mobile Phone System (AMPS) and Nordic Mobile Telephone (NMT). However, its spectrum efficiency is very low and it is not suitable for variable rate transmission. In Time Division Multiple Access (TDMA), each channel is divided into time slots which are allocated to multiple users. The TDMA system has a higher spectral efficiency and provides us with variable rate transmission. However, it needs a long guard interval and synchronization sequence. The TDMA is used in 2G systems. Global System for Mobile communications (GSM) combines TDMA and FDMA. In GSM, each user is allocated in specific bandwidth and time. In Code Division Multiple Access (CDMA), multiple users simultaneously transmit and receive data in same frequency and time. Using spread spectrum technology, each user is identified. The CDMA system has many advantages: high spectral efficiency, flexible resource allocation, soft handoff, and so on. On the other hand, some disadvantages are self-jamming, near far problem, and overall decrease in service quality when the number of user increases. The 2G system such as IS-95 used CDMA and the 3G system adopted Wideband CDMA (W-CDMA). OFDMA is based on the OFDM technique and assigns a subset of subcarriers to multiple users. The LTE multiple access scheme is based on OFDMA for downlink and SC-FDMA for uplink. Although their multiple access schemes are different, they have common frame structures. The type 1 frame structure is defined for Frequency Division Duplexing (FDD) mode. The FDD mode uses two distinct frequency bands for downlink (from a base station to mobile stations) and uplink (from mobile stations to a base station). In order to isolate the transmit/receive signals from the receive/transmit signals, a duplex-filter is used. Figure 13.4 illustrates the type 1 frame structure. The duration of the type 1 radio frame is 10 ms. The radio frame consists of 10 subframes (1 ms) and each subframe consists of 2 slots (0.5 ms). Each slot consists of seven OFDM symbols when the normal cyclic prefix is employed.

The type 2 frame structure is defined as Time Division Duplexing (TDD) mode. The TDD mode assigns a single frequency band to both a transmitter and a receiver but the signal is transmitted/received in different times. The expensive duplex filters are not required in the TDD mode so that the hardware cost of the TDD mode is substantially less than the FDD mode.

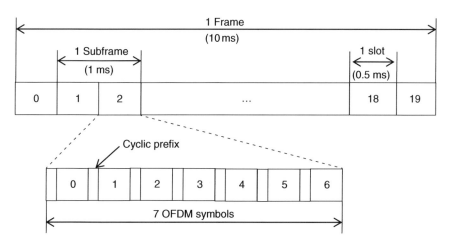

Figure 13.4 LTE type 1 (FDD) frame structure

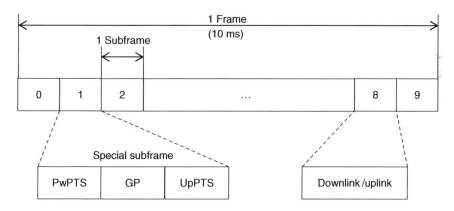

Figure 13.5 LTE type 2 (TDD) frame structure

However, the adjacent channel interferences of the TDD mode are much higher than the FDD mode. Figure 13.5 illustrates the type 2 frame structure. The type 2 frame consists of two 5 ms half frames. Subframes consist of three type transmissions: downlink transmission, uplink transmission and special subframe transmission (downlink pilot time slot (DwPTS), guard period (GP) and uplink pilot time slot (UpPTS)).

The type 2 frame configuration is determined by one of the seven difference configurations as shown in Table 13.1. In the table, D, U, and S represent downlink transmission, uplink transmission, and special subframe transmission, respectively.

The OFDMA uses time and frequency resources of the available bandwidth. A physical Resource Block (RB) is defined as consisting of 12 consecutive subcarriers in frequency domain and 7 consecutive OFDM symbols in time domain. It is the smallest element of resource allocation and contains 84 resource elements ($=12 \times 7$). Each RB occupies 180 kHz ($=12 \times 15$ kHz (the subcarrier spacing Δf)) in frequency domain and 0.5 ms (one slot) in time domain.

Table 13.1 LTE type 2 (TDD) frame configuration

Type 2 frame configuration	Switch point periodicity (ms)	Subframe number									
		0	1	2	3	4	5	6	7	8	9
0	5	D	S	U	U	U	D	S	U	U	U
1	5	D	S	U	U	D	D	S	U	U	D
2	5	D	S	U	D	D	D	S	U	D	D
3	10	D	S	U	U	U	D	D	D	D	D
4	10	D	S	U	U	D	D	D	D	D	D
5	10	D	S	U	D	D	D	D	D	D	D
6	5	D	S	U	U	U	D	S	U	U	D

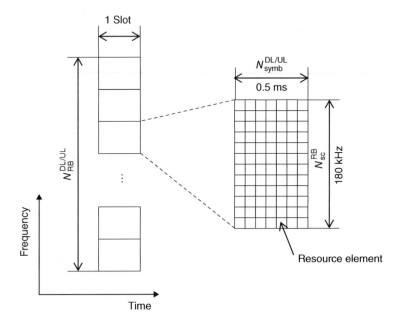

Figure 13.6 Resource grid and resource blocks

When the channel bandwidth is 5 MHz, we can use maximum 27.7 RB (=5 MHz/180 kHz). However, we use 25 RB for the 5 MHz channel bandwidth. Figure 13.6 illustrates resource blocks.

In Figure 13.6, $N_{RB}^{DL/UL}$ represents the number of resource blocks in the downlink and uplink. N_{sc}^{RB} represents the number of subcarriers and the value is 12 for standard operation. $N_{symb}^{DL/UL}$ represents the number of OFDM or SC-FDMA symbols and the value is 7 or 6 for standard operation. Table 13.2 summarizes the physical resource block parameters.

The LTE air interface consists of *physical channels* and *physical signals*. They are defined in 3GPP documentations [6]. The physical channels correspond to a set of resource elements conveying information from upper layers and they are mapped onto transport channels.

Table 13.2 Physical resource block parameters

	Downlink		Uplink	
	N_{sc}^{RB}	N_{symb}^{DL}	N_{sc}^{RB}	N_{symb}^{UL}
Normal CP ($\Delta f = 15\,\mathrm{KHz}$)	12	7	12	7
Extended CP ($\Delta f = 15\,\mathrm{KHz}$)	12/24 ($\Delta f = 7.5\,\mathrm{KHz}$)	6/3 ($\Delta f = 7.5\,\mathrm{KHz}$)	12	6

Table 13.3 LTE downlink physical channels

Downlink channels	Full name	Modulation	Purpose
PDSCH	Physical downlink shared channel	QPSK, 16 QAM, 64 QAM	Carries user data
PDCCH	Physical downlink control channel	QPSK	H-ARQ, resource allocation and UL scheduling grant
PCFICH	Physical control format indicator channel	QPSK	Indicates number of PDCCH OFDM symbols per subframe
PHICH	Physical hybrid ARQ indicator channel	BPSK	Carries H-ARQ information (ACK/NACK)
PBCH	Physical broadcast channel	QPSK	Carries cell information
PMCH	Physical multicast channel	QPSK, 16 QAM, 64 QAM	Carries multicast data

The transport channels are Service Access Points (SAP) between MAC and PHY. The LTE supports two types of *downlink physical channels*: control channels and data channels. Physical Downlink Shared Channel (PDSCH) carries the user data. Since a high data rate is required, a high order modulation (such as 64 QAM) and spatial multiplexing are used in this channel. Physical Downlink Control Channel (PDCCH) conveys the control signal to the mobile station (or User Equipment (UE)). The control signals include H-ARQ information, resource allocation, and UL scheduling grant. Multiple PDCCH can be transmitted in one subframe. Physical Control Format Indicator Channel (PCFICH) is used to indicate how many OFDM symbols are reserved for PDCCH. It is transmitted in every subframe. Physical Hybrid ARQ Indicator Channel (PHICH) carries H-ARQ information in response to UL transmission. Physical Broadcast Channel (PBCH) carries cell-specific information such as Random Access Channel (RACH). This channel is always provided with 1.08 MHz bandwidth. Physical Multicast Channel (PMCH) carries multicast data. Table 13.3 summarizes LTE downlink physical channels.

The LTE physical signals use physical resource elements but do not carry any information from/to upper layers. There are two types of the LTE *downlink physical signals*: Synchronization Signals (SSs) and Reference Signals (RSs). The synchronization signals are classified into primary and secondary synchronization signals. The Primary Synchronization Signal (PSS) is generated from Zadoff-Chu (ZC) sequence with length 62

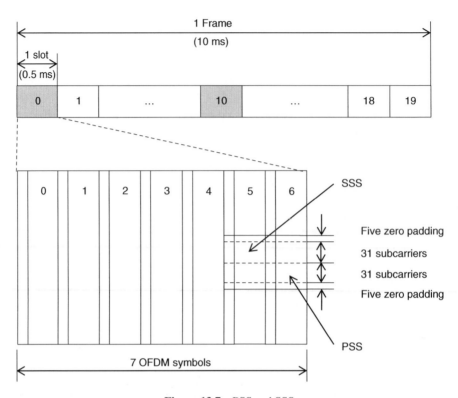

Figure 13.7   PSS and SSS

in frequency domain and mapped to the seventh OFDM symbol in slots 0 and 10. The Secondary Synchronization Signal (SSS) is an interleaved concatenation of two binary sequences with length 31 and mapped to the sixth OFDM symbol in slot 0 and 10. The concatenated sequence is scrambled with a scrambling sequence by PSS. Figure 13.7 illustrates PSS and SSS in one frame.

The LTE access procedure (cell search and synchronization) is carried out using PSS and SSS. Once a UE (or a mobile station) is switched on, it must start looking for an LTE cell and acquire all information required to register. Since the UE is unaware of the downlink channel configuration, cell information should appear on a consistent place regardless of the downlink channel configuration. Thus, PSS and SSS are transmitted in specific slots of each downlink frame. In order to search and select an LTE cell, we follow three steps. The first step is to perform symbol timing acquisition and frequency synchronization and then obtain physical layer ID using PSS. In the second step, we perform frame boundary detection, Cyclic Prefix (CP) length detection, and FDD/TDD detection and then obtain cell ID using SSS. The third step is to detect cell-specific reference signal and PBCH. Once synchronization process is finished, channel estimation and carrier offset estimation should be performed using reference signals. The reference signals are generated in physical layer as gold sequence with length 31. There are five types of downlink reference signals: cell-specific RSs, Multicast Broadcast Signal Frequency Network (MBSFN) RSs, UE-specific RSs, positioning RSs, and channel-state information RSs. The location of the RSs depends on the

MIMO configuration. Figure 13.8 illustrates downlink reference signal mapping for MIMO antenna configuration. Each antenna has a specific reference signal pattern and each reference signal is transmitted on equally spaced subcarrier (six subcarriers) and time domain spacing (four OFDM symbols). Namely, there are four reference signals per resource block at each antenna. When one reference signal is transmitted at a specific resource element from one antenna, the other antenna resource elements are null. The LTE uplink RSs play an important role in channel estimation, power control, timing estimation, and direction of arrival estimation.

The LTE physical layer receives data in the form of a certain size transport block. The LTE downlink physical layer processing scheme consists of many steps according to different physical layer channels and transmission modes. For example, the LTE-baseband signals representing a downlink physical channel (PDSCH) are processed in terms of the following

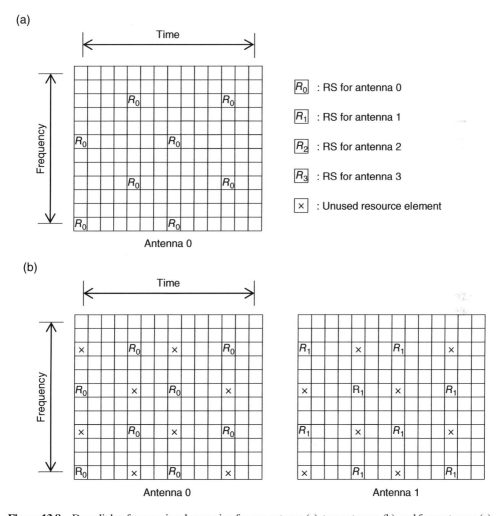

Figure 13.8 Downlink reference signals mapping for one antenna (a), two antennas (b), and four antennas (c)

(c)

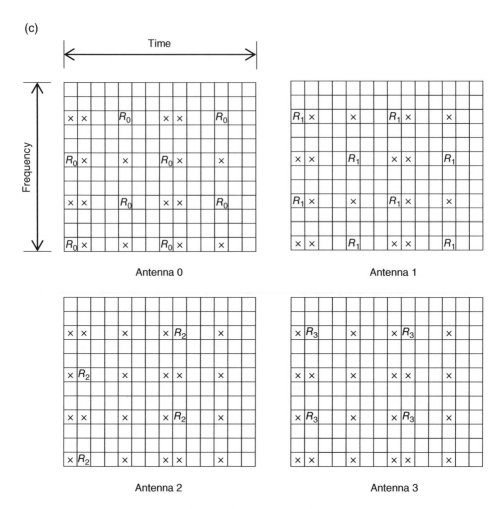

Figure 13.8   (*Continued*)

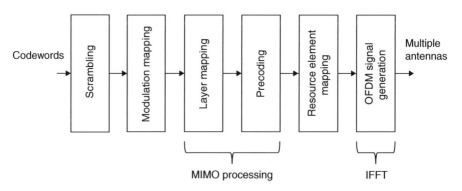

Figure 13.9   LTE downlink physical channel (PDSCH) processing (transmitter)

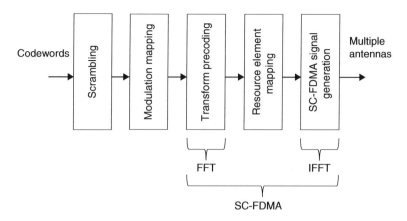

Figure 13.10 LTE uplink physical channel processing (transmitter)

steps: (i) Scrambling is applied by multiplying codewords from H-ARQ with the scrambling sequence. (ii) Modulation is performed on the scrambled bits and complex-valued modulation symbols are generated. (iii) Layer mapping and precoding is performed to transmit on MIMO antenna. LTE antenna mapping can be configured to support different MIMO schemes. (iv) Resource element mapping allocates complex-valued modulation symbols to the resource elements of the resource blocks. The resource blocks are assigned by the scheduler in MAC layer. (v) IFFT generates OFDM symbols for each antenna port. Figure 13.9 illustrates the LTE PDSCH processing.

The LTE supports three types of *uplink physical channels*: Physical Random Access Channel (PRACH), Physical Uplink Shared Channel (PUSCH), and Physical Uplink Control Channel (PUCCH). The PRACH carries the random access preambles which are generated from ZC sequences. The ZC sequences reduce the PAPR of the LTE uplink transmission. The location of the PRACH is defined by upper layer signaling. The PUSCH is the main uplink channel and carries the Uplink Shared Channel (UL-SCH). Since it carries user data, it uses a high-order modulation (such as 64 QAM) as the PDSCH did. The PUCCH carries the uplink control information such as Channel Quality Indicators (CQI) reports, ACK/NACK in response to downlink transmission, scheduling request, MIMO codeword feedback, and so on. There are two types of the LTE *uplink physical signals*: reference signals and random access preambles. The reference signal is based on ZC sequences and has two variants: Demodulation Reference Signal (DRS) and Sounding Reference Signal (SRS). The DRS is used for channel estimation and transmitted in the 4th symbol in each slot for normal CP. The SRS provides the base station (eNB) with uplink channel quality information and transmitted in the last symbol of the subframe. The random access preamble is used in various scenarios such as initial access and handover. When the UE initiates the cell search, the random access preamble is transmitted to eNB. In the LTE uplink, SC-FDMA is adopted. It is very effective to reduce the PAPR. LTE uplink physical layer processing is similar to LTE downlink physical layer processing except MIMO processing and SC-FDMA. Figure 13.10 illustrates LTE uplink physical channel processing.

Table 13.4 summarizes LTE channel bandwidth and resource configuration [6].

Table 13.4   LTE channel bandwidth and resource configuration [6]

Channel bandwidth (MHz)	1.4	3	5	10	15	20
FFT/IFFT size	128	256	512	1024	1536	2048
Number of resource blocks (N_{RB})	6	15	25	50	75	100
Number of occupied subcarriers	72	180	300	600	900	1200
Sample rate (MHz)	1.92	3.84	7.68	15.36	23.04	30.72
Samples per slot	960	1920	3840	7680	11520	1560
Subcarrier spacing (KHz)	15					
Physical resource block bandwidth (KHz)	180					

Example 13.1   LTE physical layer

Consider the following LTE downlink physical layer parameters:

Parameters	Values
Duplexing	FDD
Downlink physical channel	PDSCH
Channel bandwidth	5, 10, 15, 20 MHz
MIMO	2×2
Modulation	QPSK and 64 QAM
ECC	Turbo coding (code rate = 1/2)
Channel estimation	LS
MIMO detection	MMSE
Precoding	Codebook based
Channel model	ITU PB channel (Low mobility), AWGN and SNR = 20 dB

Compare the OFDM symbol spectrums according to the different channel bandwidth and modulation.

Solution

As we can see from Table 13.4, each channel bandwidth has different resource configuration. However, we have similar LTE downlink physical channel (PDSCH) processing as shown in Figure 13.9. Each physical layer block is developed independently and then they are integrated with flexible system parameters. Using computer simulation, we can obtain the LTE downlink OFDM symbol spectrums as shown in Figures 13.11, 13.12, 13.13, and 13.14.

As we can observe from Figures 13.11, 13.12, 13.13, and 13.14, the OFDM symbol spectrums with QPSK modulation has a lower ripple than 64 QAM because a high-order

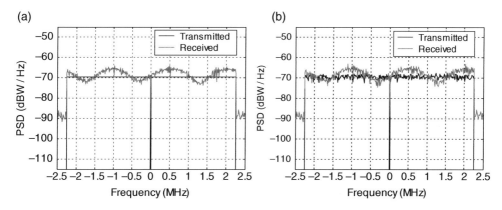

Figure 13.11   LTE downlink OFDM symbol spectrums with QPSK (a) and 64 QAM for 5 MHz mode (b)

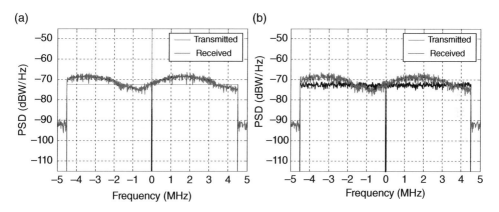

Figure 13.12   LTE downlink OFDM symbol spectrums with QPSK (a) and 64 QAM for 10 MHz mode (b)

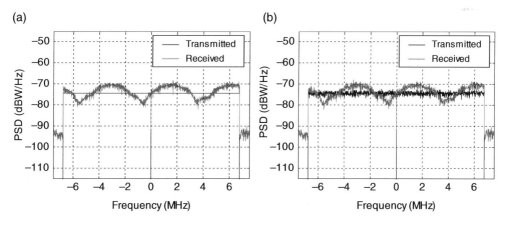

Figure 13.13   LTE downlink OFDM symbol spectrums with QPSK (a) and 64 QAM for 15MHz mode (b)

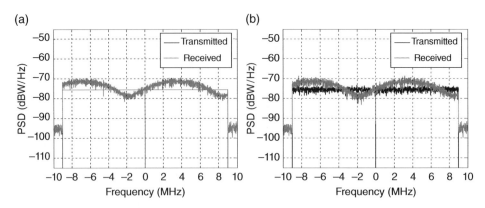

Figure 13.14   LTE downlink OFDM symbol spectrums with QPSK (a) and 64 QAM for 20 MHz mode (b)

modulation scheme exhibits a higher PAPR than a lower-order modulation scheme. These are investigated further in Section 13.4.   ∎

13.2.2   WiMAX

The IEEE 802.16 working group developed a new amendment of the IEEE 802.16 standard to meet 4G requirements by the ITU. The purpose of the standard is to develop a broadband wireless access at high speed and low cost, which is easy to deploy and provide a scalable solution in areas beyond the reach of Digital Subscriber Line (DSL). The initial IEEE802.16 standard was developed for the higher microwave band where the line-of-sight between a base station and a mobile station is required for reliable service. However, most vendors did not attempt to implement high-frequency multipoint systems based on the IEEE802.16 standard due to the poor economics of the high frequency system. In 2000, a low frequency IEEE802.16 standard was initiated and this revision supported non-line-of-sight service, OFDM, Turbo Codes, and licensed and unlicensed band implementations. Since 2007, they have embarked on the IEEE802.16m (Mobile WiMAX) which works in a low frequency (2.3, 2.5 or/and 3.4 GHz) and supports flexible bandwidths. The previous version of WiMAX supports TDD, FDD, and half-duplex FDD. However, the mobile WiMAX supports only TDD. Similar to the LTE, the mobile WiMAX is based on scalable OFDMA and has two-dimensional resource blocks in time and frequency domain. Figure 13.15 illustrates the mobile WiMAX OFDMA frame structure for TDD. As we can observe the figure, each frame is divided into the downlink subframe and the uplink subframe. They are separated by the guard interval called Transmit Transition Gap (TTG) or Receive Transition Gap (RTG). In the subframes, there are several elements to carry control information. The preamble is used for synchronization and located in the first OFDM symbol of every downlink frame. The Frame Control Head (FCH) provides the frame configuration such as usable subchannels, the length of the DL-MAP, and coding scheme. The location of the FCH is fixed in the frame. The DL/UP-MAP describes sub-channel allocation including the number of downlink bursts and length. The ranging part is allocated for mobile stations to perform closed loop adjustments. The ACK is used for downlink H-ARQ acknowledgment. The fast channel feedback/Channel Quality Information Channel (CQICH) is used for channel-state information feedback. The user data is

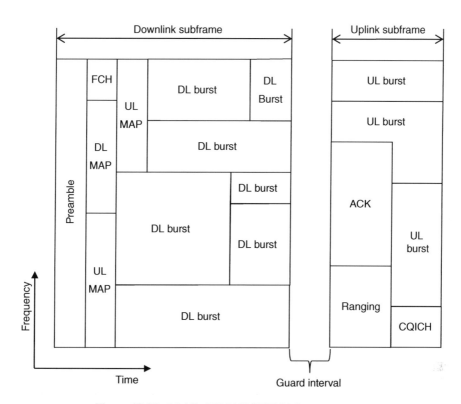

Figure 13.15   Mobile WiMAX OFDMA frame structure

assigned in a burst. In WiMAX, there are two subcarrier permutations: diversity permutation and contiguous permutation. The diversity permutation provides frequency diversity and includes Fully Used Subchannels and Partially Used Subchannels (PUSC). The contiguous permutation provides multiuser diversity by choosing the subchannel with the best frequency response. The multiuser diversity gain promotes system throughput substantially while it requests AMC scheduler to consider PHY channel state information as well as MAC QoS requirements [7, 8].

The mobile WiMAX is based on scalable OFDMA and supports various bandwidths. In order to maximize the system throughput, Adaptive Modulation and Coding (AMC) is used. The downlink supports BPSK, QPSK, 16 QAM, and 64 QAM. The high-order modulation (such as 64 QAM) is optional in the uplink. It supports both convolutional codes and Convolutional Turbo Codes (CTCs) with variable code rates. The Block Turbo Codes (BTCs) and Low-Density Parity Check Codes (LDPCs) are used as optional features. After FEC processing, puncturing is performed. The puncturing block removes a certain bits in the transmitter and replaces the deleted bits with an unbiased value in the receiver. The interleaver takes the codewords and rearranges them in a different order. The mobile WiMAX supports the open loop and closed loop MIMO. The open loop MIMO includes two mandatory MIMO techniques. The first is MIMO A (Space Time Code) proposed by Alamouti for transmit diversity. The second is MIMO B (Spatial Multiplexing) which uses two transmit antennas to transmit two independent data streams. The MIMO A was originally proposed to avoid the use of receive diversity on the downlink and keep the subscriber stations simple. In the WiMAX system, this is applied subcarrier-by-subcarrier.

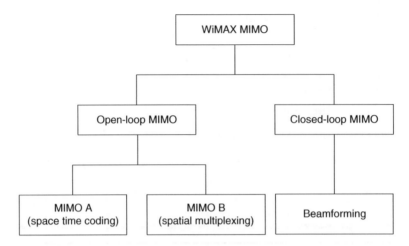

Figure 13.16   WiMAX MIMO schemes.

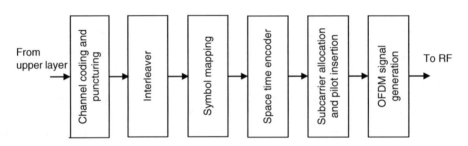

Figure 13.17   Simplified mobile WiMAX physical layer functional diagram (transmitter)

Table 13.5   Mobile WiMAX channel bandwidth and resource configuration

Channel bandwidth (MHz)	1.25	5	10	20
FFT/IFFT size	128	512	1024	2048
Sampling frequency (MHz)	1.4	5.6	11.2	22.4
Number of data subcarriers	72	360	720	1440
Number of pilot subcarriers	12	60	120	240
Number of null subcarriers	44	92	184	368
Number of subchannels	2	8	16	32
Symbol time (ms)	91.4			
Guard time (ms)	11.4			
Subcarrier spacing (KHz)	10.94			
Number of OFDM symbols (5 ms frame)	48			

The MIMO A achieves the maximum available diversity with a simple receiver structure but it does not give any spatial multiplexing gain [9]. The MIMO B supports full-rate, as it transmits one symbol per antenna use. However, it does not offer any diversity gain from the transmitter side because each symbol is transmitted from one antenna only. In the 2×2 MIMO system, it offers a diversity gain of 2 on the receiver side provided it used Maximum Likelihood (ML) detection [9, 10]. Figures 13.16 and 13.17 illustrate Mobile WiMAX MIMO schemes and simplified physical layer functional diagram, respectively.

Table 13.5 summarizes Mobile WiMAX physical layer parameters.

Example 13.2   WiMAX physical layer
Consider the following WiMAX downlink physical layer parameters:

Parameters	Values
Channel bandwidth	10 MHz
FFT size	1024
Cyclic prefix	1/8
Subcarrier permutation	PUSC
Number of users	1
Repetition rate	1
FEC	CTC
MIMO	A and B (2×2)
Channel estimation	Perfect channel estimation
Channel	AWGN, ITU-PB (3 km)

Also consider the following coding and modulation configuration:

Modulation	Code rate	Number of symbols per frame	Number of subchannels	Number of slots	Data block size (bytes)
QPSK	1/2	4	5	10	60
QPSK	3/4	4	3	6	54
16 QAM	1/2	2	5	5	60
16 QAM	3/4	2	3	3	54
64 QAM	1/2	2	3	3	54
64 QAM	2/3	2	2	2	48
64 QAM	3/4	2	2	2	54
64 QAM	5/6	2	2	2	60

Compare the throughputs in different MIMO types and coding and modulation types.

Solution
MIMO techniques have been incorporated in many wireless communications standards and provide a significant performance improvement for the Mobile WiMAX. Using computer simulation, we firstly compare MIMO A and B throughput as shown in Figure 13.18.

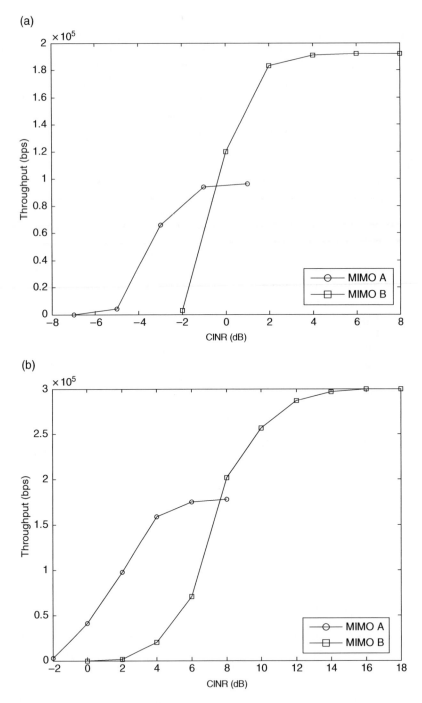

Figure 13.18   MIMO A and B throughput comparison with QPSK 1/2 (a) and 16 QAM 1/2 (b).

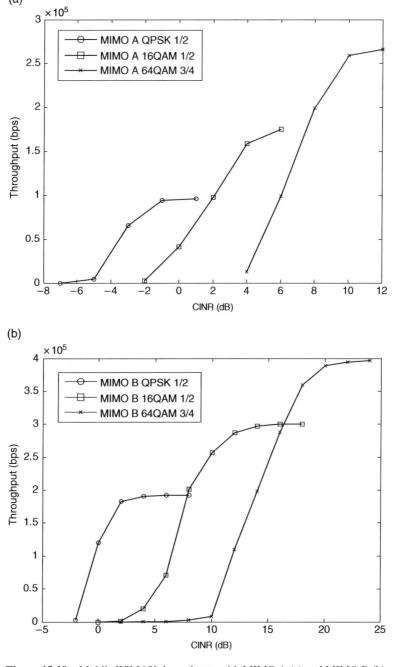

Figure 13.19 Mobile WiMAX throughputs with MIMO A (a) and MIMO B (b)

As we can observe from the figure, the MIMO A has a better throughput at low Carrier to Interference-plus-Noise Ratio (CINR) but the MIMO B has a better throughput at high CINR regardless of coding and modulation types. Secondly, we compare throughputs in different coding and modulation as shown in Figure 13.19.

As we easily anticipate, a high-order modulation and a low code rate provide us with a better throughput.   ∎

13.2.3   Comparison of LTE and WiMAX

In comparison with LTE and WiMAX, the advantages of LTE are as follows: (i) it is compatible with previous mobile technologies (GSM, UMTS, etc.), (ii) it is more suitable to high mobility users, and (iii) LTE mobile users consume low power than WiMAX mobile users due to SC-FDMA. The advantages of WiMAX are as follows: (i) WiMAX network is cheaper than LTE networks and (ii) spectral efficiency is higher. Table 13.6 summarizes the comparison of LTE and WiMAX.

Table 13.6   Comparison of LTE and WiMAX

	LTE (3GPP)	WiMAX (IEEE802.16)
Evolution	Release 8–12	IEEE802.16-2001, 16d, 16.2-2004, 16e-2005, 16m-2011, etc.
Network architecture	All IP Network	All IP network
Multiple access	DL: OFDMA UL: SC-FDMA	DL: OFDMA UL: OFDMA
Mobility	High	Moderate (started from fixed wireless communication)
Duplexing	TDD and FDD (mainly FDD)	TDD and FDD (mainly TDD)
Channel bandwidth	1.4, 3, 5, 10, 15, and 20 MHz	1.25, 5, 10, and 20 MHz
FFT/IFFT size	128, 256, 512, 1024, 1536, and 2048	128, 512, 1024, and 2048
Modulation	QPSK, 16 QAM, and 64 QAM	QPSK, 16 QAM, and 64 QAM
FEC	Convolutional code and turbo code	Convolutional code, turbo code, LDPC
Legacy network	GSM, GPRS, UMTS, HSPA	IEEE802.16 series

Summary 13.2   4G physical layer system

1. The LTE physical layer is similar to the WiMAX physical layer. Their key physical layer technologies include MIMO, OFDMA, AMC, Turbo code, and so on.
2. The LTE was initiated in 3GPP Release 8 and continues the revision of the LTE standard from release 8 to release 12.
3. The LTE multiple access scheme is based on OFDMA for downlink and SC-FDMA for uplink.
4. The LTE air interface consists of physical channels and physical signals.
5. The LTE downlink physical layer processing scheme consists of many steps according to different physical layer channel and transmission modes.

6. The purpose of the WiMAX standard is broadband wireless access at high speed and low cost, which is easy to deploy and provide a scalable solution in areas beyond the reach of Digital Subscriber Line (DSL).

7. The mobile WiMAX is based on scalable OFDMA and supports various bandwidths. In order to maximize the system throughput, Adaptive Modulation and Coding (AMC) is used. The downlink supports BPSK, QPSK, 16 QAM, and 64 QAM. The high-order modulation (64 QAM) is optional in the uplink. It supports both convolutional codes and Convolutional Turbo Codes (CTCs) with variable code rates. The Block Turbo Codes (BTCs) and Low Density Parity Check Codes (LDPCs) are used as optional features.

8. The advantages of LTE are as follows: (i) it is compatible with previous mobile technologies (GSM, UMTS, etc.), (ii) it is more suitable to high mobility users, and (iii) LTE mobile user spends a low power consumption than WiMAX mobile user due to SC-FDMA. The advantages of WiMAX are as follows: (i) WiMAX network is cheaper than LTE networks and (ii) spectral efficiency is higher.

13.3 SoC Design for 4G Communication System

In Part II, we selected key wireless communication algorithm blocks and discussed their design. However, many other algorithm blocks (e.g., scrambler, interleaver, Received Signal Strength Indication (RSSI) measurement, DC-offset compensation, I/Q imbalance estimation and compensation, automatic gain control (AGC)) are needed in a whole wireless communication system. In this section, we discuss System-on-Chip (SoC) design for a WiMAX handset. The SoC design is very common now when implementing wireless communication systems. It integrates almost all components of the wireless communication system. Thus, there are many advantages such as lower power consumption, low cost per gate, and reliable implementation. On the other hands, the disadvantage is not to replace a particular component. The other drawbacks are high fabrication cost, high system complexity, and complex verification, and so on. As we discussed in Chapter 12, the first step of HW/SW codesign is system specification and architecture definition. Using several partitioning algorithms, we can partition HW tasks and SW tasks in a unified HW/SW representation. Since the performance of microprocessors is getting more powerful and the flexibility is getting more important in wireless communications, Software Defined Radio (SDR) got the limelight in wireless communication design. The SDR is a radio communication where each block is implemented in software. In the SDR, ADCs are located nearby the antenna as closely as possible. Thus, the digitalized signals can be processed in software. However, the flexibility and the efficiency are basically trade-off relationship. The SDR technology can achieve a high flexibility but its efficiency is very low. Especially, the power consumption and overall cost are very high. Thus, the SDR technology came down from the main stage.

The SoC for the wireless communication design contains multiple processors, large memories, multiple peripherals, and multiple interfaces. Software design and hardware design are inevitably tied. It is so much complex to design a whole wireless communication system. However, we must manage all components efficiently. In terms of the target application and metric, HW/SW partitioning could be different. In many cases, a wireless communication

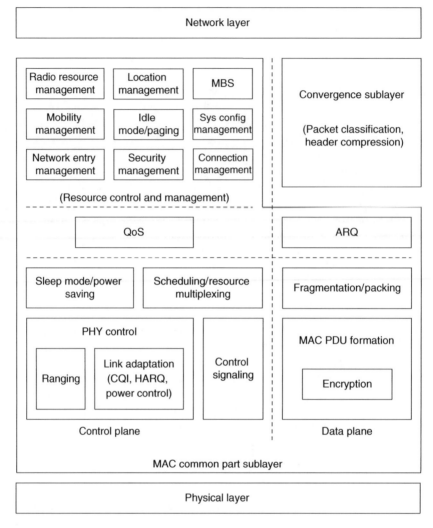

Figure 13.20   WiMAX protocol stack

system designer partitions and implements subjectively. Nevertheless, it is very common to integrate physical layer parts (RF and baseband) and MAC parts (low and high MAC) in the SoC implementation. Typically, the SoC design covers physical layer and MAC layer. The physical layer algorithms and low MAC algorithms are implemented in hardware and high MAC algorithms are implemented in software. The SoC implementation for the WiMAX system can be divided into software part and hardware part. Figure 13.20 illustrates the WiMAX protocol stack. As we can see in Figure 13.20, the WiMAX MAC layer is composed of MAC Common Part Sublayer (CPS) and Convergence Sublayer (CS). The role of CS is to classify Protocol Data Unit (PDU) into the appropriate connection and compress the header information. The CPS is classified into upper MAC and lower MAC. The upper MAC deals with radio resource control and management. The lower MAC on the control plane performs sleep mode/power management, link adaptation, ranging, and control signaling. The lower MAC on the data plane handles ARQ, fragmentation/packing, and PDU formation. These protocols are implemented in software.

13.3.1 Software Design for 4G Communication System

The SoC software is composed of three parts: management plane, data plane, and control plane. The *management plane* is composed of three parts: system management, host–target connection management and, diagnostic monitoring. The WiMAX MAC software includes many tasks based on event-driven programming. They communicate with each other through message queue or semaphore. The system management part has the highest priority so that it enables all tasks to work properly. If an error occurs or a task does not work, it initializes the tasks. The host–target connection management part deals with command and information exchange between host and target. Namely, the target device informs a mobile user of network connection information. A user sets up the configuration using the host–target connection management part. The diagnostic monitoring part is designed as software developer interface. It informs the developer of device and network information such as RSSI, frequency offset, frame synchronization, detected preamble index, and message log. The *data plane* is literally WiMAX MAC data plane software implementation. Depending on multimedia services such as voice, video, VoIP, HTTP, and email, the data packets are classified in the CS and associated them with MAC Service Flow Identifier (SFID) and Connection Identifier (CID). The data plane part is composed of device driver, CS, and CPS. The device driver enables the traffic data to transfer to display, memory, or other interface. The CS deals with packet classification and header compression and receives/sends the packets to the upper/lower layer. In the CPS of the data plane, Service Data Unit (SDU) is encapsulated by addition of header (or Protocol Control Information (PCI)) to Protocol Data Unit (PDU). In addition, it deals with scheduling, ARQ, and data encryption/decryption. When a handset powers on, the first step is to find and access a network. The *control plane* deals with the network entry process [11]. Figure 13.21 illustrates the WiMAX network entry process. In order to access a network, a mobile station scans for channels in the WiMAX frequency list. If the mobile station finds a channel and detects the periodic frame preamble, we can obtain the network parameters. Then, the mobile station sends a ranging request using the minimum transmission power and begins the initial ranging where a ranging is a dynamic time alignment process. It allows a WiMAX base station to receive transmitted signals from mobile stations in an exact time slot, which are located in different distance. Thus, ranging is necessary because mobile stations may keep moving. If there is no response from the base station, it sends the message again using a higher transmission power. If initial ranging is successful, the mobile station is ready to send data to the base station. The next step is to negotiate capability. The mobile station sends a capability request to the base station. Based on the base station's capability, it assigns a modulation order and coding rates to the mobile station. After completion of capability negotiation, the base station authenticates the mobile station. Then, the mobile station sends a registration request to the base station and the base station registers the mobile station. Then, the mobile station receives the IP address and establishes IP connection between them. After completion of registration and IP connection, the transport connections are created and the ranging is periodically performed.

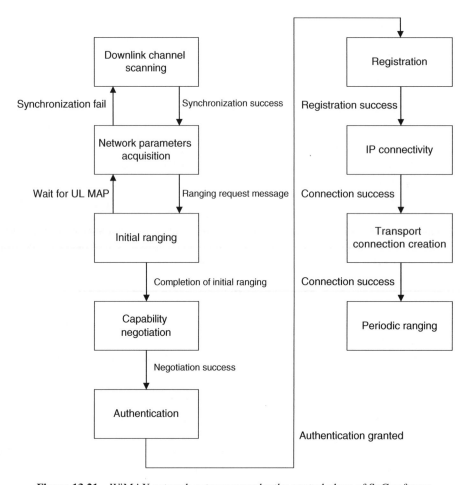

Figure 13.21   WiMAX network entry process by the control plane of SoC software

13.3.2   Hardware Design for 4G Communication System

The SoC hardware of the WiMAX handset includes a RF, PHY (baseband), and low MAC. The Advanced Microcontroller Bus Architecture (AMBA) is widely used for wireless communication SoC design. It describes interconnection and management of functional blocks in the SoC hardware. Figure 13.22 illustrates the SoC block diagram using AMBA.

In the SoC block diagram, the number of DAC/ADC depends on the number of MIMO size. The PHY and RF controller compensates RF impairments using AGC, I/Q imbalance estimation and compensation, and so on. The PHY processor controls the data flow and enables signals for the PHY block. The MAC processor controls the data flow for low MAC block as well as handles high MAC signal processing. The ARM processors [12] are widely used for both PHY and MAC processors. The Direct Memory Access (DMA) is included in order that the access time of memory can be fast enough to keep up with both processors. Two types of buses are used for the SoC block diagram: Advanced High-performance Bus

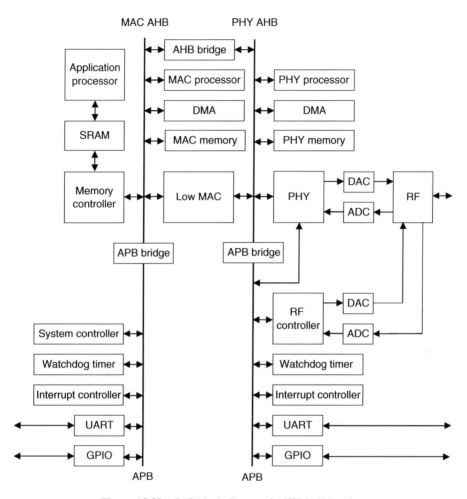

Figure 13.22   SoC block diagram for WiMAX handset

(AHB) and Advanced Peripheral Bus (APB). The AHB is suitable for high bandwidth accesses among PHY, low MAC, and application processor. The APB is suitable for low bandwidth control accesses among system controller, watchdog timer, interrupt controller, Universal Asynchronous Receiver/Transmitter (UART), General-Purpose Input/Output (GPIO). In the SoC block diagram, the low MAC deals with PDU formatting, encryption/decryption, Cyclic Redundancy Check (CRC) checking. The data stream is transferred to/from high MAC (or PHY) using MAC and PHY AHB. Figure 13.23 illustrates the low MAC architecture.

As we discussed in Section 13.1, the baseband is composed of the inner and outer transmitter/receiver. The inner transmitter/receiver includes analogue interface, synchronization, FFT/IFFT, channel estimation, and MIMO encoder/decoder. The output transmitter/receiver includes convolutional code, turbo code, interleaver, and symbol mapping. The analogue

MAC AHB
(high MAC –low MAC)

PHY AHB
(low MAC –PHY)

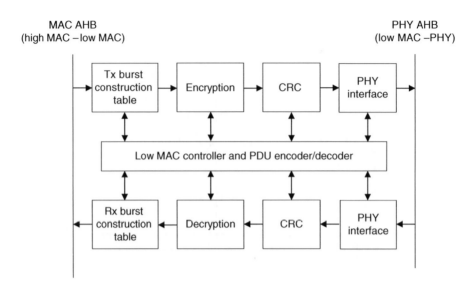

Figure 13.23   Low MAC block diagram

interface needs to support multiple bandwidths because WiMAX channel bandwidths are 1.25, 5, 10, and 20 MHz. The synchronization block detects a preamble and estimates a frequency and time offset. Then, it calibrates system clock frequency and sampling rate using the RF controller. The FFT/IFFT supports multiple-points for multiple channel bandwidths. The channel estimation block calculates the channel state information using the WiMAX pilots. The MIMO block supports the MIMO A and B. In the outer transmitter/receiver, the convolutional encoder/decoder and the turbo encoder/decoder are implemented. Figure 13.24 illustrates PHY block diagram.

In the analogue interface, the Automatic Gain Control (AGC) measures the signal level for each frame and adjusts the RF device gain. Thus, it maintains input signal level as much as the baseband desires. The imbalance between in-phase and quadrature-phase such as the power imbalance results from a nonideal RF component. The IQ mismatch is unavoidable in the direct conversion receiver. The IQ mismatch causes severe degradation of demodulation performance. Especially, when the WiMAX system uses a high-order modulation, it becomes the major channel impairment. Basically, an equalizer can compensate the IQ mismatch but it is a big burden on the equalizer. Thus, the IQ mismatch compensation block is designed. It estimates difference between the in-phase signal path and the quadrature-phase signal path and compensate the imbalance. The LPF in both the transmitter and the receiver is the digital polyphase filter. The LPF in the transmitter performs filtering and upsampling so that it satisfies spectrum masking. Then, it transfers the signal to RF front-end. The LPF in the receiver rejects sidelobe signals and perform decimation. The resampler adjusts sampling rate and the Tx/Rx signal comparison block compensates the I/Q imbalance. Figure 13.25 illustrates analogue interface block diagram.

As we discussed in Chapter 10, the role of synchronization block is to detect a preamble and estimate the frequency and time offset. Using the autocorrelation and crosscorrelation

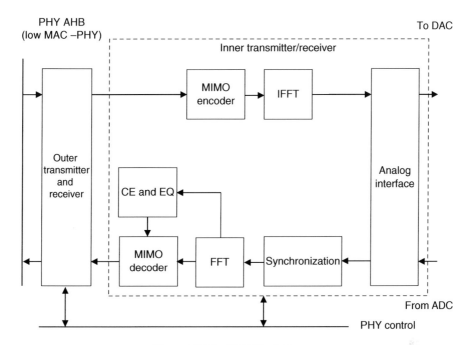

Figure 13.24　PHY block diagram

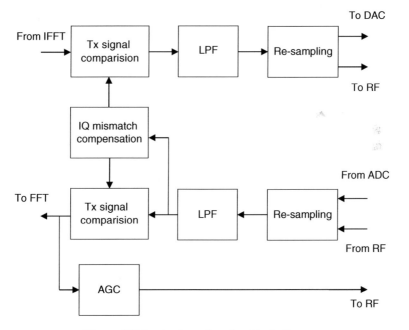

Figure 13.25　Analogue interface block diagram

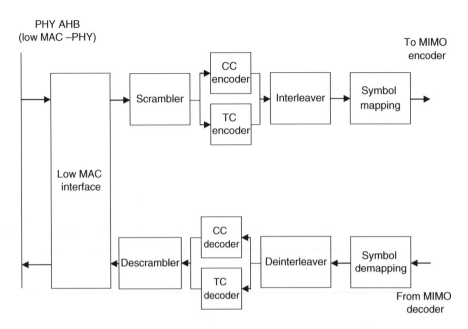

Figure 13.26   Outer transmitter and receiver block diagram

techniques, it synchronizes a frame in support of different bandwidths. As we discussed in Chapter 7, the FFT and IFFT are implemented. In this block, we should consider different sizes of cyclic prefixes and different frame structures. In order to assemble and disassemble a frame, the PHY controller provides the FFT/IFFT block with the frame information such as FFT/IFFT size, subcarrier allocation, segment number, cell ID, and frame structure. As we discussed in Chapter 9, channel estimation and equalization can be implemented using ML, LS, and MMSE algorithm. The pilot-based channel estimation technique is used because the pilot signals are periodically transmitted. In this block, pilot signals should be stored and interpolation should be performed. Thus, memory management is one of the important issues. The WiMAX system supports the MIMO A (Space time coding) and B (Spatial multiplexing). Depending on MIMO modes, both MIMO encoders/decoders can be implemented as we discussed in Chapter 8. The outer transmitter/receiver includes the low MAC interface. It receives PDU data and appropriately reassembles it for the FEC blocks. Scrambling/descrambling is performed in order to randomize signals using a Linear Feedback Shift Register (LFSR). Convolutional Codes (CC) encoder/decoder and Turbo Codes (TC) encoder/decoder are implemented as we discussed in Chapters 5 and 6. Both FECs supports multiple code rates. Depending on transmission modes, one of FECs is used. The WiMAX PHY includes two interleavers. One interleaver is used in order to overcome a burst error and the other is used as one component of Turbo encoder. The codewords from FECs are mapped into signal constellation plane as QPSK, 16 QAM, and 64 QAM. Figure 13.26 illustrates the WiMAX outer transmitter and receiver block diagram.

Summary 13.3   SoC design for 4G communication system

1. Although key wireless communication blocks are designed in Part II, many other algorithm blocks (e.g., scrambler, interleaver, Received Signal Strength Indication (RSSI) measurement, DC-offset compensation, I/Q imbalance estimation and compensation, and automatic gain control (AGC)) are needed in a whole wireless communication system.
2. Typically, the SoC design covers physical layer and MAC layer. The physical layer algorithms and low MAC algorithms are implemented in hardware and high MAC algorithms are implemented in software.
3. The SoC software of the WiMAX handset can be composed of three parts: management plane, data plane, and control plane.
4. The SoC hardware of the WiMAX handset includes a RF, PHY (baseband), and low MAC. Advanced Microcontroller Bus Architecture (AMBA) is widely used for SoC design.

13.4   Problems

13.1. Describe LTE and WiMAX protocol stacks.

13.2. The Evolved Universal Terrestrial Radio Access (E-UTRA) interface provides connectivity between the UE (mobile station) and the eNB (base station). The interface can be split into control plane and user plane. Describe the interface and signaling in E-UTRA.

13.3. In Evolved Universal Terrestrial Radio Access Network (E-UTRAN), S1 interface links among E-UTRAN elements, eNB (base station), and Evolved Packet Core (EPC). Describe S1 signaling.

13.4. In E-UTRA, X1 interface connects between eNBs (base stations). Describe X1 signaling.

13.5. The Application Programming Interface (API) can provide an interface between PHY and MAC. It supports data transfer, buffering, error checking, and interpreting between PHY and MAC. Design the API for WiMAX PHY and MAC.

13.6. A cellular system suffers from inter-cell interference. In order to mitigate this interference, frequency reuse scheme is very useful. Explain the Fractional Frequency Reuse (FFR) technique of WiMAX and Soft Frequency Reuse (SFR) technique of LTE.

13.7. Describe the handoff procedure of LTE and WiMAX.

13.8. Compare RF discrete device design with RF SoC design.

13.9. Beyond SoC, a phone-on-a-chip is actively investigated but not yet realized. What is the main obstacle?

13.10. Describe the pros and cons of Multi-Chip Modules (MCMs), System-on-a-Chip (SoC), Systems-in-Package (SiP), and Systems-on-Package (SoP).

13.11. As SoC technology improves, digital noise and thermal effect become important research challenge. Describe implementation techniques to suppress them.

13.12. As SoC design is getting more complex, SoC testing is getting more difficult. Survey the SoC testing methods.

13.13. Consider the system configuration of Example 13.1. (i) Compare spectral efficiencies between the 2×2 MIMO and the 4×4 MIMO. (ii) Compare BER performance according to different code rates of Turbo coding. (iii) Find the suitable LTE downlink physical layer configuration for a high mobility (ITU VA 100 km channel).

References

[1] V. Kawadia and P. R. Kumar, "A Cautionary Perspective on Cross Layer Design," *IEEE Wireless Communications*, vol. **12**, no. 1, pp. 3–11, 2005.

[2] F. Foukalas, V. Gazis, and N. Alonistioti, "Cross-Layer Design Proposals for Wireless Mobile Networks: A Survey and Taxonomy," *IEEE Communications Surveys & Tutorials*, vol. **10**, no. 1, pp. 70–85, 2008.

[3] V. Srivastava, "Cross-Layer Design: A Survey and the Road Ahead," *IEEE Communications Magazine*, vol. **43**, pp. 112–119, 2005.

[4] The 3GPP Documentation. Overview of 3GPP Release 8 V0.3.3. http://www.3gpp.org/specifications/releases/72-release-8 (accessed April 4, 2015).

[5] H. Myung, J. Lim, and D. J. Goodman, "Single Carrier FDMA for Uplink Wireless Transmission," *IEEE Vehicular Technology Magazine*, vol. **1**, no. 3, pp. 30–38, 2006.

[6] LTE: Evolved Universal Terrestrial Radio Access (E-UTRA). Physical channels and modulation, 3GPP TS 36.211 version 10.0.0, January 2011.

[7] WiMAX Forum Whitepaper, Mobile WiMAX – Part I: A Technical Overview and Performance Evaluation, August 2006. http://www.wimaxforum.org (accessed April 1, 2015).

[8] L. Wan, W. Ma, and Z. Guo, "A Cross-Layer Packet Scheduling and Subchannel Allocation Scheme in 802.16e OFDMA System," *Proceedings of IEEE Wireless Communications and Networking Conference (WCNC2007)*, pp. 1865–1870, Hong Kong, China, March 11–15, 2007.

[9] S. Sezginer and H. Sari, "A High-Rate Full-Diversity 2x2 Space-Time Code with Simple Maximum Likelihood Decoding," *Proceedings of IEEE International Symposium on Signal Processing and Information Technology*, pp. 1132–1136, Giza, Egypt, December 15–18, 2007.

[10] D. Tse and P. Viswanath, *Fundamentals of Wireless Communications*, Cambridge University Press, Cambridge, 2005.

[11] G. Nair, J. Chou, T. Madejski, K. Perycz, D. Putzolu, and J. Sydir, "IEEE 802.16 Medium Access Control and Service Provisioning," *Intel Technology Journal*, vol. **08**, pp. 50–65, 2004.

[12] AMBA Specification (Rev 2.0), May 13, 1999. http://www.arm.com/ (accessed April 1, 2015).

Index

Note: Page numbers in *italics* refer to Figures; those in **bold** to Tables

Printed and bound by CPI Group (UK) Ltd, Croydon, CR0 4YY